Continuity and Change
in Rural Russia

Continuity and Change in Rural Russia

A Geographical Perspective

Grigory Ioffe and Tatyana Nefedova

Routledge
Taylor & Francis Group

NEW YORK AND LONDON

First published 1998 by Westview Press

Published 2018 by Routledge
605 Third Avenue, New York, NY 10017
4 Park Square, Milton Park, Abingdon, Oxon OX14 4RN

Routledge is an imprint of the Taylor & Francis Group, an informa business

Library of Congress Cataloging-in-Publication Data
Ioffe, G. V (Grigoriĭ Viktorovich)
   Continuity and change in rural Russia : a geographical perspective
/ Grigory Ioffe and Tatyana Nefedova.
      p. cm.
   Includes bibliographical references and index.
   ISBN 0-8133-3634-1
   1. Agriculture—Economic aspects—Russia, Western. 2. Russia,
Western—Rural conditions. I. Nefedova, T. G. (Tat'ĭana
Grigor'evna) II. Title.
HD1995.15.164 1997
338.1'0947'2—dc21   94-756
                                                                                        CIP

   ISBN 13: 978-0-8133-3634-3 (pbk)

Contents

Acknowledgments

Funding for this book was provided by the National Council for Soviet and East European Research (NCSEER) and in part by Radford University.

George Demko's (Dartmouth College) encouragement was critical to the entire undertaking. He blessed both our research project and book proposal and enthusiastically recommended them to the NCSEER and to Westview Press. George made us believe we could do it. Don Dahmann (U.S. Bureau of the Census) joined him in setting us on our course. It was one thing to set out—and another to navigate. Glenn Embrey's (Radford University, professor of English) patient, generous help was crucial in this regard. Without him, the two of us, born and raised in Russia, would certainly have failed to produce an intelligible English-language book.

There are actually many more individuals both in the U.S. and in Russia without whom this book would not have appeared. In the U.S. we are grateful to Robert Randolph and Christopher Bowe for their smooth and flexible administration of our project on behalf of the NCSEER, and to Janet Hahn, Rhea Epstein, and Sarah Underwood from Radford University's Office of Research and Sponsored Programs. We greatly appreciate Lori LeMay's (Radford University, professor of Geography) self-sacrificing assistance in adjusting our computer maps to the standards of the publisher. We also thank George Rogachev of RADVA Corporation for his inestimable help with our living arrangements in his native Yaroslavl.

In Russia we wish to express our gratitude to the editorial board of the magazine *Vash Vybor* and individually to Alexander Mineyev and Olga Glazer. They provided us with the opportunity to use extremely valuable data previously gathered for other purposes and with logistical support for our field trip in the summer of 1995. Sergei Safronov and Lena Oleshkievich have been of great help in producing maps. Finally, Andrei Trievish's erudition combined with his magnanimity and kindness assisted us in our historical analysis, as did books and articles kindly supplied by our friends and colleagues Alexander Akhiazer and Zhanna Zayonchkovskaya.

Grigory Ioffe and Tatyana Nefedova

St. Petersburg
Moscow
1
2
3
4
5
6
7
8
9
10
11
12
13
14
15
16
17
18
19
20
21
22
23
24
25
26
27
28
29
30
31
32
33
34
35
36
37
38
39
40
41
42
43
44
45
47
48
49
50
51
52
53
54
55
56
57
58
59
60
61
62
63
64
65
66
67
68
69
70
71
72
73
74
75
76
77
78
79
80
81
82
83
84
85
86
87

Tallinn
St. Petersburg
Riga
Vil'nus
Minsk
L'vov
Kiev
Moscow
N.Novgorod
Perm'
Khar'kov
Voronezh
Krasnodar
1
2
3
4
5
6
7
8
9
10
11
12
13
14
15
16
17
18
19
20
21
22
23
24
25
26
27
28
29
30
31
32
33
34
35
36
37
38
39
40
41
42
43
44
45
47
48
49
50
51
52
53
54
55
56
57
87
88
89
90
91
92
93
94
95
96
97
98
99
100
101
102
103
104
105
106
107
108
109
110
111
112
113
114
115
116
117
118
119
120
121
122

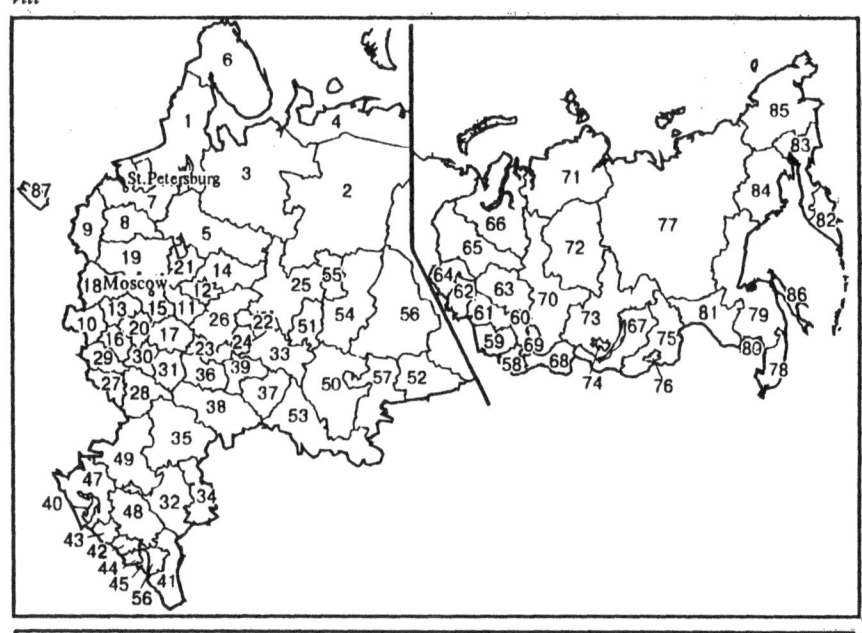

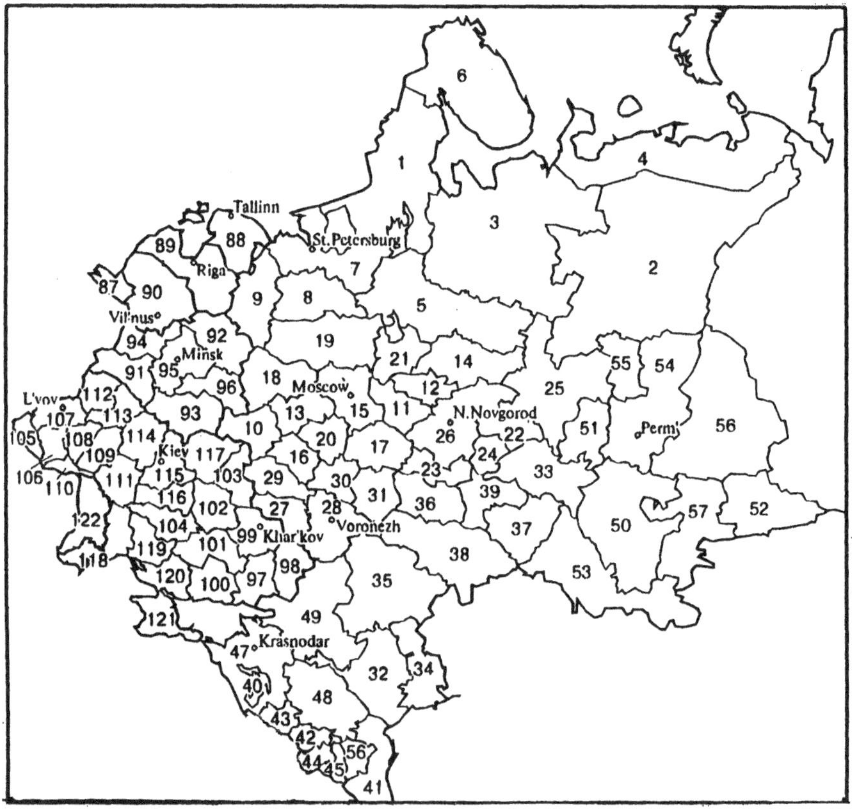

Territorial Sub-Divisions of the Russian Federation (upper map) and of the European Part of the Former USSR[1] (lower map)

1: Karelia	42: Kabardino-Balkar	83: Koryak
2: Komi	43: Karachay-Cherkess	84: Magadan
3: Arkhangelsk	44: North Ossetia	85: Chukotka
4: Nenets	45: Ingush	86: Sakhalin
5: Vologda	46: Checnya	87: Kaliningrad
6: Murmansk	47: Krasnodar	88: Estonia
7: Leningrad	48: Stavropol	89: Latvia
8: Novgorod	49: Rostov	90: Lithuania
9: Pskov	50: Bashkortostan	91: Brest
10: Bryansk	51: Udmurt	92: Vitebsk
11: Vladimir	52: Kurgan	93: Gomel
12: Ivanovo	53: Orenburg	94: Grodno
13: Kaluga	54: Perm	95: Minsk
14: Kostroma	55: Komi-Permyak	96: Mogilev
15: Moscow	56: Sverdlovsk	97: Donetsk
16: Orel	57: Chelyabinsk	98: Lugansk
17: Ryazan	58: Gorno-Altay	99: Kharkov
18: Smolensk	59: Altay	100: Zaporozhye
19: Tver	60: Kemerovo	101: Dniepropetrovsk
20: Tula	61: Novosibirsk	102: Poltava
21: Yaroslavl	62: Omsk	103: Sumy
22: Mari-El	63: Tomsk	104: Kirovograd
23: Mordva	64: Tyumen	105: Zakarpatye
24: Chuvashia	65: Khanty-Mansi	106: Ivano-Frankovsk
25: Kirov	66: Yamalo-Nenets	107: Lvov
26: Nizhni Novgorod	67: Buryatia	108: Ternopol
27: Belgorod	68: Tuva	109: Khmelnitsky
28: Voronezh	69: Khakas	110: Chernovtsy
29: Kursk	70: Krasnoyarsk	111: Vinnitsa
30: Lipetsk	71: Taymyr	112: Volyn
31: Tambov	72: Evenk	113: Rovno
32: Kalmyk	73: Irkutsk	114: Zhitomir
33: Tatarstan	74: Ust-Ordyn Buryat	115: Kiev
34: Astrakhan	75: Chita	116: Cherkassy
35: Volgograd	76: Aginsk-Buryat	117: Chernigov
36: Penza	77: Sakha (Yakutia)	118: Odessa
37: Samara	78: Primorsky	119: Nikolayev
38: Saratov	79: Khabarovsk	120: Kherson
39: Ulyanovsk	80: Jewish	121: Crimea
40: Adyge	81: Amursk	122: Moldova
41: Dagestan	82: Kamchatka	

[1]Names of non-Russian subdivisions are transliterated from Russian.

In Place of an Introduction

Occasionally, a Russian writer heads a preface as this one is headed, in the hope, no doubt, of getting it read.

--G.T. Robinson[1]

To get a feel for the real significance of events in Russia, both recent and remote, has ever been a mind-boggling task for a Westerner. First because, as G. T. Robinson put it in 1932, "living along from year to year in a world of warm breakfasts and trains-on-time, it is easy to believe that history has no place, nowadays."[2] While in America warm breakfasts have given way to cereal and juice, and trains have retreated against the onslaught of planes and cars, Russia is now in the warm breakfast era and following some of the West's own history. However, it is doing so in a very serpentine and often confusing way. Secondly, the average American is not much interested in foreign countries, and since it ceased to be the West's principal military foe, Russia is no longer the exception to the rule. Perhaps to justify their ongoing invasion of Russia since Gorbachev's perestroika, Western media are prone to hyperbolize even small and inconsequential developments in Russia, reporting them as stunning successes or, more often than not, stunning failures. Therefore, to media watchers in America Russia appears intermittently as a place where either nothing is happening at all or something outright fateful is going on. Thirdly, even when cleansed of hyperbole, news from Russia is difficult to penetrate since it often appears in a symbolic guise designed by Russians themselves for domestic consumption, so that foreigners, even those with PhDs in Russian studies, often associate those figures of speech with similar phraseology of Western vintage and so fail to understand Russia on its own terms.

This last shortcoming has been one of American sovietology's most vivid birthmarks. One of our undergraduates unwittingly provided a suggestive parody of it. In reference to a lecture comment that in West Europe roots of democracy are much deeper than in Russia, the student enriched her essay with the following statement: "Russia is pretty much different from Europe. Whereas Europe is democratic, Russia is republican." Though not as entertaining, many assessments of things Russian we have read or heard since coming to the US echo that student's remark. To be sure, there have been a number of exceptions to this general rule -- but not many -- which we will be referring to with gratitude.

Just recently, upon completing this book, we came across yet another sign, this one more personal, of the incongruity between Western and Russian views of Russia. Long considered the Bible of late 19th century village life, *Letters from the Country* by Alexander Engelgardt has been published seven times in Russia, including four Soviet editions. The bulk of Engelgardt's exceptionally rich letters consists of descriptions of everyday life. A few paragraphs, however, are distillations of the author's philosophy and reveal the quintessence of his views. One of us borrowed one such statement for an epigraph to a Russian language book[3] and we set aside few more for future use. Imagine how glad we were when we found out that Engelgardt's Letters are available in English translation.[4] That, we thought, would spare us the translation work. However, it turned out that all the pieces we deemed quintessential are missing from the American English edition, apparently winnowed out as less important or perhaps too didactic.

Here is one that bears directly on this study: "A plant grows in Russia in exactly the same way as in England; both here and there it would require, for example, phosphoric acid for its development; a bone in Russia consists of calcium phosphate as is the case in England; in a nondescript village like Sikorshchina [a toponym fairly typical for Engelgardt's native Western Russia] you could breed Indian corn in a water solution just as you would do it in Eldena [in Mecklenburg, Germany]; but when it comes to practical application of a bone-meal fertilizer or to cultivation of wheat, one cannot always apply methods that are used in England or France. Sciences have no Fatherland, but to the applied science of agriculture cosmopolitanism is alien. There is no chemistry that is Russian, English or German, there is just one chemistry for the whole world, but agricultural science can be Russian or English or German."[5]

Given the abundant difficulties of understanding Russia, we must nevertheless caution against retreating into the agnosticism implied in Tyutchev's famous verse: "Not with the mind is Russia comprehended,/ The common yardstick will deceive/ In gauging her: so singular her nature -- /In Russia you must just believe."[6] The message of this verse has long been, for some Russians, a disguise for a deeply ingrained inferiority complex vis-a-vis the West. But the same message is oftentimes embraced by Westerners falling prey to intellectual laziness.

Surely one can and ought to employ the yardstick of reason in explaining things Russian. Especially now when Russia is once again at a crossroads. Will developments in Russia be along lines followed by other European post-communist nations or will Russia again resort to something unique? It seems to us that life experience in both Russia and the West is an extremely valuable asset in any attempt at answering such questions.

This book has been produced through the collaboration of two co-authors trained in the same school of spatial analysis in Moscow, Russia, but living in different parts of the world since 1989.

After sixteen years of professional life devoted to rural research in Russia, Grigory Ioffe emigrated to the US and is currently living and working in Radford, Virginia. Inspired in part by his research background and in part by new data now gushing out of Russia, Ioffe proposed to the National Council for Soviet and East European Research a project "Rural Restructuring in Central Russia." The project was funded by the Council and was the springboard for this book.

Tatyana Nefedova continues to live in Moscow and to work for the Institute of Geography of the Russian Academy of Sciences. She is also affiliated with the Research Center of Russian Lands founded under the auspices of *Vash Vybor* magazine. This Moscow-based Russian-language magazine receives its major funding from provincial Russian administrations and addresses their interests. Information gathered by Nefedova both personally and with the assistance of her colleagues from *Vash Vybor* has been crucial for this book.

It has been written both in Moscow and in Virginia, with the huge help of e-mail and other means of regular, sometimes daily communication. Given the brief face-to-face encounters between the two co-authors, the book would have been impossible without the mutual understanding already established through years of working together in the same research lab in what is now the Russian Academy of Sciences.

We not only write about one of the most neglected and underrepresented Russian subjects, we do so with insiders' knowledge. Together we have over eighty years of life experience in Russia, including what amounts to several years of time spent in rural field research trips in Russia. Since 1985, the start of Gorbachev's reforms, which did away with Russian isolation from the rest of the world, we have also been communicating with scholars in Western Europe and the USA.

Research Premises

The relationship between continuity and change has always been a subject of culture-oriented research. For the Russian culture *change* has been routinely and accurately for the most part associated with borrowing, while Russia's tenacious domestic tradition has been the basis of its *continuity*.

What Russians have borrowed from the West throughout their history includes many of the underpinnings of modern civilization: technologies (from crop rotation in agriculture to steam engines in transportation and to production lines in manufacturing; and from laundries and vacuum-cleaners to the Internet); health care (from major vaccines to routine drugs); food (from sausage and candy to beer and wine); fashion design; leisure and entertainment activities; and patterns of land-use and beautification (from landscape architecture to principles of industrial aesthetics). Added to the above list should also be: major ideologies (from Karl Marx adapted by Lenin to Milton Friedman adapted by Gaidar), architectural styles (classicism, baroque, etc.), and institutions of power. In most cases, however, borrowing neither resulted from nor led to blind or mindless imitation. All the above appropriations were substantially altered or adjusted by the adoptive country. One may say they were "Russified." "The patriots who preach that Russia is losing its original ways and its historical peculiarities are worrying in vain," writes a Russian political scientist. "No matter what is going on in the country and what innovations are being introduced from abroad, they are being transformed so completely on our soil that, not infrequently, an original source or a foreign model, from which these innovations were copied, is hard to recognize."[7] And as an historian of Soviet architecture once pointedly stated, there is an intimate connection between the questions "Is Russian communism really Russian?" and "What does the so-called Naryshkin baroque have to do with 17th century Italian baroque?"[8] To comprehend the culture of any historic Russian period it is more essential to explore the nature of the *transformation* the borrowed ideology, organization, or style has undergone than to focus on the primordial nature of what was borrowed.

This assertion is related to the first of the two major research premises: to truly understand Russia we must understand it on its own terms. If our first premise concerns the subject of our research, the second one concerns our research perspective. Andrew Gilg, one of the few Western authorities in *rural geography,* once lamented that "general texts on population geography tend to devote only a few pages to rural population...which may show how rural population is inferior to and dependent on an urban civilization."[9] The second part of this observation strikes a familiar chord. In the 1980s the Russian rural research community experienced a split, the forerunner of a deeper ideological rift yet to come. Certain scholars maintained that the wretchedness and the overall "idiocy of rural life" in Russia results entirely from an urbocentric investment policy: countryside was deemed inferior by the investment planners and so given only the leftovers of urban spending. Strange as it may seem, the opponents of this view did not discard its presumption about investment strategy. They simply downplayed the consequences of this presumption by arguing that the root cause of rural backwardness lay elsewhere. Namely, many urban areas themselves were too weak to become sustaining focal points for rural hinterlands, which *naturally* depend upon those urban entities in many ways. If a town is itself a god-forsaken place, what, if anything, could one expect of surrounding villages?

This rhetorical question may sound (though not necessarily *be*) out of place in a modern America or Britain long impacted by the forces of counterurbanization, population deconcentration, and, in America, downtown decay. But it has always been and still is an appropriate question in most other world regions, and in Russia in particular. In fact, drawing on our own understanding of the unresolved problems haunting Russia's countryside, we have never believed that our research subject, this countryside, is in any way demeaned because of its subordinate position with respect to urban cores. As in the human body, many organs that are vitally dependent upon brain are nevertheless important in and of themselves because of their own functions.

That the countryside's problems today derive from some essential features of urban civilization (just as the reverse was true once upon a time) forms the *second* major premise of our work. In fact, we prefer to talk about rural settlement and agriculture as unfolding in *inter-urban space*.

Three Inter-Urban Spaces

Visual impressions derived from travel and living experiences in what may be called the Northern Hemisphere's European Culture Sphere, point to the existence of three distinctive inter-urban social spaces (niches) that contain rural settlements and agriculture. They are: *West European, Russian,* and *North American.* Russian and West European spaces are, in essence, poles apart, while the locus for North American rural space seems to be somewhere in between, perhaps leaning somewhat to the West European pole.

Trying to sort out our holistic impressions we hit upon two essential features that actually underlie them: *scale* or *magnitude of enclosures* (i.e., landholdings with separate ownership and/or economic function); and the *physical condition of a cultural landscape* (built environment, roads, and arable land). In terms of the latter criterion, West European space appears exceedingly *groomed,* while Russian space is *unkempt.* By the former criterion West European space is filled with *small-sized* enclosures and *crisp* border lines, while Russian space has *vast* landholdings with *fuzzy* outer limits and an abundance of *no-man's land.*

In fact, the amount and visual variety of land not put to any use whatsoever are the most pervasive marks of Russia's rural and small-town landscape. These pockets of under-utilized, idle space may contain semi-demolished structures or abandoned fields, or may be derelict littered gaps between sparsely settled residential and office buildings. Many dumps, both officially designated and improvised, add to this impression. In contrast, in West Europe there seems to be no land at all exempt from broadly defined productive use, and dumps are not readily visible.

The differences in the aesthetics of land use can be striking around pockets of West European landscaping *inside* Russia, for example, around the former royal estates in the vicinity of Saint Petersburg that are currently maintained as museum grounds. Unless you are among Western tourists, who are normally whisked away in a luxurious coach, upon exiting the manicured parks you will find yourself at a commuter bus stop leading to the nearest railway station. The impression of dishevelled surroundings, contrasting with what you have just seen, may overwhelm you, even though these suburban communities (Pushkin and Pavlovsk) are actually rather compact and tidy compared with most other Russian small towns. A similar dramatic contrast occurs upon exiting Yasnaya Poliana, Leo Tolstoy's estate, and Spasskoye-Lutovinovo, the estate of Ivan Turgenev, and other remarkable estates described with virtuosity by Priscilla Roosevelt.[10]

One may, of course, reply that any rank-and-file area pales in contrast to royal holdings or those of brilliant celebrities. True enough. However, in Russia the breadth and depth of the contrast rather than the contrast itself, is what captures imagination. A visit to Versailles, Windsor, Aranjuez, or Blenheim castles alongside their Russian counterparts would highlight the differences. The surroundings of West European landmark estates are, aesthetically speaking, not nearly as remote from those estates as is the case in Russia. Is it only because, as Ms. Roosevelt put it, few ways of life have vanished as completely as those of Russian nobility? Or is there another reason, one which might also suggest why these habitats of Russian gentry's life were doomed?

The cultural dichotomy at the heart of Russian society may offer at least a partial answer to this question. This split was addressed most emphatically by Count Trubetskoy who wrote in 1921 that the top and the bottom of Russian culture, its grassroots and its upper crust, "gravitated to different ethnographic zones."[11] More recently Alexander Akhiazer[12] devoted to this cleavage at the heart of culture two of the three volumes of his fundamental work *Russia: A Critique of Historic Experience*. In the words of Nickolas Riazanovsky, "one hardly needs to be reminded that the division between the elite and the masses in Russia paralleled similar divisions in other countries. Still, the Russian split was not quite like the others, or at least it represented a more extreme species of the same genus."[13]

Thus, the sharp impressionistic differences that exist between Russia and West Europe imply more profound socio-cultural differences.

If North American rural and small town scenery shares some common features with that of Russia -- and it seems that it surely does -- the commonality is primarily the enormity of open spaces. In North America they are not as groomed as in West Europe but not as unkempt as in Russia. Some more objective indicators of rural land use, like yields per unit of land, also place North America firmly in between the two poles.

Russia Vs. North America: Exurbia and Periphery

The point of Russian-American similarity, however, is too important to mention only in passing. Quite a few researchers of the past believed that the cause of many Russian woes was simply that Russia does not cope with the challenge of land superabundance.

America confronted the same challenge of land enormity, and "the Westward movement in the annals of the United States" is "no less important than the Eastward movement in the history of Russia."[14] And yet "though the Atlantic is broad ... there is no ocean of difference between the social development of [West] Europe and that of [North] America."[15]

What especially sets Russia apart from both West Europe *and* North America is the history of land ownership. In America *free proprietorship* began at the outset of colonization, as in New England, or, not unlike West Europe, evolved from semi-manorial tenure -- a system of land ownership under which the best and the largest parcels of land still belonged to a select few, but obligatory agricultural services on the lord's domain ceased to be performed by country folk.[16] In Russia, on the contrary, free proprietorship has never had any substantial persistence. Nor is it in place today, when political obstacles, at least in theory, seem to have been removed.

Denis Shaw drew a comparison between Russian and American frontier experiences[17] and Mark Bassin attempted to compare the roles that the abundance of open spaces has played in Russian and in American history. Although Bassin's line of argument occasionally[18] springs from some notions that are alien to us,[19] his comparison of Frederick Jackson Turner's and Sergey Mikhailovich Solovyev's frontier hypotheses is illuminating.

Bassin's discoveries are essentially as follows:

1. Long before Turner authored his frontier hypothesis, the Russian historian Solovyev (1820-1879) had advanced a similar one concerning Russia.

2. There is a profound affinity between the two theories in that both asserted that restless colonization disrupts the nation's natural evolution. Historically, "the United States and Russia represented the product of European expansion into geographical realms that...were not European...In a sort of cultural reflection of this geographical situation, the societies that have resulted from this process could be seen as peripheral offshoots or extensions of European civilization which in important respects were no longer European."[20] Rough-hewn frontier life, the vastness of space, and the paucity of population reduced both societies to a less civilized order.

3. The latter assertion, however, leads to the major interpretative divergence between Turner and Solovyev. "For Solovyev there was

essentially nothing to add once the above point was made, for with it he had effectively accomplished his explanatory task of accounting for the origins of Russia's initial and enduring backwardness vis-a-vis the rest of Europe."[21] Turner, on the other hand, called on the so-called recapitulation hypothesis (the notion that in the course of development from an embryo to a mature specimen, an individual organism repeats the evolutionary ascent of the entire species) to account for the rapid evolution by American society from the very bottom to the very top of the social-evolutionary scale.

Bassin clearly sees this view of American evolution as an intellectual artifice by Turner: "If the civilization that the New World had created were truly to be pronounced a superior society, this would have to be done in terms of the generally accepted and 'scientific' scheme of social evolution. American society, in other words, would somehow have to be moved -- and moved quickly -- from the lowest to the highest of its stages."[22]

Some historical and other geographical observations are useful to make at this point. They can, we hope, reconcile Turner's frontier hypothesis with actual facts at an appreciably lower intellectual cost and perhaps cast some additional light upon the role of open spaces in Russia as compared to North America.

Time is a critical parameter in the evolution of any natural and social entity. No less important is whether this evolution is a spontaneous result of the entity's interaction with the environment alone, however harsh it may be, or whether interbreeding and cultural mutations have intervened as well.

The North America that we know today was initially colonized by West Europeans. In its infancy and formative years it was a direct transplant of West European civilization, its accumulated knowledge, instincts, values, and traditions including the skill of working land in a densely settled space. This knowledge and these traditions evolved throughout a millennium or so of existence in a relatively uniform, compared with East European, cultural milieu. No matter how harsh the frontier environments of North America turned out to be, 300 years was too short a period for all those historical accumulations to be stripped away. But in fact, the period has actually been much shorter than 300 years. As recently as 1800, there were only about 5.5 million people in the territories now covered by the US, and some half million in Canada.[23] Only later did the majority of immigrants arrive: over 30 million in 1815-1914 in the US, while in Canada the main immigration came even later.[24] Like the original European settlers, most of these immigrants also came from West Europe. It was not until the 1890s that migration from the eastern part of Europe started on a large scale; yet even in 1980, after many years of increased immigration from Asia and Latin America, more than two thirds of all American residents born in the US had a British, Irish, or German background.[25] Immigrants from other West European nations make the prevalence of West Europe's cultural ferment in N. America even greater. What is also important is that international migration has always been a highly selective process which typically involved the most enterprising, energetic, robust, and self-sufficient.

By 1800, the Russian civilization had endured about 1,000 years of autochthonous existence largely unaffected by migration from abroad. German colonists arriving in Russia during the reign of Catherine the Great (1762-1796) could be considered virtually the only exception to the rule. They, however, have never become an ethnic intrusion of a truly national scale; having created ethnic enclaves in some Russian regions (predominantly in the Northern Caucasus and the Volga)[26] they have persisted doggedly opposed to assimilation. In fact, Russians have never succeeded in cultural and linguistic assimilation of any ethnic group from an area west of the East Slavic realm. Poles, who are in many ways closer to Russians than Germans, have managed to hold on to their own identity despite the long-term absence of statehood. But what did not materialize west of the Russian heartland abundantly succeeded east of it. Contrary to its westward expansion, the eastward one has rarely "resembled a bold foray of colonial conquest into a foreign realm identical to those of West European empires."[27] Instead there was a progressive settling of the lands east and southeast of the Volga and subsequently of Siberia which continued for centuries of gradual and unhindered sprawling by ethnic Russians. While the newly obtained lands were not empty of indigenous settlement, they lacked any significant nuclei of national or even pre-national consolidation, although in some cases such embryos had earlier effectively degenerated, thus clearing the way for unobstructed Russian colonization. Only the subjugation of Central Asia proper (not including Northern Kazakhstan) bore signs of colonial conquest, but that area has never been culturally assimilated despite the fact that substantial numbers of ethnic Russians settled there.

Integral to the assimilation processes unfolding virtually everywhere eastward of the Russian heartland, except but in Central Asia, has been their reciprocity. While linguistically predominantly one-way traffic occurred, with the Russian language steadily taking over, culturally traffic was clearly two-way. Centuries of intermixing in this gigantic oikumene of Asiatic nomads not only changed European facial types,[28] it altered the settlers' mentality producing what the Slavophile Lamanski defined as the "middle world of Eurasia."[29] Alexander Blok put it in a highly emphatic, poetic form challenging the alleged Western misunderstanding of Russia: "You are the millions, we are multitude/ And multitude and multitude./ Come, fight! Yea, we are Scythians, /Yea, Asians, a slant-eyed, greedy brood./ For you the centuries, for us -- one hour./ Like slaves, obeying and abhorred,/ We were the shield between the breeds/ Of Europe and the raging Mongol horde."[30]

In the 1920s, the so-called Eurasianists (*Yevraziytsy*), Russian immigrant geographers, ethnologists, and philosophers, dwelled on the same subject. "Are there indeed many people in Russia, in whose veins the blood of Khozars, Polovtsy, Tatars or Bashkirs, Mordvins or Chuvash does not flow?" asked Piotr Savitsky rhetorically. "Are there many Russians who are alien to the imprint of oriental spirituality: its mysticism, its love of contemplation, and, eventually, its meditative laziness? Russian common folk is noticeably drawn to the common folk of the East and in its organic brotherly association of the Christian Orthodox with the nomad and pariah of Asia, Russia appeared to be a truly Orthodox-Muslim and Orthodox-Buddhist nation."[31]

"Speaking of ethnography," wrote Nikolay Trubetskoy, "Russian people do not represent just a Slavic realm. Together with Ugro-Finns and Volga Turks they make up a specific cultural zone related both to Slavs and to the Turan [i.e. Turan lowland of Central Asia, the ancestral land of Turks] East, whereby it is hard to say to which of the two Russians are attached more firmly ... Links between Russia and the Turan East have been solidified not only ethnographically but also in terms of anthropology since in Russian veins not only Slavic blood flows but...that of Ugro-Finns and Turks as well. In the Russian national character there are definitely some points of affinity with the Turan East. A kind of mutual understanding and brotherly association that sets itself so easily between us and these 'Asians' is based upon those invisible threads of racial sympathy."[32] Several years later, Nikolai Berdiaev went further attributing the most essential features of Russian communism to Russia's oriental affinities and associations rather than to Marxism: whereas the latter had come to be the source of communist symbolism, the former provided human capital susceptible to social engineering and made it possible to muster energy of the whole nation.[33] The second paragraph of Berdiaev's book, whose real significance was underrated by his contemporaries, reads: "By their spiritual make-up, Russians are Oriental people. Russia is the Christian Orient, which in the course of two centuries has been subjected to the strong Occidental influence and assimilated all the Western ideas in its upper cultural crust."[34]

It is thus safe to say that the expanding frontier similarity between North America and Russia actually conceals a profound difference: in the course and in the wake of its continual sprawl, Russia managed to distance itself from its European roots far more than America ever dreamed of doing. That Russia had actually done so even before most European settlers started to flock to America only exacerbates the difference. Thus, although colonization might have had the effect, in both Russia and America, of lowering the level of civilization, the historical backgrounds against which this effect occurred are so dissimilar as to dwarf any similarity between the two nations.

Such a conclusion is further supported by the fact that *geographically* the outcomes of the colonization processes are also different. Whereas in America and in Canada the densely settled heartlands and sparsely settled hinterlands are spatially disunited, in Russia they interpenetrate. In other words, while in both North American nations population densities may be deceptively low on average, within the areas where the vast majority of people live, the actual densities are not as far from the West European level, as they are in Russia. That Russia's only hinterland is Siberia is, in fact, a spurious notion. To be sure, Siberia is Russia's *classic periphery,* i.e., an area that not only is perceived but actually *is* remote from the heartland. But many Westerners do not realize that a *sense* of remoteness is far more acute in areas barely 150-250 km from Moscow and Saint Petersburg, for example, where the perception of living in the middle of nowhere belies the actual proximity to those population centers.

This sense of remoteness is poignantly captured in the poetic Russian diminutive *glubinka* (from *glubina* = depth) reserved for outlying areas. Of course, a combination of factors, including a sparse network of vibrant urban cores and the primitive condition of roads and other means of communication, creates this perception. In the mid 1970s in Poshekhonye, a district in the Yaroslavl province almost directly between Moscow and Saint Petersburg, we came across the term "*zona nieuverennogo pryioma teleperedach*" (a zone of unsteady TV broadcast reception). It turned out to be a set phrase of technical jargon and implied that the district was too remote from the retransmission towers. We then also discovered that electricity itself had come to villages of Poshekhonye, located close to a district of Volga-based hydroelectric power stations, only in the late 1960s. Districts with similar backgrounds abound in Central Russia.

Many such revelations lie in wait for a stranger to the Russian "glubinka," and sometimes urban-based Russian researchers confess that they feel more out of place there than during a stay abroad.

In *European Russia,* which this book is largely about, a second-order (non-classic) periphery, a kind of hinterland within a heartland, has always existed. It appears that the area of old colonization has shed its spatial continuity and dissociated into original nuclei, big and small. Looking like oases in a rural vastness these nuclei have been able to cast the web of intense social interactions upon only a relatively small part of the inter-urban space. This stands in contrast to North America, and especially to West Europe.

A transitional zone between country and city where both overlap and inter-penetrate naturally exists both in Russia and in the West. In the West this area resulted primarily from urban sprawl and is labelled "suburb." A standard English-Russian dictionary would tell us that the Russian equivalent of "suburb" is *prigorod*. However, the connotation of this Russian word is somewhat different. While "suburb" implies the outgrowth or extension of an *urban realm,* its intrusion into the countryside and the ensuing reimplantation of typically urban amenities in a more rarified residential setting, "prigorod" suggests the *countryside* itself, but a countryside impacted by its close proximity to a city. For urbanites "impacted" would mean ennobled and gentrified, but for country people, it often means the countryside has lost its innocence. Needless to say, this semantic difference has its backing in real life: American-style suburbia is all but non-existent in Russia. Consequently, in this book we have been reluctant to use the word "suburb" when in fact what we mean is *prigorod,* and have resorted to the more seldom used word "exurb" as its more fitting English language equivalent.

In the early 1970s, merely 4% of the vast expanses of European Non-Chernozem (Non-Black-Earth) Russia (2.8 thousand square km) were within two hours accessibility by existing public transportation, to large cities (over 100,000 residents).[35] Since then the situation has changed only slightly for the better, but in the meantime intensive out-migration drained the outlying areas of population, producing more "open" spaces, sharper population density gradients, more abandoned fields, and correspondingly fewer incentives for allocating road construction investments to revitalize the land. Needless to say, these abundant "godforsaken" places have never been conducive to a "European" level of civilization that depends on on-going innovations, a strong labor ethic, etc.

In fact, the socio-cultural cohesion of disparate Russian communities has traditionally advanced less through direct coordination or cooperative action than through being presided over by the same national center with similar lines of command extending from it and transmitting orders down the urban hierarchy. Vertical, centralist, unitarian traditions took precedence over horizontal, neighboring links. It is likely that the presence of an "internal" periphery at the heart of the Russian oikumene and the aesthetics of land-use are related to what Leonid Smirniagin refers to as "the vague concept of space" which he believes is integral to Russian culture. "Maybe it seems paradoxical," writes Smirniagin, "that a people with so vague a concept of space were nevertheless able to command such gigantic territories. But in fact there is no paradox here at all: it was precisely that vagueness which allowed Russians to penetrate such monstrous expanses, with no particular detriment to their own cultural essence. The Russians covered these enormous distances as if in a trance -- they absorbed them, consumed them, while they themselves were wholly absorbed and consumed with their own being, their inner selves: to space itself they gave little of their attention."[36]

The above, once again, stands in contrast to the way the colonization process shaped North America. However, this is not to say that comparing Russia to other nations is futile. Not at all. Comparison is essential for any geographic research whatsoever. Yet, out of the two comparison options, with North America and with the nations of Europe, we prefer the latter and it is more frequently used in this book. Our feeling is that this line of comparison has a richer potential in general and specifically in the context of this book. This is because Russia and Europe share autochthonous development, contiguous land mass, and have been relatively closely interacting throughout history.

Agriculture and Countryside

To the Russian mindset, the notions of *agriculture* and of *countryside* appear as all but perfect synonyms. The convergence of these notions disappeared long ago in the advanced nations of the West. Its persistence in Russia may be seen from economic and cultural perspectives:

Economically, the two concepts are closely related because of the de facto significance of agriculture in rural life and in rural employment. Slightly over half of the rural labor force in Russia works in agriculture. However, if we include those who work only in personal auxiliary farming (subsidiary plots), the proportion is higher, about two-thirds. About a quarter of the rural population are retirees, the overwhelming number of whom work subsidiary plots too, and in the past the bulk of them were affiliated with collective and state farms. Those officially employed in agriculture account for 38% of the rural population as a whole; when retirees are added this percentage becomes 61%. If, however, we exclude children below 15 years of age from the total population, the percentage is as high as 70%, or 20,300,000 out of 29,597,000. This percentage includes only those whose *main* source of income derives from working the land. But many more people engaged in rural industry, forestry, and in social services also depend, at least in part, upon their private plots. So in reality the economic significance of agriculture in the countryside is enormous even if one disregards urbanites whose second dwellings in the country are in large measure devoted to subsidiary farming as well.

From a *cultural* perspective, every country of recent urbanization is more rural in its way of life than formal records of the urban/rural population split would suggest. Historically the words "rural" and "agricultural" meant the same thing, and they still retain more in common in Russia than in the West. Many Russians believe that their economic troubles are rooted in the fact that their genuinely Russian village has been victimized by a not quite genuinely Russian city. This protracted victimization has reportedly undermined the Russian sense of self and caused all sorts of economic problems. This belief has been elevated to a position of national myth by ultra-patriotic political currents. It has echoes in predominately conservative attitudes toward land ownership and toward allotting land to non-agricultural purposes. In fact, as recently as ten years ago it was illegal for an urbanite to buy an abandoned house in a village primarily because locals were opposed to outsiders and opposed to transferring any land, however idle, to use that was not strictly agricultural. The Russian agricultural minister Victor Khlystun's statement about the establishment of what he calls a "civilized market in land" is part of this attitude. To him "civilized" appears to mean (a) regulated by the state; and (b) "a market in which agricultural land cannot be sold for other purposes."[37] As clearly irrational as it is, considering the enormous amount of abandoned and wastefully used land in Russia, the same attitude looms behind current attempts by a heavily communist Russian Duma to eventually do away with private ownership of land.

Research Questions

This book attempts to address the following questions:

- Why, despite the vast and variable land at its disposal, does Russian agriculture remain the most wretched, backward, and the least reformable sector of the Russian economy even against the backdrop of its other crisis-ridden sectors?
- Why, despite a great demand for farming produce, cannot Russian agriculture make ends meet?
- Which contemporary features of the Russian countryside are really new and which only seem so while, in fact, deriving from some deeply ingrained traditions, that have survived many shocks of history?
- What has set the Russian agrarian scene far apart from its West European counterpart?
- What are the major factors of Russian agriculture's spatial development and what is the rural population's role in this regard?
- What are the major characteristics and dimensions of national, macroregional, provincial,[38] and local rural space in Russia?

In the following chapters we answer these questions by discussing developments that have left a distinctive imprint on Russia's inter-urban spaces. The first seven chapters concern Russia as a whole with an emphasis on European Russia; Chapters 8 through 11 deal with specific case studies at the regional and provincial level; and Chapters 12 through 16 focus on specific *issues,* without which the picture of the Russian countryside would be far from complete.

Notes

1. Robinson, Geroid Tanquary. *Rural Russia Under the Old Regime*. New York: Macmillan 1932, p. VII.

2. Ibid., p.2.

3. Ioffe, G.V. *Selskoye Khozyaistvo Niechernozemya: Territorialnye Problemy*. Moscow: Nauka 1990.

4. *Aleksandr Nikolaevich Engelgardt's Letters from the Country, 1872-1887*. Translated and edited by Cathy A. Frierson. New York: Oxford University Press 1993.

5. Translated by the authors from Engelgardt, A.N. *Iz derevni. 12 Pisem.* Moscow: Mysl 1987, pp. 189-190.

6. *A Treasury of Russian Verse*. Edited by A. Yarmolinsky. New York: Macmillan 1949, p. 86.

7. Bondarev, Victor. "Russkaya Democratiya: Bespredel Protiv Bespredela." *Rubezhi* No 6, 1995, p. 124.

8. Paperny, Vladimir. *Kultura "Dva."* Ann Arbor: Ardis 1985, p. 14.

9. *Progress in Rural Geography*. Edited by Michael Pacione. London: Croom Helm 1983, p. 75.

10. Roosevelt, Priscilla. *Life on a Russian Rural Estate*. Yale University Press 1995.

11. Trubetskoy, N.S. "Verkhi i Nizy Russkoi Kultury." In *Iskhod k Vostoku*. Sofia: Balkan 1921, p. 103. Trubetskoy wrote : "As long as the edifice of Russian culture was being crowned by the Byzantine cupola, the whole construction was stable. However, ever since this cupola began to be replaced with an upper floor of Roman-Germanic descent, any stability and harmony of the edifice's parts were being diminished the top tilting more and more until it finally collapsed while we Russian intellectuals, who have spent so much time in effort to prop up the Roman-Germanic roof sliding down Russian walls, are now standing stunned stiff before this gigantic wreck and still thinking how to build a new roof, yet again of Roman-Germanic style." (Ibid., p. 101).

12. Akhiazer, Alexander. *Rossiya: Kritika Istoricheskogo Opyta*. Moscow: FO SSSR 1991.

13. Riasanovsky, Nickolas. V. "Afterword: the Problem of the Peasant." In *The Peasant in Nineteenth-Century Russia*. Stanford University Press 1968, p. 263.

14. Robinson, Geroid Tanquary. Rural Russia, p. 8.

15. Gras, Norman Scott Brian. *A History of Agriculture in Europe and America.* New York: F.S. Crofts & Co 1925, p. 252.

16. Ibid., p. 258.

17. Shaw, Denis, J. B. "The Frontier Experience in Romanov Russia." In Pallot J. and Shaw, D. J. B. *Landscape and Settlement in Romanov Russia 1613-1917.* Oxford: Clarendon Press 1990, pp. 13-22.

18. Bassin, Mark. "Geographical Determinism in Fin-de-siecle Marxism: Georgii Plekhanov and the Environmental Basis of Russian History." *Annals of the Association of American Geographers.* Vol. 82, No 1, 1992, p. 3-22.

19. We specifically mean the popular assertion that environmentalism is a fallacy advanced by reactionary conservatives to prove the pre-eminence of Western civilization. As we understand it, in the West this cliche has been part of a certain ideological portfolio on the one hand, and on the other it has been sustained by the cult of political correctness. During our formative years communist agitprop in Russia, whose inventions could not be taken at face value, widely used the same cliche; however, the moral suasion inherent in Western-style political correctness was missing, so we came to believe (and still do) that the above assertion is an ideological myth not worthy of being rebutted by a respectable scholar, and, as well, it is self-destructive for a geographer to take it seriously. What may be merely a pet peeve or a dead-end for a philosopher or historian is for a geographer a primary professional prism; in a way all geographers espouse *environmental* determinism the day and hour they choose their profession.

20. Bassin, Mark. "Turner, Solov'ev and the 'Frontier Hypothesis': the Nationalist Significance of Open Spaces." *Journal of Modern History,* Vol. 65, September 1993, p. 485.

21. Ibid., p. 505.

22. Ibid., pp. 505-506.

23. Paterson, J.H. *North America. A Geography of the United States and Canada.* Ninth edition. New York: Oxford University Press 1994, p. 43.

24. Ibid., p. 44.

25. Ibid., p. 45.

26. About agrarian experiences of German colonists in Russia see: Pallot, J. "Agricultural 'Culture Islands' in the Eastern Steppe." In Pallot, J. and Shaw, D. J. B. *Landscape and Settlement in Romanov Russia 1613-1917.* Oxford: Clarendon Press 1990, pp. 79-111.

27. Bassin, Mark. "Russia between Europe and Asia: The Ideological Construction of Geographical Space." *Slavic Review,* Vol. 50, No 1, p. 11.

28. One of the authors well remembers how surprised he was back in the early 1970s when during a kayak trip in Latgali (East Latvia) he came in touch with ethnically Russian Old Believers whose ancestors had migrated from the Pskov and Novgorod lands (least of all Russian lands affected by Asiatic invasions) in the 17th century. The strongest impression was produced by their faces. Many of them resembled saints on Christian Orthodox icons; they lacked the prominent cheek bones, flat noses, and squinting eyes so typical of the majority of people in Central Russia. By the same token, when a couple of years ago Boris Yeltsin met with a leader of Mongolia it was hard to distinguish on a TV screen which of them was actually more Mongolian-looking. Had we not been familiar with Boris' rugged face, we would have had a hard time guessing, who is who.

29. Lamanski, V. I. "Tree Mira Azyisko-Europeyskogo Materika." In *Slavyanskoye Obozrenye* 1892, Vol. 1, No 1, pp. 19-41.

30. *A Treasury of Russian Verse,* p. 167.

31. Savitsky, Piotr. "Povorot k Vostoku." In *Iskhod k Vostoku.* Sofia: Balkan 1921, pp. 1-2.

32. Trubetskoy, N.S. "Verkhi i Nizy Russkoi Kultury." In *Iskhod K Vostoku.* Sofia: Balkan 1921, p. 100.

33. Berdiaev, Nikolai. *The Origin of Russian Communism.* London: G.Bles, The Centenary Press 1937. This book is a must for anybody striving to understand Russia. Later Nickolas Vakar attempted to adapt a more diluted and lay-oriented version of Berdiaev's views on Russian communism for American audience not versed in Russian history. However, Vakar's book -- *The Taproot of Soviet Society.* New York, Harper and Brothers 1962 -- was deemed eccentric and fell into oblivion. Although more artfully designed than many later sovietological writings, it apparently was released at the wrong time. Touting some shadowy cultural roots for communism was apparently too much for politically agitated readers. The cold war was in full swing. Communism was not only supposed to be ominous, Western "agitprop" viewed it entirely as a product of rational political indoctrination. The fact that Marxism enjoyed the support of quite a few people in the West aggravated ideological excitement. It was only much later, after communism had retreated from the forefront of international affairs, that Berdiaev's idea began to gain ground and in fact now carries the day in the Western-oriented intellectual circles of Russia itself.

34. Our translation from the original Berdiaev, N.A. *Istoki i Smysl Russkogo Kommunizma.* Moscow: Nauka 1990, p. 7.

35. Lola, A.M., Savina T.M. "Zakonomernosti i Perspectivy Preobrazovanyia Selskogo Rasselenyia Niechernozemnoy zony RSFSR." In *Izvestia AN SSSR, Serya Geograficheskaya* 1978, No 1, pp. 77-89.

36. Smirniagin, Leonid. "Russia is Going Out of Her Mind." *Geograffity,* Vol.1, No 2, 1993, p. 1.

37. Wegren, Stephen. "Rural Reform and Political Culture in Russia." *Europe-Asia Studies* 1994 Vol. 46, No 2, p. 230.

38. We prefer the term "province" over "*oblast,*" "*kray,*" and "*gubernia*" for two simple reasons: first, it captures all three without distorting their common meaning of a Russia's civil division; secondly, it does not grate upon the ears of our English-speaking readers. For the same reason Russian *rayon*s or the lower-level (minor) civil divisions are referred to as "districts."

1

Historical Construction
of Russia's Inter-Urban Space

All of Russian history is the history of a country which is being colonized.
--Vasily Kliuchevsky[1]

Social scientists disagree on how deep the roots of issues in contemporary Russia lie buried. "There is an opinion," writes Alexander Akhiazer, "that for today's Russia and its build-up, only those events are significant that occurred in the last fifty years; others believe that only events since 1917 are meaningful, and so on. Everything that happened prior to any arbitrarily selected date, therefore, is merely academic history. Such a viewpoint is profoundly delusive. History is not just events that pass by leaving no trace. History is a permanent forming of cultural layers which never disappear entirely but continue to exist under new strata and make it to the surface in times of crisis."[2] In fact, Akhiazer's point has gained ground in Russia since the beginning of Gorbachev's perestroika. Upon this writing it seems to be accepted by scores of Russian intellectuals. We share it as well. In particular, we are convinced that the current state of the Russian countryside and of its principal employer, agriculture, cannot be meaningfully examined, let alone diagnosed, without taking a closer look at their historical antecedents.

This chapter and the following two address some phenomena of the Russian past that are still evident in events and developments of today. Those phenomena have already been fairly well studied; we will simply single out the most formative of them, relying on sources, some familiar and some less so to the Western reader.

A popular way of thinking holds that economic activity and life in general are conditioned primarily by some fundamental traits of the existing political order and by the cataclysms that set that order in place. From that perspective communism in Russia and the lack thereof in, say, France would be regarded as far more important to life in the two countries than, say, the dissimilar patterns of settlement and of spatial development in the two countries. In fact, such "minutiae" as where communities live with respect to one another or whether land is settled sparsely or densely are themselves thought of as consequences of the political order, not the other way around.

In Russia's intellectual circles this mode of thinking was never so pervasive as during the early years of Gorbachev's perestroika. At that time the formerly dissident idea that Russian problems could be solved only by deep, "revolutionary" reforms was espoused by the national leadership itself. This belief led some to associate solutions to the country's social and economic problems with radical departures from the established political and economic system.[3] Over the course of Russian history the following appeared to be such departures, alternately positive and negative: the abolition of serfdom; the communist revolution; the collectivization of agriculture; the New Economic Policy; repressions; political thaw; its fiasco; and now its modern offshoot, perestroika.

While this view of history as a sequence of cataclysms has some truth, it is not without shortcomings. If it were entirely true, one sure sign of its truth would be that political cataclysms would set the nations they befell far apart from others in terms of economic activity -- and not just for the short term. However, in such a vital economic sphere as agriculture we do not come across major discontinuities of that sort. For example, time series of grain outputs and areas under crops published by Mitchell[4] in the US and by Mironov[5] in Russia show that the yields' ratio between Western Europe and the European part of the Russian Empire (the USSR to come) was relatively stable, fluctuating for roughly 100 years (1880s-1980s) only between 2.5 and 3.8. According to calculations based on these and other sources,[6] at the very beginning of this century one weighted average hectare of arable land in France outproduced its Russian counterpart by a factor of 2.1; in 1986-90 the proportion was 2.4, that is, roughly the same. Germany, where at the turn of the century farming efficiency exceeded that in France, outproduced European Russia by a factor of 3.8; whereas in the late 1980s East and West Germanies combined outproduced the European part of the USSR only by a factor of 2.5, or almost the same as France.[7] Livestock density comparisons, beginning with data collected long before the advent of communism, show an even more striking stability. As early as the 1870s Germany exceeded European Russia in livestock density by a factor of 3.3; 110 years later the proportion was exactly the same. The superiority of France over Russia in this regard (a factor of about 1.6) also survived all the storms of the century[8]. Note that while farming productivity has been, of course, growing on both flanks of Europe, the ratios between countries have either remained stable or -- despite the supposedly damning influence of communism, for which agriculture used to be the first stumbling block -- changed somewhat in the Soviets' favor.

TABLE 1.1 Length of Railroads in km per 100 sq. km, for Selected Countries in 1874 and in 1982.

Countries	1874	Countries	1982
Germany	3.0	FRG & GDR	120.1
France	2.0	France	62.5
European Russia	0.3	European USSR	10.9

Source: Ioffe, G.V. Selskoye Khoziaistro Nechernozemya. Moscow, Nauka, 1990: 15.

We are not claiming that these observations shatter the cataclysmic view of history (our observations in fact are incomplete and warrant a separate inquiry). Still, it would be logical to assume that if some kind of stability persists through "shocks" of the century, then evidently not only the shocks themselves ought to command attention but also the specific characteristics that remain *stable*. The most wide-ranging characteristic in this regard is the pattern of spatial development, and one of its correlates is the density of reliable transportation routes, a factor which necessarily influences the intensity of any form of land use. Since Europe, and Russia in particular, is still in the railway era, we can compare, for example, the recent railroad density with that over 100 years ago when the rail network was not yet complete. As can be seen from Table 1.1, the change in ratios between Germany, France, and European Russia/USSR has been quite insignificant.

Evidently, then, the extrinsic characteristics of any economic activity are created not only by social upheavals (leading to short-term fluctuations in the indicators of the productivity of an area), but also by certain features of socio-economic space that are resistant to change. In fact, at times those features may be more important than the upheavals.

Distinguishing between the inherently territorial (*spatial*) and generic (*aspatial*) factors of economic development and comparing their impacts have long interested us.[9] And agriculture has always seemed to be a good probing ground for such an analysis. While commanding an ever-decreasing share of labor, agriculture still involves more land in both developing and developed countries than any other form of activity. Because land is of variable quality agriculture remains the most site-specific form of land use, that is, agriculture adjusts to local conditions of natural environment.

However, spatial variation in a *social* environment, that is, in the quality of life, in local traditions, and in the demographic situation, is able to create differences in farming activity even more than nature does. Several crucial circumstances have impacted Russia's social environment: the nation's protracted expansion with backward forms of land tenure and delayed economic development (including delayed but then landslide urbanization); a sparse urban network and vast inter-urban distances; and a pronounced east-west gradient of land-use intensity on the Great European and Russian plains. These inter-related circumstances highlight the specific features of Russia's space viewed as oikumene (literally, receptacle or niche) for agriculture and agrarian communities very distinct from the West.

We will now elaborate on each of these circumstances.

Development Implications of Russia's Colonization Zeal

A pithy saying of Piotr Semionov is a key to understanding social consequences of Russia's irrepressible spatial expansion: "For Russia's premature striving for colonization, its peasantry paid its freedom."[10] The immediate reference was to Russian peasants' 17th-century migrations from what was later labelled the Central Chernozem region to the newly opened lands in European Russia's south-east; the nobility's complaint about losing laborers led to Russian peasants' becoming legally attached to their lords' lands, thus introducing full-fledged bondage. But Semionov's statement runs deeper. Whereas in West Europe internal and overseas *migrations* were in large measure conditioned by agrarian *overpopulation,* in Russia these two factors became linked only in the mid-1800s. Throughout the previous centuries lower and high social strata alike were mobile, at least partly for the simple reason that in Russia there had always been room for new settlements. Peasants could switch from one lord to another of their choosing once a year. *Boyars* (the nobility) in the 14th-16th centuries also often switched from one prince to another at will and owned land apart from where they lived while serving their prince. In places where a prince's entourage was located, *boyars* were assigned temporary landholding (*kormleniya*). Many areas in the Russian heartland experienced intermittent periods of relatively dense settlement and then abandonment. What one can see today in many outlying non-Chernozem districts is in many ways similar to what the Englishman Fletcher came across in the second half of the 16th century when he travelled between Vologda and Moscow: scores of totally abandoned villages with plow-land taken up by forest.[11] Such mobility was unknown in West Europe during that period. Only much later, in modern times, did mobility become considered a feature of advanced societies, and that is what Semionov means by his adjective "premature." The irony of Russian history is that the high and unpredictable pace of resettlement eventually began to threaten the existing economic order and Russian peasants became rivetted to land at a time (1649) when most of their West European counterparts were being set free because their sedentary life had already created economic preconditions for the release of farm labor. Belated bondage exacerbated Russia's social backwardness. In the late 1700s when on the Western flank of Europe the advent of industry took advantage of the recently unleashed mobility of peasant masses, in Russia peasant mobility became rigorously restricted. According to Kliuchevsky, "Up till the end of the first half of the 19th century, the territorial expansion of the [Russian] state proceeded in inverse proportion to domestic freedoms."[12]

Figure 1.1 depicts the level of bondage/serfdom in different parts of Europe at different times. The figure is derived from a variety of literary sources[13] but not based on any strict criteria. East and West Europe have been far apart in terms of serfdom's actual life span. In the Roman Empire *coloni* (serfs) became ultimately attached to land in AD 332 under Constantine. Barbaric invasions and the free settlement of Germanic communities within the fallen empire loosened bondage, but did not eliminate it; only in the 10th-12th centuries was it squeezed out from the hinterlands of commercial towns. In the Frankish Kingdom serfdom took root under the Carolingians (AD 750-887); later it made its way to England and northern Germany; in Scandinavia serfdom did not develop at all. By the 12th-13th centuries classic serfdom (with servitude and personal dependency) had already begun to decline. At local markets peasants became more active than their senhors (lords): they expanded cropland and began to rent land for money or crops, redeeming it as their own and freeing themselves from their lords. But while serfdom had left the scene in the West in the 16th and 17th centuries, in Prussia, Hungary, and Russia it entered it. In the 18th century the east-west differential in terms of serfdom becomes even steeper as the feudal East becomes the granary of the newly-emerging capitalist West. The later serfdom was introduced, the more abrupt was its subsequent decline, because this decline occurred due not to gradual replacement but to drastic one-time reforms (manifested by steep slopes on the right side of the curves in Fig. 1.1). However, while decline by gradual replacement in the West paralleled and in many ways derived from the loosening economic bonds between serfs and their lords, bondage's abrupt formal annulment in the East left these bonds intact for decades.

Another effect of the super-abundance of land (an effect exacerbated by Russia's social backwardness) was the delayed assimilation of farming innovations, for example, crop rotation systems, a key to higher yields. A three-field system (Fig. 1.2), that is, a system of three adjoining unhedged fields sequentially assigned to winter grain, spring grain, and fallow, was practiced in West Europe already in the 7th century, reached its apogee in the 11th-15th centuries, and then gave way to more sophisticated crop rotation patterns. The three-field system penetrated East-Central Europe in the 12th-14th centuries, reached full swing in the 15th century and then faded into the background, although in some places (eastern Poland and the Balkans) it persisted up until this century. However, the three-field system came to Russia only in the 14th-15th centuries, that is, 200 years later than to East-Central Europe and 600 years later than to West Europe. But even "at the

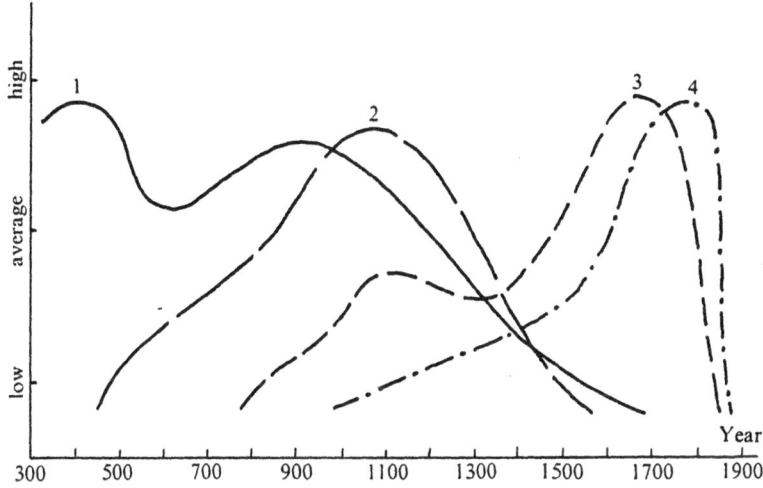

FIGURE 1.1 Serfdom in Europe: Estimate of the Degree of Adoption. 1: South-western Europe; 2: Northwestern Europe; 3: East-Central Europe; 4: European Russia.

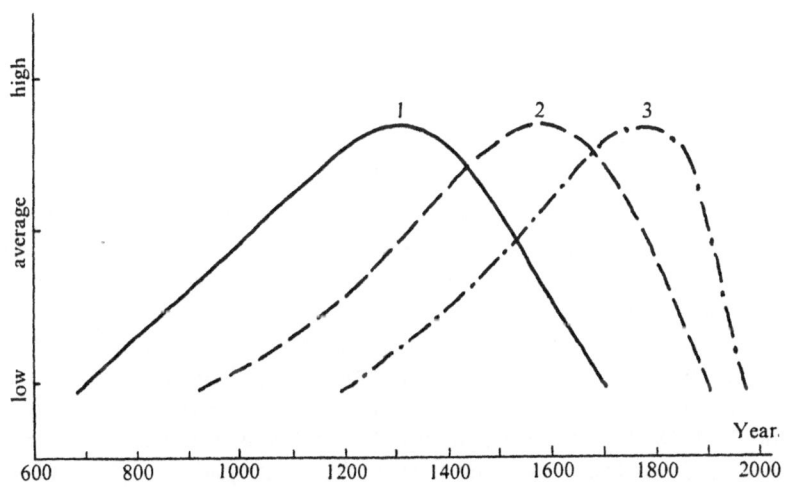

FIGURE 1.2 The Three-Field System of Crop Rotation: Estimate of the Degree of Adoption. 1: Western Europe; 2: East-Central Europe; 3: European Russia.

TABLE 1.2 Percentage of Urban Population in Western Europe, USA, and European Russia.

	1800	1900	1950	1970	1990
West Europe	12	41	59	67	80
USA	5	42	57	66	74
European Russia	6	13	42	62	74

Sources: Bairoch, Paul. *Cities and Economic Development,* The University of Chicago Press, 1988: 221, 290; *1992 World Population Data Sheet,* Population Reference Bureau, Washington D.C., 1992; Davies, K. *World Urbanization 1950-1970.* V.1. Greenwood Press, 1976:75-79; *Demografichesky Yezhegodnik* 1991. Moscow, Goskomstat 1992:7.

the beginning of the twentieth century a more primitive routine of cropping the land year after year to the point of exhaustion, and then leaving it unseeded for ten years or more, was still extensively employed in the far north and in the newly developing regions of the southeastern steppe."[14] Elsewhere in Russia, however, the three-field system had come to dominate cropland in the 18th-19th centuries, and "only" about one-third of the plowland was annually left fallow, much more than in West Europe at the same time.

Russia's Delayed and Yet Landslide Urbanization

Nineteenth-century Russia was a land of peasants. Population statistics for 1811 reveal that 95% of its inhabitants were rural.[15] By the end of the century, in 1897, the countryside of European Russia contained 87.2% of the country's population; in 1916, 79%.[16]

While the current percentage of urban population in the European part of Russia is quite similar to the West European and North American levels (Table 1.2), there are striking differences in the *longevity* and in the *quality* of the urban experience.

According to Paul Bairoch, at the beginning of the 17th century, in West European countries the level of urbanization ranged from 10-12% to 35-45% of the total population, with the average for the region as a whole around 14-17%. "To the East," writes Bairoch, "was a scattering of small towns spread out thinly over the face of the countryside, where levels of urbanization ranged between 4-5% and 7-8%... Many of the cities of vast Russia were not only small, but also at the fringe of the rural world, because of the dominance of agricultural activities of their inhabitants."[17] The first surge of urbanization in Russia fell between 1861 (the abolition of serfdom) and 1914 (the start of World War I). Whereas the total population of European Russia had grown 89% during that period of time, its urban population almost tripled.[18] The second major peak of urban growth came in the late 1920s. Triggered by the so-called socialist industrialization, urban growth at this time took off: between 1926 and 1939 the proportion of urban population in the Russian Federation almost doubled, increasing from 17.7% to 33.4%, at a time when there was a very high natural increase of rural population. By 1959 urbanites already exceeded one half of Russia's total population.[19]

In the USSR, industrialization was pursued as a *Sturm und Drang* campaign, its flip side being the outright plunder of the countryside. Only decades later could villages themselves take advantage of the industrial transformation of Russian society and of nearby urban centers. Not all such centers, though, few in fact, could exert a modernizing influence on their surroundings.

Although cities all across the world owe their development to industries, urban settlement in the West long ago acquired a self-contained *social* quality (an *externality*, in the jargon of econometrics) that in turn exerts its own influence upon the economy. While the same, of course, occurred to cities and towns in Russia, one qualification is in order. The numerical threshold of population concentration required to support a mature urban environment is incomparably higher in Russia than anywhere in the West; it is even much higher than in West Europe's eastern fringe, the Baltics. While this is an informal observation rarely (if at all) subjected to empirical verification,[20] it is not unimportant. On the contrary. In Russia only large (that is, in excess of 200,000 residents) and very large cities are in possession of a reasonably advanced tertiary job-base including higher education, modern health facilities, attractive restaurants, modern retail facilities, and a more or less manicured downtown. A handful of security-closed cities used to be the exceptions that only serve to confirm the rule.

Smaller towns have existed either as mere appendages to big industrial enterprises or -- in the absence of such enterprises -- as local centers distinguishable from villages only by their size. In both cases the *social quality* of such population concentrations does not fit their urban status. Partially because of this but also because urbanization in Russia occurred recently and during a fairly short period of time, many Russian urbanites are virtually villagers in transition. A plethora of writings in Russia has been devoted to their painful acculturation,[21] a process, however, that did not unfold linearly. According to Alexander Akhiazer, during the socialist industrialization of the 1920s-1930s, and later after the war, Russian cities were becoming more "ruralized" than villages were being affected by the intellectual urban mold.[22] Such observations suggest that viewing urbanization merely as an increase in the percentage of urbanites (rather than an actual spread of the urban way of life) would be short of meaningful in the Russian context.

One of the phenomena that escapes attention in such a numerical approach to urbanization is the relationship between the countryside and emerging towns. In Russia this relationship is shaped differently than in the West. George Schoepflin studied this difference; his description of that relationship in East-Central Europe actually overstates the case and, in our view, fits Russia even more than it does, say, Hungary or Poland. Whereas the West exemplifies a fairly dense network of towns, these towns having the ability to develop independently of the national authorities and to *integrate surrounding rural areas,* Schoepflin notes, in the East "the gap between town and country was significantly wider than in the West... The cities that did grow in the Nineteenth Century tended to acquire a certain alien quality in the minds of the bulk of the population; they were all but colonial intrusions in the countryside."[23] While this insightful observation cannot be entirely transferred to current conditions, it still fits remarkably well the cases of Moscow and Saint Petersburg as viewed by other areas of Russia, and not only rural areas. But there is even more to this point. The perception of cities as alien intrusions underwent a peculiar twist: as urbanites, formally speaking, began to dominate numerically, they began to look upon the countryside as inferior, and the word *derevnia* (village) assumed a derogatory meaning in Russia's urban parlance. In our view, the infamous urban bias in the Soviets' investment policy is more rooted in this cultural disposition than in any kind rational thinking or in the nature of central planning.

Sparse Urban Network and Vast Inter-Urban Distances

Russia is a sparsely settled country. The macrogeographical contrasts in land-use intensity revealed by the simplest indicators are substantial. Whereas the average population density of Western Europe is about 100 people per square km, it is merely 27 in European Russia. Only one civil division, the province (*oblast*) of Moscow (47,000 square km), with its 334 residents per square km (which includes the city of Moscow), exceeds the West European average. An analysis of transportation network densities (railway lines and paved roads) by Andrew Treivish revealed that the European part of the former USSR as a whole was less dense than the former Comecon countries of East Europe by a factor of 3. Compared to West Europe this factor would be even higher. Out of all the civil divisions of the former USSR, only Estonia, with 600 km of transportation routes per 10,000 square km, exceeds the level of East Europe and is close to the former GDR (East Germany). The three Baltic states combined, along with the Ukraine, Moldova, and the Moscow province of Russia, are the only entities that are on a par with some of the least developed nations of East Europe (Romania

TABLE 1.3 Spatial Distribution of Large (over 250,000 population) Cities in Western Europe, USA and European Russia in 1990-91.

Area	Number of Cities	Land Area, 1000 sq. km	Average Inter-City Distance, km
Western Europe	144	3598	158
USA: Northeast, Midwest, and South	124	3087	158
European Russia	42	4142	314

Sources: Western Europe: Geographical Perspectives. Essex, Longman 1994:131; *Statistical Abstract of the United States 1993*:42-43 (in case of the US data for MSA were used because cities within their corporate limits are ill-suited for comparison with their European and Russian counterparts); *Demograficheskiy Yezhegodnik SNG.* Moscow, 1991:47-54.

and Bulgaria) where there are 300-500 km of railway lines and paved high-ways per 10,000 square km of territory.[24]

It is not just the population and transportation network densities that are low in Russia. The network of urban settlement is also sparse. In this regard especially it stands in striking contrast with West Europe. However, given the inferior quality of roads and the limited availability of private vehicles, the average *accessibility* of mature urban cores is actually much lower in Russia compared with the West than Table 1.3 might imply. Note that according to 1991 estimates of inter-urban transportation mobility, Russia is inferior to Western Europe by a factor of two and to North America by a factor of three.[25] It would be reasonable to assume that Russia's rural population mobility would be at even more of a disadvantage.

In a major piece of comparative research Hall and Hay revealed that by 1950 already as many as 86% of the population of 15 West European nations could be classed as metropolitan, i.e., living in "daily urban systems in the American sense," while by 1970 the percentage had grown to 88.3%.[26] The equivalent percentage in European Russia is certainly much lower. Yevgeny Leyzerovich, whose expertise in regionalization schemes for Soviet physical planning is second to none, distinguished four types of microregions with an ascending order of density of economic activities. In his classification scheme only the fourth type -- areas with "intensive land use and dense clustered settlement" -- may be thought of as identical to the core part of Western Europe, while the third type -- areas with "medium and relatively evenly distributed intensity of land use" -- is similar to West Europe's hinter-lands (north-central Scandinavia, southwestern France, Spain) and some segments of the American Midwest. In the late 1980s both types accounted for 22% of European Russia's territory and hosted 71% of its population, with the fourth type accounting for only 7% and 39% respectively.[27]

Andrew Treivish has provided a more explicit Russian/European com-parison. He set out to designate the so-called core areas in the European part of the former USSR, in which the density of land use is on par with urbanized regions of West Europe and with the North American megalopo-lis. With this in mind he delineated his so-called "demo-economic base structure" of the European USSR. It contains urban agglomerations with populations in excess of 100,000, including rural areas within one hour accessibility range to urban centers. The demo-economic base structure also includes other important industrial nodes that do not meet the above cri-teria, as well as 5-20-km-wide strips along major highways. The base struc-ture thus consists of nodal and linear components. It will be shown later in Chapter 5 that the emerging spatial concentration of Russia's rural popula-tion is especially consistent with the nodal components, as an ever-increasing number of rural dwellers have been living in the city's countryside. However, strips along major highways have also been demographically stable. For example, in 1959-79, the rural population of the Udmurt republic had con-tracted by only 4.3% within a 3-km accessibility range by bus transportation; by 6.3% within a 5-km range; but by 24.1% within the remaining territory.[28] Alexeyev's analysis of the Toropets district of the province of Tver revealed that in 1970, 52.4 % of the rural population lived within a 2-km accessibility range by bus, 35.3% within 2-8-km range, and 12.3% be-yond 8 km. By 1982 the respective percentages were: 76.6, 22.9, and 0.5.[29]

Treivish's map of the demo-economic base structure is reproduced in Fig-ure 1.3, and its distribution across macroregions of European Russia and of what Russians now call "the near abroad" is given in Table 1.4. This table shows that although in European Russia alone the overall land area dimen-sion of the base structure (535 thousand square km) exceeds the total land area of France, it accounts only for 13% of European Russia's total land. However, it contains more than a half of its population, including three fourths of its urbanites. The average population density within the base structure is close to 200 people per square km or about eight times as high as beyond it.

Not only has the rural population been gravitating towards nodal (see Chapter 5) and linear components of the base structure; the output of agriculture has as well (see Chapter 10 for a detailed treatment). This trend seems to be a more radical version of the spatial shifts in agriculture that the West had been experiencing prior to the 1960s. As our earlier research

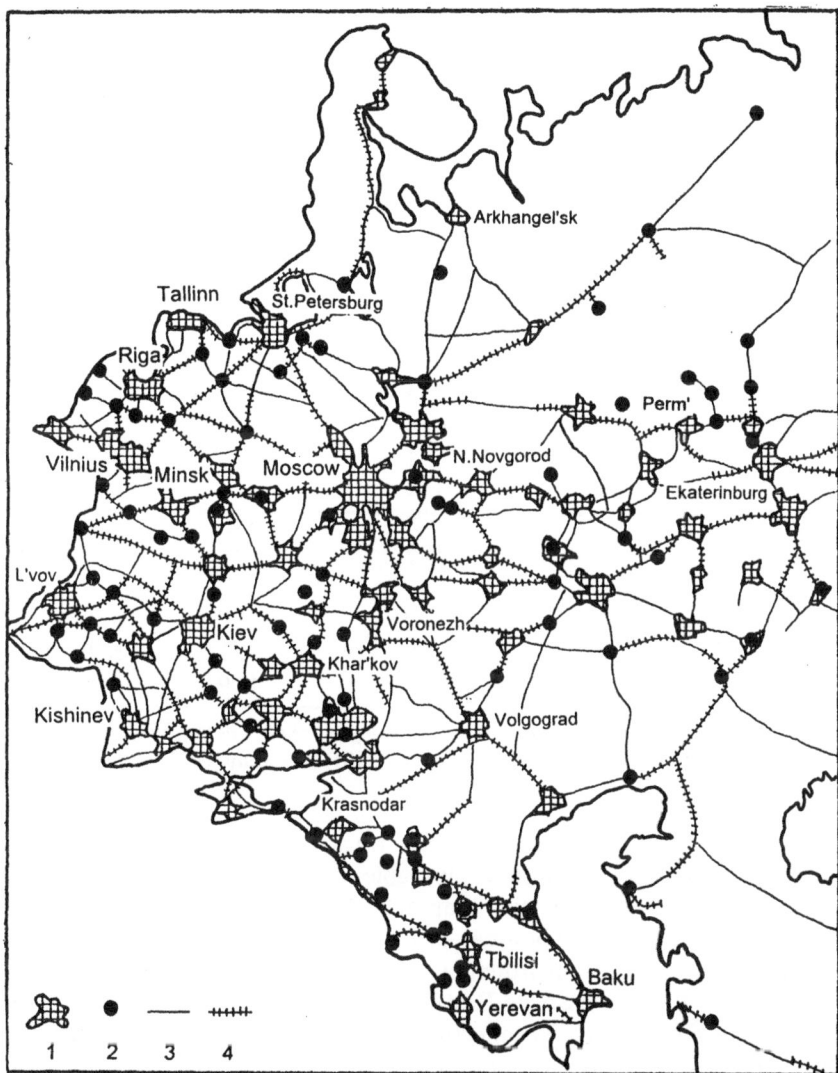

FIGURE 1.3   Demo-Economic Base Structure of European USSR. 1: Urban agglomerations; 2: Free-standing urban centers; 3: Solitary transport routes; 4: Infrastructural bands composed of two or more supplementary transport routes. *Source:* Treivish, A.I. *Osvoyeniye Territorii i Territorialnaya Kontsentratsiya Proizvoditelnykh Sil.* In: *Territorialnaya Organizatsia Khoziaistva kak Factor Economicheskogo Razvitiya.* Moscow, IGAN SSSR 1987:66.

TABLE 1.4 Base Structure of Macro-Economic Regions and Union Republics of the European USSR.

Territory	Land Area of Base Structure in 1000 sq km	Land Area of Nodal Base Structure Only, in 1000 sq km	% of Base Structure in Total Land Area	% of Nodal Base Structure Only, in Total Land Area
North	56.3	25.0	4.1	44.4
Northwest	43.0	20.3	21.9	47.2
Center	125.0	86.5	25.8	69.2
C. Chernozem	43.5	32.0	25.9	73.6
Volgo-Viatka	38.1	25.4	14.5	66.7
Ural	87.8	61.4	11.7	69.9
Volga	79.4	53.1	14.8	66.9
N. Caucasus	62.0	39.4	17.5	63.5
Baltics	65.8	40.8	34.8	62.0
Belarus	57.3	38.2	27.6	66.7
Ukraine and Moldova	202.0	141.2	31.7	69.9
Transcaucasus	39.3	21.6	21.1	55.0
Total	899.5	584.9	16.9	65.0

Source: Territorialnaya Organizatsiya Khoziaystva kak Factor Ekonomicheskogo Razvitiya. Moscow, Institute of Geography, USSR Academy of Sciences 1987:68.

showed, in Russia only the exurban rings of major centers have had agricultural land use intensity comparable to that in Western Europe.[30] As was pointed out earlier in this chapter, in the rarefied social space of the East European Plain, only areas in the immediate vicinity of large cities offer the type of social environment that is widespread in Western Europe.

In the 1980s, in then Soviet Estonia a major factor in the variance in net profits earned by state and collective farms was the farms' proximity to urban centers; soil quality differentials were significantly less important.[31] Note that nowhere in the former USSR were rural roads better or the quality of rural amenities higher than in that Baltic republic. So if tiny Estonia provided evidence of the crucial role proximity to urban centers played in agricultural output, what may one expect of the spacious provinces of Russia, in which no more than 20% of the land area lies within two-hour public transportation-based accessibility to major urban centers?[32]

Despite all the investment directed to Soviet rural areas since the mid 1960s and the resulting improvement in living conditions, nowhere in the former USSR, least of all in Russia, have rural landscapes been transformed to the point that the quality of life differentials between exurban and outlying districts have ceased to exist, let alone been reversed. The levelling processes in the countryside in Western Europe between 1950-1990 described by Fielding[33] have never occurred in Russia.

East-West Gradient of Land Use Intensity

Nothing has contributed to our understanding of European macrogeographical proportions quite as much as our repeated train voyages from Moscow to Warsaw, Prague, Cologne, Paris, and Hoek-van-Holland. Given the various departure times of our business trips, undertaken in different years,[34] these trips created a clear impression in us of a continent-wide regularity or pattern.

For one travelling within central Russia beyond the western reaches of the Moscow urban agglomeration, open fields and forests are the only landscapes that meet the eye. They are occasionally interspersed by clusters of dwelling structures and other buildings. But urban clusters teeming with life are far apart, and road crossings are infrequent. By Belarus the scene from the railway has changed. Fields and forest enclaves visibly shrink, while man-made structures occur more frequently, and the distances between their concentrations become shorter. Adjoining roads and intersections appear with increasing frequency. In Poland this drift toward densely-packed space picks up appreciably and reaches its culmination at the crossroads of Germany and the Benelux.

What was especially surprising for us, raised as we were in the Soviet Union, where deference to and an outright awe of national borders used to be instilled from early childhood, was that the borders, those allegedly formidable discontinuities in socio-political space, did not seem to impact the smooth trend toward densely settled space in any noticeable way, neither accelerating nor inhibiting it. So the scene had an altogether natural, spontaneous streak to it not unlike what one sees sliding along the gentle slope of a hill.

Figures 1.4 and 1.5 corroborate this impression, and there is a definite similarity between the spatial patterns of road density and of grain yields.

Has this east-west gradient been around for a long time? Reliable sources[35] answer "yes" tracing the above mentioned trend to as early as the Roman times.

According to data compiled by Yanson, in 1870, 89.9% of European Russia's land area had a population density that did not exceed 2 persons per square mile and 70% of European Russia's total population lived within this sparsely settled area. The equivalent estimates for 1870 Germany were 9% (of land area) and 3.5% (of population); for Austria-Hungary, 20% and

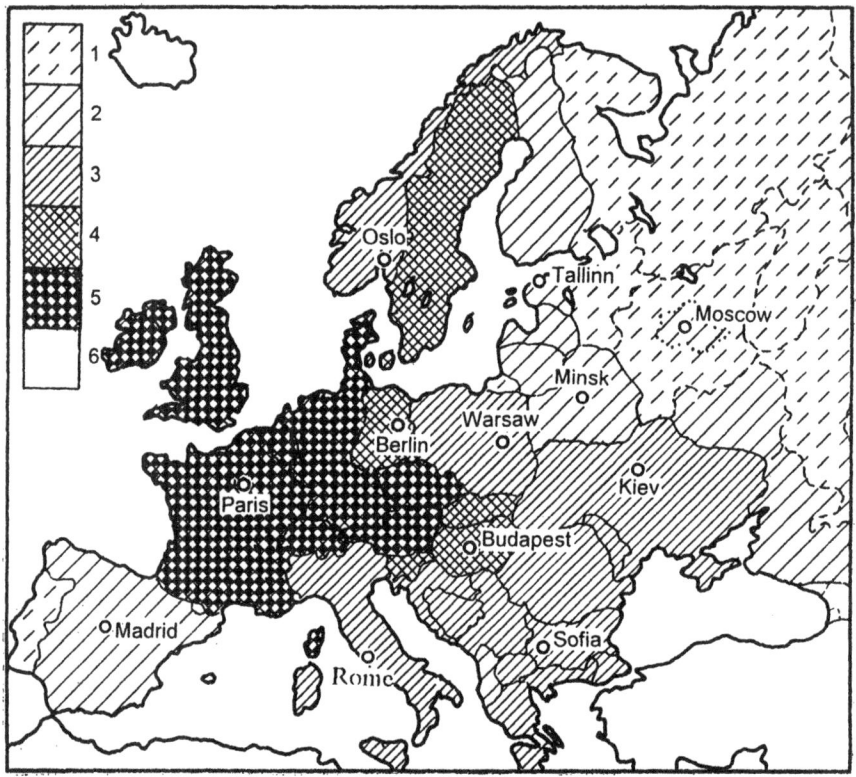

FIGURE 1.4 Average Grain Yields in Europe in Centners per Hectare (1986-1990).
1: Below 20; 2: 21-30; 3: 31-40; 4: 41-50; 5: 51-65; 6: Data unavailable.

8.8%; France, 6.8% and 3.0%; Great Britain, 22.0% and 4%. "In Russia," Yanson writes, "there are no areas at all with a population as dense as that which occurs in up to 15% of the named nations' land; on the contrary, on nine-tenths of Russia's land, the population is as thin as the population on merely one-fifth to one-fifteenth of West Europe's land. Even localities with a medium population density, which in Europe account for 63 to 91% of various nations' land, account only for 9% of European Russia's land. In European Russia the *most densely* populated area, which would match only the *average* population density of West Europe, stretches as a semi-circular band widening westward, from Moscow through Kursk, Kiev, and Warsaw. Outward from this band, the population gets thinner more or less quickly and reaches its lowest ratio to land area in the extreme North, in Zavolzhye [east of Volga], and in the Caspian lowland."[36]

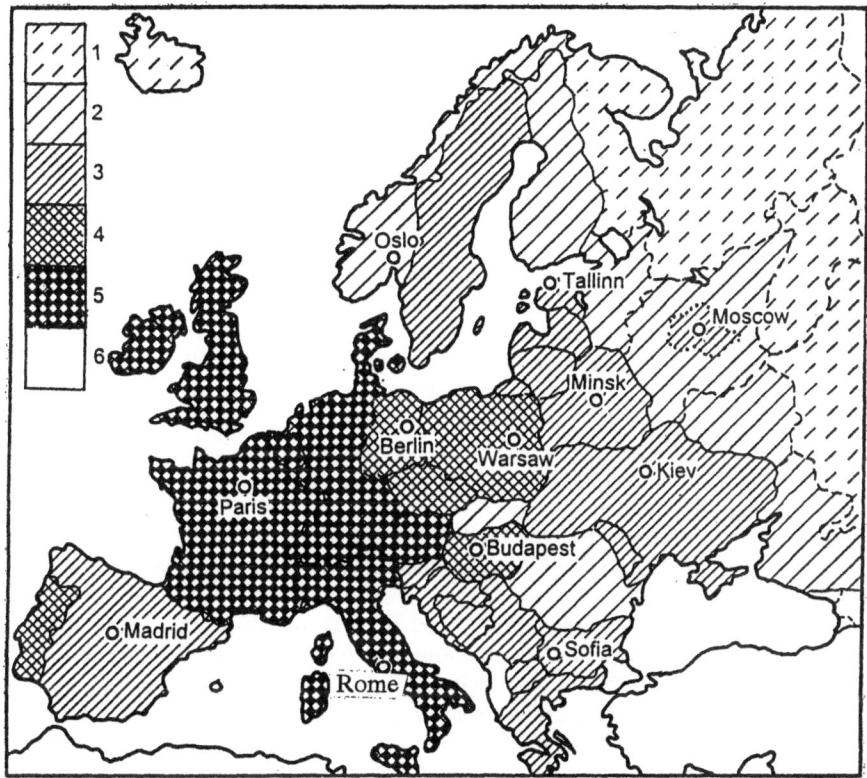

FIGURE 1.5 Length of Paved Roadways in Kilometers per 1000 Square Kilometers (1990). 1: Below 100 km; 2: 100-200 km; 3: 200-400 km; 4: 400-1000 km; 5: 1000-4100 km; 6: Data unavailable.

While the same geographical proportions (though not the values) have survived to this day, European Russia itself exemplifies this continental regularity: note that it is its easternmost regions that have the lowest densities. It stands to reason, then, that in the West the cities' countryside or exurban rings command a higher share of inhabited space than they do in the East, while in the East the outback areas claim the lion's share of land.

In summary, we believe that the four inter-related features of Russia's spatial development converge in support of the premise that inter-urban spaces in Russia shaped up differently than they did in the West. Yet, however substantive, evidence so far provided to this effect is external with respect to the major subject of our research as it does not reflect specific processes unfolding in the countryside. We will now move to fill this gap.

Notes

1. Kliuchevsky, V.O. *Russkaia Istoria. Polny Kurs Lektsii.* Moscow: Mysl, Kniga Vtoraya, pp. 128-129.

2. Akhiazer, Alexander. "Dialectika Transformatsii." *Rubiezhi*, 1995, No 6, p. 96.

3. In the Russia of the mid-1980s, articles by Nikolai Shmeliov conveyed this thought most emphatically and captured the imagination of many.

4. Mitchell, B.R. *International Historical Statistics. Europe 1750-1988.* Third edition. New York: Stockton Press 1992.

5. Mironov, B.N. *Istoriya v Tsifrakh.* Moscow: Nauka 1991, p. 149.

6. Yanson, Y. *Sravnitelnaya Statistika Rossii i Zapadno-Yevropeyskikh Gosudarstv.* Two volumes. Saint Petersburg: 1878-1880; *Selskoye Khozyaistvo SSSR.* Moscow: Finansy i Statistika 1988; *Russia 1913.* Saint Petersburg: Blitz 1995, p. 80.

7. Comparisons in crop harvesting took into account all cereals, potatoes, and sugar beets; their yields per hectare were attached weights equal to their respective percentage shares in sown area.

8. For animal unit mouths, cattle were given a statistical weight of 1.0; horses, 1.5; pigs, 0.25; sheep and goats, 0.1.

9. Ioffe, G.V. and Shoyfer, A.I. "The Transport Accessibility of a Town and the Differentiation of the Factors of Farm Production." *Soviet Geography* 1981, No 2, pp. 94-98.

10. Semionov, P.P. (Tian-Shanski). "Znacheniye Rossii v Kolonizatsionnom Dvizhenii Yevropeiskikh Narodov." *Izvestia Russkogo Imperatorskogo Geograficheskogo Obshchestva,* 1892 Vol. 28, No 4, p. 360.

11. Kliuchevsky, Vasily. *Russkaya Istoriya. Polnyi Kurs Lektsii v Triokh Tomakh. Kniga Piervaia.* Moscow: Mysl 1993, p. 53.

12. Ibid., p. 128.

13. Blum, Jerome. "The Rise of Serfdom in Eastern Europe." *The American Historical Review* 1957, Vol. 62, p. 807-836; Boesch, H. *A Geography of World Economy.* New York: Wiley 1974; Earle, Peter. (ed.) *Essays in European Economic History.* Oxford: Clarendon Press 1974; Makkai, L. "Neo-Serfdom: Its Origin and Nature in East and Central Europe." *Slavic Review* 1975, Vol. 2, pp. 225-238; Kliuchevsky, V. *Russkaia Istoriya. Polnyi Kurs Lektsii v Triokh Knigakh.* Moscow: Mysl 1993.

14. Robinson, Geroid Tanquary. *Rural Russia Under the Old Regime.* New York: Macmillan 1957, p. 98.

15. Drobizhev, V., Kovalchenko, I., and Muravyev, A. *Istoricheskaya Geografiya SSSR.* Moscow: Vysshaya shkola 1973, p. 196.

16. Ibid., p. 198.

17. Paul Bairoch. *Cities and Economic Development from the Dawn of History to the Present.* The University of Chicago Press 1988, pp. 181-82.

18. Drobizhev, V; Kovalchenko, I., and Muravyev, A. *Istoricheskaya,* p. 199.

19. *Demographic Yearbook of the Russian Federation* 1993. Moscow: Goskomstat 1994, Vol. 13, p. 16.

20. One exception known to us is the work of Vadim Miachin: where explicit Soviet/American comparisons are made in regard to such thresholds. They were elaborated upon in greater length and detail in V. Miachin's unpublished doctoral

thesis *Territorial Organization of a National Settlement System (Case Studies of the USSR and the USA)*. Moscow: Institute of Geography 1988. For a brief discussion of that work see: Gritsai, O., Ioffe, G., and Treivish, A. *Tsentr i Peripheria v Regionalnom Razvitii.* Moscow: Nauka 1991, pp. 94-96. Of interest in this regard would also be a structural typology of the largest cities by Agafonov and associates (in: Aga-fonov, N. T. (ed.) *Krupny Sotsialisticheskiy Gorod.* Leningrad: Nauka 1987). It shows that even the largest Soviet cities, with the exception of Moscow and republican capitals, used to deemphasize services.

21. See for example: Zaionchkovskaya, Z.A. *Novossioly v Gorodakh.* Moscow: Statistika 1972; Perevedentsev, V.I. *Plachu Dolgi, Dayu Vzaymy.* Moscow: Sovetskaya Rossiya 1983.

22. Akhiazer, A.S. *Rossiya: Kritika Istoricheskogo Opyta.* Moscow: FO SSSR 1991, Vol. 2, p. 209.

23. Schoepflin, George. "Political Traditions of Eastern Europe." *Daedalus* 1990, Vol. 119, p. 69.

24. Treivish, Andrew. "Territoryalnaya Kontsentratsya Khozyaystva i Rasselenya." In *Geograficheskye Aspekty Vzaymodeystvyia Khozyaystva i Okruzhayushchey Sredy.* Moscow: Institute of Geography, USSR Academy of Sciences 1987, pp. 31-50.

25. Shlichter, S.B. and Tarkhov, S.A. "Transport Rossii Posle Raspada SSSR." In *Rossiya i SNG.* Moscow: Institut of Geography of the Russian Academy of Sciences 1995, p. 71.

26. Hall, P., Hay, D. *Growth Centers in the European Urban System.* University of California Press 1980, p. 85.

27. Our calculations are based on: *Recommendatsii Po Rayonirovaniyu Territorii SSSR Dla Tseley Rasselenyia i Rayonnoy Planirovki.* The author (shown only on the fly-leaf) - Yevgeniy Leizerovich. Moscow: Stroyizdat 1988.

28. *Geograficheskiye Problemy Transporta.* Kazan: KGU 1984, pp. 32-45.

29. *Voprosy Sotsialno-Ekonomicheskoi Geografii Verkhnievolzhya.* Kalinin: KGU 1983, pp. 51-63.

30. Ioffe, G.V. "Different Perspectives on Changes in Rural Areas." *Soviet Geography* 1991, No 5, p. 335.

31. Bronshtein, M. L. (ed.) *Otsenka Zemli i Yeio Ispolsovaniye v Planirovanii i Economicheskom Stimulirovanii.* Moscow: Economica 1984, pp. 74-78.

32. A comparison of this sort, i.e., Estonia vs. Russia, may appear to be somewhat discredited by the fact that even a small free-standing town in the Baltics could serve the surrounding rural area better than a medium-sized town in Russia could and therefore that urban accessibility might matter more for the former than for the latter. However, while this difference in the integrating capacity of central places seems to be quite real, it can lead to a corresponding difference in the threshold value of population concentration; in other words, an urban center with 20 thousand people in Estonia may "weigh" as much as one with 80 thousand in Russia. But a counterclaim could be made as well. It would point to unequal expectations on the side of rural dwellers: the average Estonian may well be more demanding of the social role that a nearby town plays in his life than the average Russian.

33. Fielding, A. J. "Counterurbanisation: Threat or Blessing?" In Pinder, John (ed.). *Western Europe: Challenge and Change.* London: Belhaven 1990, pp. 226-227.

34. Such trips became possible for us with the commencement of the changes brought about by Gorbachev's perestroika. Prior to 1985/86 none of us travelled beyond Comecon.

35. See, for example: Gottmann, Jean. *A Geography of Europe*. Fourth edition. New York: Holt, Reihart & Winston 1969, p. 44.

36. Yanson, Yu. E. *Sravnitelnaya Statistika Rossyi i Zapadno-Yeuropeyskikh Gosudarstv*. Saint Petersburg 1878, Vol. 1, pp. 21-23.

2

Russian Agriculture Early in the 20th Century: Social Peculiarities and Spatial Distinctions

Why, we have enriched the world with a capacious, untranslatable term. Marc Kramer, a "village" writer from the USA, now a Boston University professor who graduated from nearby Harvard, when asked to render "raskulachivaniye" in English, uttered, after conversing in whispers with himself: "dekulakization," and spread his hands, "no notion -- no term."

--Yuri Chernichenko[1]

Daniel Thorner distinguishes five criteria of agrarian or peasant economy[2] that fit 19th-century *fin de siecle* Russia quite well:

- According to the 1897 census, 87% of Russia's population lived in the countryside.
- Of the economically active population, 82% were entirely or partially employed in agriculture.
- Feudal land tenure had been weakened: in 1905, out of the total of 395 million *desiatinas* of land in European Russia (1 *desiatina* = 1.09 ha = 2.7 acre), communal lands accounted for 35.1%, lands belonging to the state treasury accounted for 39.1%, and private landholding for only 25.7%, of which landlords in the countryside, the so-called *pomeshchiki*, owned only 52%, that is, about 13% of all land.[3]
- Sharp differences existed between towns and country whereby the peasantry was considered a "junior" or subordinate estate.
- Finally, peasant families produced the bulk of Russian agricultural output, including 80% of the wheat and predominant portions of flax and potatoes; only sugar beets were being cultivated on relatively large landed estates.

TABLE 2.1 Per Capita Production of Grain in Selected Countries, in kilograms.

	1887-88	1913	1921	1929	1939
Russia	475	727	502	422	448
Austria	405	317	153	256	252
Denmark	840	865	753	893	951
Germany	314	468	283	387	355
France	420	434	382	432	442
USA	1109	980	1026	875	784

Source: Mironov, B.N. Istoriya v Tsyfrakh, Leningrad, Nauka 1991:150.

The overwhelmingly agrarian nature of the Russian economy resulted in a narrow spectrum of employment opportunities and in equally narrow local markets. These in their turn caused the preservation of the subsistence character of the economy. According to some calculations, the commercialization (that is, the proportion that was sold of what was produced) of peasant farms did not exceed 20% of their output.[4]

In 1909-1913 agriculture contributed 60% of the country's total economic output and 90% of its national export.[5] Food, mostly grain, made up two-thirds of that export; the remaining third was non-food crops, mostly flax. The emphasis on grain export was by no means the result of overproduction. By the end of the 19th century Russia produced 475 kg of cereals per capita, whereas Denmark produced 840 kg, Austria 405 kg, France 420 kg, etc. (Table 2.1). Large-scale export of non-food crops was related to industrial backwardness: only one-fifth of Russian flax was processed domestically; shipping coarse wool abroad, Russia imported fine flax and woolen fabric.

Russian grain export at the turn of the century was without precedent in the world: in 1880-1889 it amounted to seven million metric tons a year, and in 1909-1913 to over ten million tons. The US grain export of four million tons a year was a distant second. Most European countries purchased grain although their domestic per capita production was almost on a par with Russia's. In 1909-1913 even France imported one million tons a year, while Germany did about seven million tons.[6]

The principal grain-producing regions of Russia in 1913 were the Middle Volga, the Central Chernozem, the Ukraine, and the Eastern Urals (Fig. 2.1). The Industrial Center was the most grain-deficient region. Overall, the grain surplus in the nation, according to various estimates, amounted to 1.6-3.2 million tons or 5-10% of the total output. However, since yields fluctuated within a range exceeding 10%, mostly due to unstable precipitation, this "surplus" was essentially crop failure insurance.[7] This is another indication that Russia's image as the breadbasket of Europe was largely achieved at the expense of poverty and malnutrition of Russia's own people.

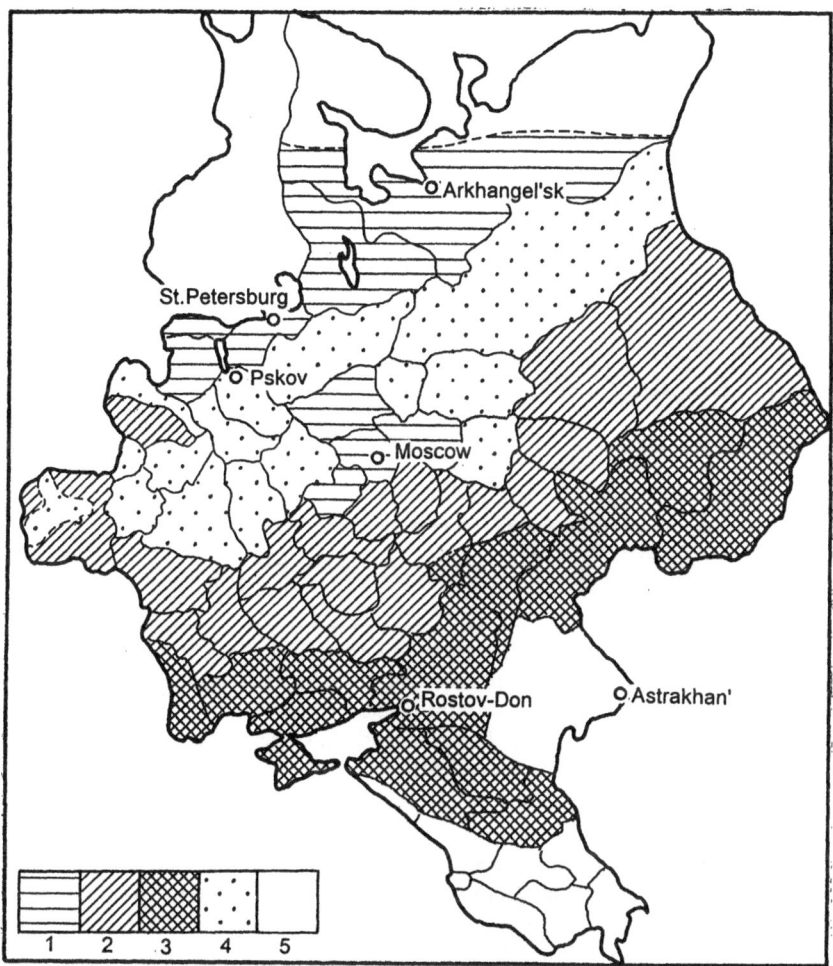

FIGURE 2.1. Grain-Deficient and Grain-Surplus Regions (1913). 1: Production is far short of consumption; 2: Production exceeds consumption; 3: Production is far ahead of consumption; 4: Production-consumption parity; 5: Data unavailable. *Source: Ekonomicheskaya Geografia.* Volume 2. Moscow, Sverdlov Communist University 1929: 359.

Agrarian Overpopulation

By the end of the 19th and the beginning of the 20th century, the countryside of the central and south-western regions of European Russia had become overpopulated (Figure 2.2). This resulted from the combination of the following factors: the mostly agrarian character of the national

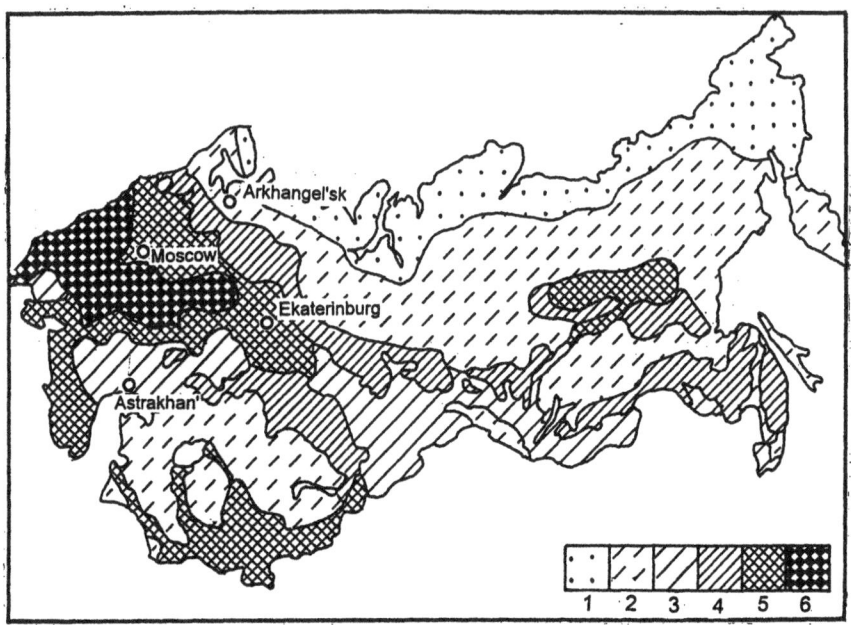

FIGURE 2.2. Rural Population's Pressure on Land (circa 1913). 1: Open spaces of tundra; 2: Scattered, resource-linked colonization; 3: Sparse settlement with pockets of idle land; 4: Sparse settlement without pockets of idle land; 5: Equitable population; 6: Overpopulation. *Source: Ekonomicheskaya Geografiya.* Moscow, Sverdlov Communist University 1929; Volume 2:61.

economy; bondage, which restricted the mobility of rural folk, and the demographic explosion. The 1896-97 total fertility rate in European Russia was as high as 7.06,[8] with a crude birth rate of about 48 per 1000 population (the same as in countries like Ethiopia, Somalia, and Uganda today); at the same time the crude death rate was slightly over 30 while its distribution among peasantry was in inverse proportion to the size of landholding.[9]

Exodus to the cities and to the European south-east and Siberia could have been a safety valve. While both did in fact take place, their rate was highly insufficient. As a result, in 1891-95, the cropping area per one peasant of European Russia was only 65% of its 1861-65 level (56% in the Central Chernozem region). Since progress in farming technology did not quite make up for this reduction in available lands, the overall output of grain in the fifty provinces (*gubernias*) of European Russia fell by 12% during the same period (by 27% in the Central Chernozem region). The shortage of bread was compensated for, in part, by the growing consumption of potatoes, but the average calory intake was reduced.[10] According to calculations by Lenz, in 1899 excessive labor per 1000 *desiatinas* (1090 hectares) increased by 176 persons as compared with 1887.[11] Most of these people, however, stayed put tremendously aggravating rural poverty and creating an environment ripe for upheavals, the response to which was a belated splash of reform activity by the Russian government.

According to Alexander Chayanov, about one-third of Russia's rural dwellers were so-called excessive hands. The main preoccupation of a rural household and of a peasant commune composed of them was to employ all their members. In his 1924 book *Organization of Peasant Economy*, Chayanov narrated a theory of family agriculture that was based on Russian experience and thus departed in many ways from dominant Western views on the issue. According to Chayanov, a peasant family strove as a unit to *jointly* maximize its overall income and employment; thus, income per laborer was not considered important. Overpopulation and the communal character (see later in this chapter) of Russian peasantry were the reasons for this orientation. To maximize employment one could either rent more land or introduce more labor-intensive crops. Chayanov illustrated the latter option with the following example: in Volokolamsk uyezd (a district of the Moscow province, 100 km northwest of Moscow) one *desiatina* sown with oats required 22 person-days of labor input and resulted in 46 rubles of income. Should oats be replaced with flax, the labor input would increase to 83 person-days and the income would rise to 91 rubles. Such replacement, therefore, would double the total income and allow more people to be employed. That the daily per person income would diminish by half would not upset the peasants at all. This explains why the high labor intensity and heightened overall profits associated with potatoes and flax in the north, and with sunflower and hemp in the south, made them so attractive for Russian peasants, regardless of low individual pay.[12] A peculiar paradox thus existed: overpopulation and unemployment went hand in hand with working overloads and extremely low pay. This, however, was a paradox only from the viewpoint of Western-style economic rationality, not to Russian peasants.

Agrarian overpopulation was at its highest in the most fertile Chernozem regions immediately south of the Russian heartland and in the Ukraine.

Early in the 20th century, per capita rural incomes in Russia were between one-half and one-fifth of those in the rest of Europe (Table 2.2). Aside from the fact that maximization of (communal and family) income *and* employment was only achievable at the expense of individual income, the low level of the latter was also related to low labor productivity. While a Russian peasant had fewer cattle than many of his European counterparts and did not use as many machines, this situation was to some extent rooted in overpopulation. According to Chayanov, many peasants consciously rejected machines because they only extended seasonal unemployment. But their incomes were impacted by backward village economy in the first place.

TABLE 2.2 Per Capita Income of Agricultural Workers in Selected Countries in Roubles.

	England	France	Germany	Austria-Hungary	Russia
1894	273	233	184	127	67
1913	463	355	292	227	107

Source: Prokopovich, S.N. *Opyt Ischisleniya Narodnogo Dokhoda.* Saint Petersburg, Soviet Vserossiiskikh Kooperativnykh Syezdov 1918:11.

Whereas in Western Europe land-use intensity had picked up appreciably by the beginning of the 20th century due to the elimination of fallow and enhanced use of fertilizers, in Russia the three-field system still prevailed. For all these reasons, yields per unit of land in Western Europe were considerably higher. As Table 2.3 shows, one German peasant supported twice as large a cropping area as his Russian counterpart in the Central Chernozem region and produced 2.5 times higher grain yields, despite the fact that nowhere in Germany were soils as fertile as in Central Chernozem. Milk yields per cow differed in a similar way: in 1913 in Russia the yield was 850 kg, whereas in the US it was 1413 kg.[13]

At the beginning of this century there were 213 hectares of agricultural land per 100 peasants in European Russia; the actual regional variation ranged between 117 hectares in the south-west (Ukraine) and 146 hectares (Central-Chernozem Region) to 345 hectares in the southeast. At the same time, in France the national average was only 197 hectares per 100 peasants, and in Germany 191 hectares.[14] However, given the actual productivity of plow-land in Russia, because of the techniques at the disposal of a Russian peasant and in part because of low natural fertility and short growing period

TABLE 2.3 Population per One Hectare of Arable Land and Average Yields in Centners per Hectare in Selected Countries and Regions (1901-1910).

	Population	Wheat Yields	Rye Yields
Denmark	76	27.5	18.0
France	78	13.5	10.5
Germany	134	19.4	16.5
Belgium	198	23.0	16.7
Russia:			
Chernozem Center	65	7.5	7.8
Little Russia	72	8.0	8.3
Middle Volga	40	6.2	7.5

Source: Rybnikov, A.A. *Osnovnye Voprosy Plana Selskogo Khoziaistva.* Moscow, Novaya Derevnia 1928:66.

in many areas, the land available to peasant households in Russia was highly insufficient. The following words of a Russian economist uttered in 1880s were even more on target at the turn of the century: "Under such allotments, using the term 'benefactor' to the land appears allegorical and devoid of any practical meaning."[15] *Malozemelye* or land hunger became a by-word in the Russian agricultural writings of the time. The fact that about one-fourth of peasant households held less than 5.5 hectares of land -- given an average household size of about six members this is equivalent to less than one hectare per person -- created an explosive social environment, the incubator for the political shocks of the beginning of the century.

Regional Disparities

The Russian countryside has never been uniform. Alfred Hettner in 1905 distinguished the following agricultural regions of European Russia: 1) the Baltics which had a humid climate, the highest (in Russia) skills in working land, the most commercialized and profitable farming, and yields much higher than on comparable lands in the Russian heartland; 2) Western Russia which had quite favorable climatic conditions but lower land working skills, a shortage of capital and of agricultural literacy; 3) Central Russia which had unstable weather conditions, primitive land skills, and low profitability of farming; 4) the eastern part of European Russia which had a severe climate and underdeveloped agriculture.[16] Hettner also distinguished between two different parts of the Chernozem belt of Russia: its west with a more intensive land-use and its east with low productivity and low standards for working land.

Another authority, Chelintsev, also described the east-west gradient in Russian agriculture. Proceeding westward results in growing intensity of agricultural land-use in both the Chernozem and non-Chernozem belts, and a concurrent rise of population density and of profitability of peasant farms. Chelintsev also pointed to an enhanced intensity of exurban agriculture.[17]

Pre-revolutionary Russia exhibited one more "paradox": on the fertile Chernozem soils, yields were no higher and, more often than not, lower than in the urbanized Center and the north-west (see Table 4.6 and the associated text for a detailed treatment of the subject). These yields were related to the variable depth of agrarian crises in different regions of Russia.

Rybnikov addressed these regional disparities using a time series of local rye and meat prices. In 1890-1913 Russia was deeply involved in the world's agrarian market, so prices were at their highest near the "windows on Europe." These were Saint Petersburg, Archangelsk, and, occasionally, Black Sea ports. In central Russia, price variation formed concentric rings in which prices increased the closer the rings were to Moscow and Nizhni Novgorod, the principal market centers. Areas remote from them experienced the deepest crises. In 1917, Russia, cut-off from external markets, came to resemble the von Thunen isolated state. In the early 1920s, when internal links were severed as well because of the Civil War, several independent systems of procurement and price regulation took shape.[18]

Rybnikov's attempts at monitoring Russia's agrarian developments early in this century revealed both a substantial correlation between agricultural prices and peasant incomes and some departures from that correlation. Both prices and incomes were at their highest in Central Russia and in the vicinity of sea ports. At the same time, because of overpopulation heightened prices in Byelorussia and in the Pskov province did not result in higher incomes. It appears, therefore, that Russian agriculture was affected not so much by soil fertility as by the combination of the following: location with respect to markets, overpopulation, and the historical east-west development (colonization) gradient. Figures 2.3 and 2.4 reflect the principal disparities in agriculture's profitability on the eve of the communist revolution. While gross output per unit of land (Fig. 2.3) looks like a function of two variables, population density and reliance on more (vs. less) progressive farming systems, wages (Fig. 2.4) reflect the degree of rural overpopulation standing in inverse proportion to it. Comparison of Figures 2.3 and 2.4 shows that in fact nowhere but in the Baltics did higher gross output mean higher wages. However, there were regions where wages were higher than output per unit of land would imply. Among such regions were the Industrial Center and the North.

One reason for relatively favorable farm wages in the Industrial Center was highly developed seasonal labor migration. This alleviated the burden of overpopulation. Peasantry living close to big cities used to seek seasonal employment in those cities. Retaining their peasant status in statistical records, these people in fact used to live in cities for months, even years. There were over 200 of these temporary migrants (*otkhodniki*) for every 1000 persons in the rural populations of the Central and Saint Petersburg regions.[19] Figure 2.5 reflects the spatial variation of this seasonal migration.

Peasant Communes

By the beginning of this century about one-third of all European Russia's land area was in peasant communal use; subtracting the northern lands (which were owned by the State Treasury and were predominantly non-agricultural) from the total would increase the above proportion to one-half.

Communal arrangements in rural Russia have been the subject of keen interest in the West and in fact have been thoroughly studied, though more as a self-contained peculiarity than as a tenacious tradition prone to persist

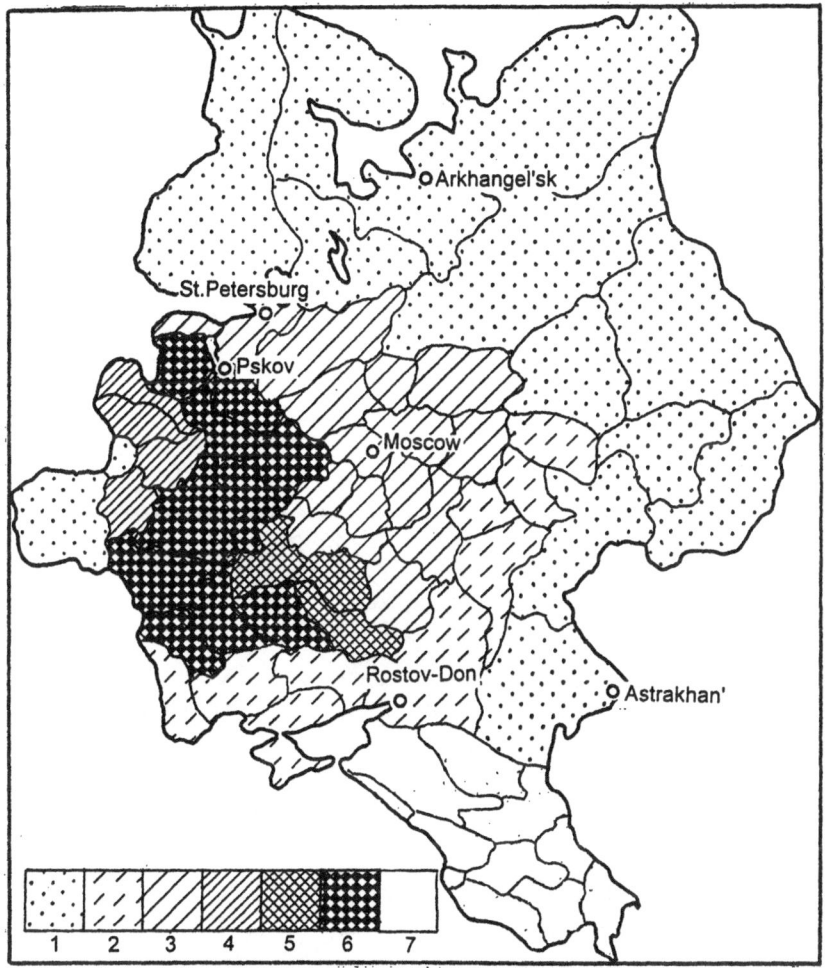

FIGURE 2.3 Gross Farming Output in Roubles per Desiatina (1.09 hectares) of Agricultural Land (1916). 1: Below 36; 2: 36-47; 3: 47-55; 4: 55-65; 5: 65-70; 6: Over 70; 7: Data unavailable. *Source:* Studcnsky, G.A. *Ocherk Selskokhoziaistvennoi Ekonomiki.* Moscow, Novaya Derevnia 1925.

indefinitely through numerous reincarnations. In the very beginning of his superb collation of pre-revolutionary Russian writings about a peasant commune, Geroid Tanquary Robinson shrewdly contrasted the two microcosms of agrarian colonization: that in Russia and that in North America: "Ordinarily...Russian colonists did not live the lonely life of an American frontier family. In the number and in the close inter-relationship

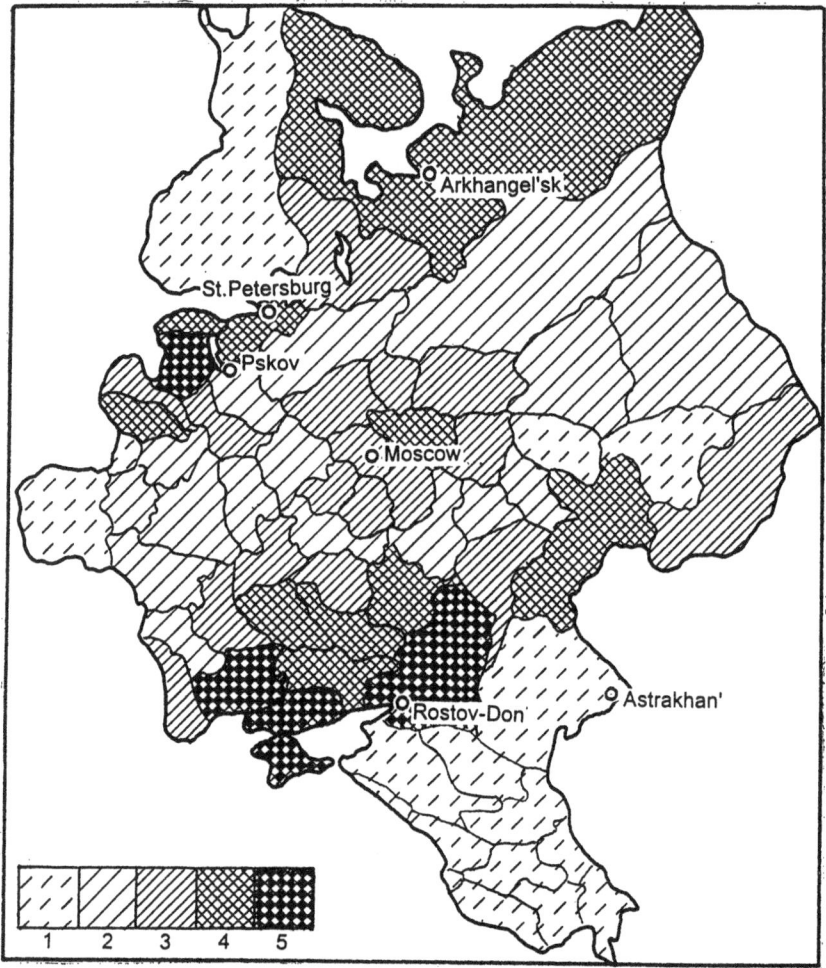

FIGURE 2.4 Average Earnings of Agricultural Laborers (1909-1913) in Kopecks per Day. 1: Below 67; 2: 67-80; 3: 81-99; 4: 100-119; 5: Over 120. *Source:* Rybnikov, A.A. *Osnovnye Voprosy Plana Selskogo Khoziaistva.* Moscow, Novaya Derevnia 1928; map #4.

of its inhabitants, the smallest Russian settlement had usually the double character of a village and a family."[20] When a village included several households, each usually had one or more strips of land in each of the several fields. And in each of these fields all the households were obliged to abide by the same cycle of cropping and fallowing. It was the custom to redivide the meadows each year before the mowing, while the pasture-lands and

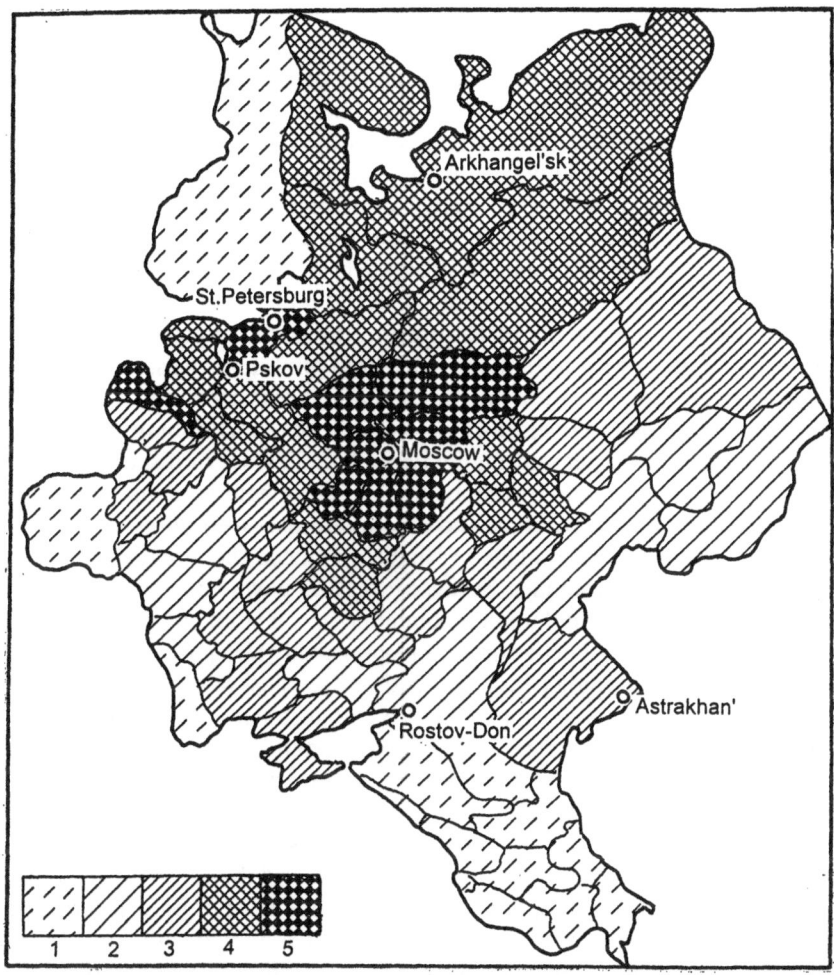

FIGURE 2.5 Temporary Rural Out-Migration (1906-1910) Based on Short-Term Internal Passports per 1000 Rural Population. 1: Below 19; 2: 20-49; 3: 50-99; 4: 100-199; 5: Over 200. *Source:* Rybnikov, A.A. *Osnovnye Voprosy Plana Selskogo Khoziaistva.* Moscow, Novaya Derevnia 1928; map #9.

forests were generally used in common. Among the powers of the commune were thus "the distribution of land and the regulation of its use, the apportionment of the *tiaglo* (feudal obligations), the election of village authorities (elders)..., the collection of funds for expenditures of the *mir* (the peasant commune in its administrative capacity), the organization of mutual aid, and the resolution of civil and minor criminal affairs... During the 17th and 18th centuries equalizing repartitions of land became the norm."[21] They were conducted in response to demographic changes (births and deaths) at the household level and to the variable quality of land. Instead of having the permanent use of a particular parcel, each household received a particular share of communal land. This strengthened the commune's fiscal and administrative rights as a collective with regard to its members. A Russian peasant commune thus involved many manifestations of collectivism. It was through periodical repartition and mutual social control that a commune provided for peasant schooling in group action. Although not all peasant communes practiced re-partitioning, about three-quarters of all peasant households with allotments in European Russia did practice it in 1877-1905.[22] Communes with hereditary household allotments existed mostly in the western and southwestern provinces, that is, areas predominantly populated by Ukrainians and Byelorussians. Yet in both repartitioning and hereditary communes each peasant family typically held a set of dispersed strips of plow-land and the whole composition of strips held by a commune involved a common crop rotation and a common pasture-right.

The 1861 abolition of bondage in Russia enhanced the already conspicuous role of peasant communes, because taxes and payments to redeem land from landlords were levied upon a commune as a whole, not upon individual households. This collective or joint responsibility (*krugovaya poruka*) was a strong motive in keeping peasants together. Needless to say, each member could count on communal support in dire circumstances. Communal bonds thus inhibited rural out-migration and could extend the life-span of crisis-ridden economic units. Even temporary migration was supposed to be subject to communal endorsement.

In a peasant commune the possibility always existed of displacing a household head and appointing another member of the same family in his place. A peasant, therefore, depended on a commune not only in terms of access to land but also in terms of social status. Combined with land repartitioning, such social organization did not facilitate initiative and motivation. Communal leadership, the elders, meddled in the property relationships of peasant families. A head of a household could not arbitrarily bequeath his property to one of his children, let alone transfer it to other hands without communal endorsement. Every son was to inherit a part of his late father's land allotment and a part of his implements in order to fulfill his duties to the commune. However, family members did not have full sovereignty over their interpersonal relationships. Everything they earned, including earnings in a city during their leaves of absence from the commune, was supposed to replenish the family budget. With family members numbering six on average it meant that, say, a childless brother earning decent money as a temporary migrant in a city was working to support the family of his brother who remained in the village sharing his father's home. The tradition of mutual material support by family members was very persistent, outliving even instances of a patriarchal family virtually splitting up when some members moved to cities or towns for good.

In the peasant communes in the Central Industrial and Central Chernozem regions and in Volgo-Viatka mutual support was especially strong because of low yields and inferior skills in working the land in these regions. However, in the newly colonized regions of the south and east, communal arrangements did not take root.

Addressing an issue intensely debated in Russian historiography, Robinson concluded that it would be "forever impossible" to resolve whether communal arrangements in Russia were official or popularly inspired or whether they drew their origin chiefly from above or from below.[23] It seems that much depends upon one's view on an historical process. We would subscribe to the notion that in the final analysis every enduring custom is conditioned from below, popular legitimacy being the only cause of a custom's persistence and longevity. Conversely, if and when that legitimacy ceases a custom is doomed to extinction, either gradually or abruptly. However, according to Yuri Samarin, an outstanding Russian historian, "the issue of a commune is perhaps the only issue where the interests of treasury and landlords [that is, of the upper crust] precisely match a peasant custom."[24] Thus, the question is another version of the chicken and the egg paradox. Whatever the answers, peasant communes are apparently a pivotal part of Russian cultural heritage.

Among other things, Russians' own introspective analysis produced knowledge that similar communal arrangements once existed in almost every European country.[25] Elsewhere, however, by the 7th and 8th centuries, under the influence of Roman law, the right of voluntary withdrawal from a commune with or without one's own land had begun to emerge. This right was a product of a social genotype different from that dominant in Russia. Starikov, a Russian historian, described the basis of social behavior in Western Europe with the following simple maxim: "It is shameful to be poor. If you are in charge of the fruits of your own labor, nothing but your own laziness could explain your misery."[26] So the miserable gave up their land and found something else to apply their skills to.

Western historians will have to judge the appropriateness of this simplification, but it clearly does not apply to developments in Russia. While the challenge of overpopulation that Russian peasantry faced was by no means unique, their reaction to it was different from that in the West. Peasant ownership of land was being gradually restricted, and the institute of private property itself never commanded respect. Intolerance of any departure from equitable distribution of land reigned supreme. According to Starikov, the communal mentality was a result of land development in Russia by colonization, an allusion to the views of the premier Russian historian, Kliuchevsky, reflected in his famous phrase: "All of Russian history is the history of a country which is being colonized."[27] In the ongoing piecemeal expansion of the vast Russian countryside, the right to own land was based on first-come-first-grab rule and land could be held only while it was actually being worked by the possessor. As land itself was not a fruit of labor, it was considered appropriate for it to be repartitioned repeatedly. The moral commandments that prevailed under these conditions were: to base one's judgments not on law but on conscience, to bestow land on the poor by taking it away from the rich.[28] This philosophy proved so tenacious in Russia that no "shock" has been quite able to wean Russian thinking away from it. While later we will show its echoes in developments today, for now it is important to emphasize once again that to the challenge of overpopulation the Russian peasant commune responded not by social stratification, as did its counterparts in the West, but by the equalizing repartitioning of land.

Reforms at the Turn of the Century

The long-lasting harmony between the interest of landlords and the popular communal mentality, the harmony upon which the stability of communal arrangements in rural Russia had long hinged, began to fade in the last quarter of the 19th century. This change was caused by a deepening rural poverty exacerbated by a land hunger. The latter was aggravated by the increased rate of population growth: from 73 million people in 1861, Russia's population had expanded to as much as 125 million by 1897 and to 170 million by 1917.[29]

Responding to the worsening situation, Russian reforms had long acquired the peculiar wavy pattern elaborated on by Alexander Akhiazer in particular, who incorporated this pattern into his cyclic theory of Russian history from Kievan Rus to today.[30] That each successive reform was followed by a rollback suggests to some commentators that the reforms were linked to external impulses and that Russian society as a whole was not quite prepared for those reforms.[31]

The reform and rollback pattern had been apparent in the inconsistencies in the crucial 1861 Peasantry Emancipation Act. According to it, the bond of personal dependence of a peasant upon a landlord was severed, but Russian peasants were supposed to redeem their land from their former lords. Since that time rural communes had been relied upon as watchdog arrangements to make sure that peasants would not renege on their financial obligations.

The issue of voluntary withdrawal from a commune was raised for the first time in the 1880s. Since withdrawal had been thought of as contingent upon receiving land-redemption payment in full, the Finance Minister Bunge presented a blueprint of a concessionary redemption deed.[32] The principal idea behind it (which Witte and Stolypin began to implement only twenty years later) was to transform peasants into private landowners, to instill in them respect for the property of others, and to resettle a part of them in newly colonized lands. But as the landed gentry saw in this idea an assault on its own socio-economic status, Bunge's blueprint was rejected even though he had succeeded in establishing a Peasant Bank in 1882 and in cancelling the poll-tax. By and large, though, the 1880s and 1890s were marked by counter-reform: the reinforcement of peasant communes, in which joint responsibility for tax and redemption payments was legally upheld and the restriction to withdrawals stiffened. In these reform rollbacks the tsarist bureaucratic establishment made a misguided attempt to re-attach peasants firmly to the land, thus thwarting their exodus to towns and the formation of an urban proletariat. The rollbacks, however, could hardly hold urbanization in check, especially around the industrial cores of Central Russia. Scores of seasonal migrants were becoming de facto urbanites even though they still formally belonged in the peasant estate. In 1889 the institute of land captains (*zemskiye nachalniki*) was introduced and peasants were subordinated to them; land captains had the right to penalize peasants without a trial, including the issuance of an exile order for the most unreliable; even corporal punishment was legal.

Peasant discontent showed up in 1902 in revolts in the Central Chernozem region, and then more strongly in 1904-1906. Many estates of nobilities were assaulted and robbed in a general rehearsal of the grand expropriation yet to come in 1917. And already by very beginning of the century many began to share the view (that Russian *narodniks,* or populists had been holding onto since the 1870s) that repartitioning peasant communes were not only custodians of Russian tradition but also hotbeds of Russian- style socialism. Now both the Socialist Revolutionaries and the conservative monarchists adopted this view,[33] but whereas the former pinned their hopes on communes, the latter saw them as a threat to the existing order. In 1903-1904 the authorities were obliged to outlaw corporal punishment. They also outlawed joint responsibility in communes, and promised to ease the procedure of withdrawal from them. Subsequently the arrears on redemption payments were cancelled. Reform-related activities were thus resurrected and Bunge's abortive blueprint returned to the forefront of public life.

This however belated serious reaction to peasant discontent may mislead as to what the discontent was primarily about. While communal arrangements were viewed by "enlightened" Russians as shackles on the economy because they inhibited the labor market, mobility, and capital accumulation, most peasants held on to their communes ferociously. What they did not want to put up with was their deepening land hunger-related poverty vis-a-vis the affluence of landlords or rampant governmental interference with their self-contained communal "worlds." Worshipping the tsar did not prevent peasants from perceiving all the institutions of the state as alien. That Russian peasants could not comprehend and accept the regulatory role of laws and especially their official enforcers was universally recognized in pre-1917 Russia.[34] In the words of Lurye, Russian peasants in the second half of 19th and the beginning of the 20th century "tried their best to evade any encounters with representatives of state power, used to be terrified at the a prospect of going to court if only as a witness, did not trust institutions of state, had strong doubts about their legitimacy, and upon the arrival of any official in the village used to hide in their huts."[35] A commune thus tended to ignore any world outside itself "and lived in a 'state' that was itself purportedly developing while complying with its own rules and entering into a conflict with representatives of actual power. And, naturally, the state created by the upper crust of Russian society did not present itself to Russian peasants as legitimate."[36] So the stage for mutual alienation had been set long before the communist revolution, in which ex-peasants eventually toppled from power the alien, Europeanized upper crust.

It is clear from the above that Russia entered the 20th century with a political environment steeped in vestiges of peasant bondage and relying on archaic communal arrangements. New attempts at reform meant that the 1861 liberation of peasants was no longer deemed to have gone far enough and the reform activities of the 1880s to have been unsuccessful. New reform measures reflected the growing alertness of Russia's ruling circles, not the inclinations of peasant masses, which wanted only more land and to be left in peace. Throughout all of 1905 the government of Sergei Witte was preparing future agrarian reform, which since 1906 had become linked with the name of the new Prime Minister Piotr Stolypin.[37] In 1910-11 laws on peasant land property and land-tenure regulations were enacted, and regional land-tenure commissions formed.

The gist of the so-called Stolypin reform was to clear the way for Russia's capitalism; specifically, it attempted to transfer land from communal to personal/household property, to accord peasants the opportunity to colonize peripheral idle lands, and to facilitate an "amicable" sale of land by gentry to peasants through the mediation of the State Treasury. Rural over-population in the central provinces of Russia was thus expected to be alleviated while peasants' spontaneous segregation into "strong" and "weak" would be sped up. Liberated from communal shackles, the strong would become Western-style farmers and the weak would be forced to leave overpopulated areas in the countryside and replenish the pool of industrial workers. "The government has placed its wager, not on the needy and the drunken, but on the sturdy and the strong -- on the sturdy individual proprietor who is called upon to play a part in the reconstruction of our Tsardom on strong monarchical foundations"[38] -- these words of Prime Minister Stolypin addressed to the Third Duma not only highlighted the major policy goal but, in hindsight, also exposed a lingering dilemma of Russian agrarian (and not only agrarian) policy at large for the remaining part of the 20th century. Integral to Stolypin reform was the final cancellation of redemption payments by peasants in 1905 and easing the procedure of withdrawal from their communes. A Chief Administration for Land Settlement was created with local committees whose task was to help volunteers consolidate their landholdings. Fearing the mass ruin of poor peasants, the reformers, however, set an upper limit to land purchases (no more than six personal allotments in one *uyezd*, which restricted one peasant landholding to 100 *desiatinas*). And yet the emphasis was clearly on giving a chance for economic advancement to the strongest households, that is, according to some estimates, to 10-15% of the peasantry.[39] The rest, however, were being deprived of commune-based security in the future.

As any action directed from outside the peasant world, the Stolypin reform was perceived by many peasants as coercion aimed at the destruction of their communes, without which the majority of Russian peasants could not conceive of their lives. Conscious of such a hostile response, the government cancelled all reform provisions upon the commencement of First World War when scores of freshly recruited peasants received arms and ammunition.

Though once again inconsistent, the implementation of the Stolypin reform brought about some tangible results. In particular, 25% of all peasant households in European Russia had taken out title to their land by 1914, although only 8.5% were actually consolidated,[40] that is, either established residences apart from a communal village on a hamlet (*khutor*) with land parcels adjoining the household property, or simply acquired land (*otrub*) apart from the communal holding while staying in the original village. The former type of enclosure was especially popular in western and northwestern Russia, where the Stolypin reform thus contributed to making the settlement structure more scattered or dispersed.

Secondly, the reform-triggered resettlement changed the spatial pattern of migration. Before the reform major migration flows were directed *to* Moscow, Saint Petersburg, and Russia's south. The reform altered this pattern to some extent diverting migration to the east.

Thirdly, out-migration *from* the most depopulated areas (Fig. 2.6) got a boost. Annual rural out-migration was highest on Chernozem lands and offset up to 30-50% of the natural increase of population.[41]

Migration to the east, though, faced many problems. The resettlement loans were insignificant; many peasants could not take root in new areas and returned home. In addition, the annual growth of agricultural output slowed from 2.4% in 1905-1905 to 1.4% in 1909-1913.[42] Stolypin himself believed

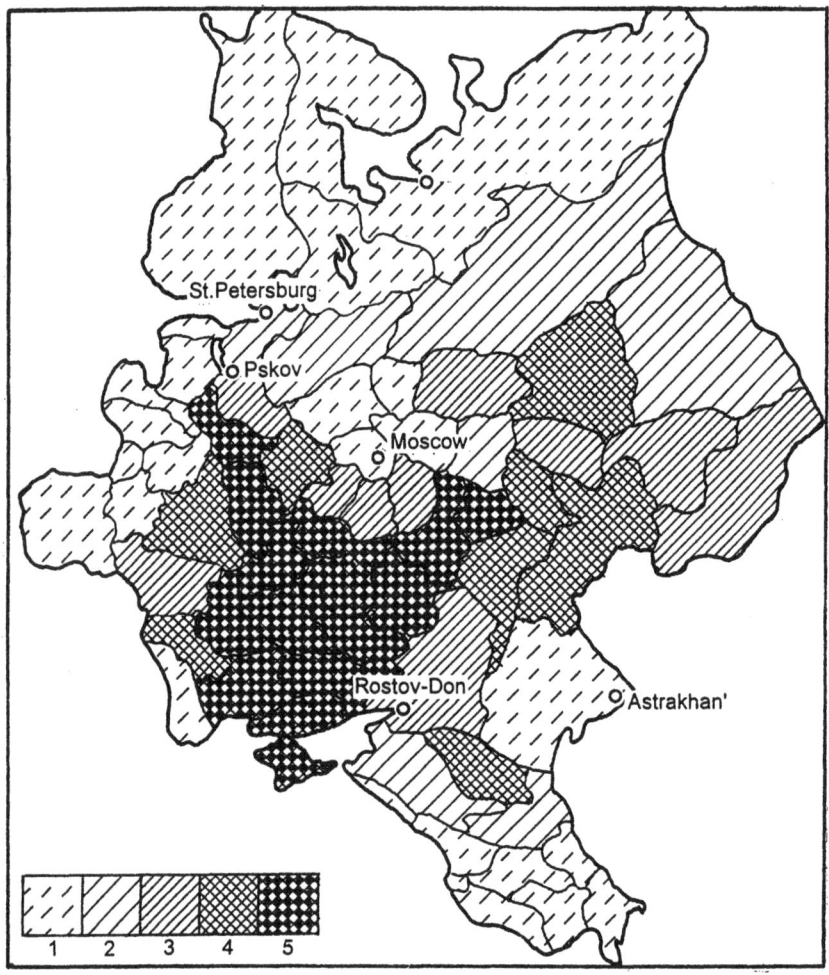

FIGURE 2.6 Annual Average Out-Migration to Newly-Colonized Lands per 1000 of Rural Population of Origin Regions. 1: Below 0.3; 2: 0.3-1.0; 3: 1.1-2.3; 4: 2.4-3.9; 5: over 4.0. *Source:* Rybnikov, A.A. *Osnovnye Voprosy Plana Selskogo Khoziaistva.* Moscow, Novaya Derevnia 1928; map #15.

that the reform would have been successful had it been given at least twenty years of peace at home and abroad.[43] This condition was not met. And the administration's coercive implementation of the reform stirred the country-side up and radicalized it.

Two Schools of Agrarian Thought

The outcomes and side effects of the Stolypin reform are still being debated.[44] Contrasting views about social stratification in the countryside and whether it proceeds at all have frequently served as reference lines for these debates. Early in this century opposite stances were taken by Russian Marxists including Lenin and by representatives of the so-called organization-production school which included Chayanov. This latter cohort of social scientists was at the time the world's leader in peasant studies.

Lenin elaborated on the issue of rural stratification in what is perhaps his most scholarly work, *Development of Capitalism in Russia,* first published in 1899, and in *The Agrarian Question in Russia at the End of the Nineteenth Century* (1908). Having analyzed the wealth of static (one-time) data concerning income, personal consumption, and cattle ownership disparity inside regional groups of Russian peasantry in the 1880s and 1890s, he wrote: "As to whether the disintegration of peasantry moves ahead and how fast, we do not have detailed statistical data. However, all the general data on our village economy testifies to on-going and fast disintegration."[45] Consequently, Lenin focused on the alleged erosion of the peasant "middle class" and the two emerging antagonist classes; rural bourgeoisie or well-to-do (*zazhitochnye*) peasants, and rural proletariat (*batraki*). He supported his argument by the following one-time data: in 1890s, 20% of the peasants held 35-40% of land, 40-45% of the area under crops and cattle, 56% of all horses, and 75-80% of new farm implements; on the other hand, 50% of the peasants held 20-30% of land, 15-30% of cropping area, and 5% of new implements."[46] We can boldly assert," wrote Lenin, "that out of the 7,500,000 desiatins of land which have been acquired by peasants as their private property in the period from 1877 to 1905, from two-thirds to three-fourths are in the hands of an insignificant minority of wealthy households."[47]

It was only natural for anybody following in Lenin's footsteps to embrace the idea that Stolypin reform would lead to an even deeper stratification of the peasantry, especially since the reform aimed at precisely that. Moreover, it would replicate developments in Russia's urban realm, for which Marxists used to have better skills of analysis. Indeed, such academic authorities as Kritsman[48] and Niemchinov[49] later upheld Lenin's view.

Chayanov, on the other hand, argued with these authorities. His detailed 1906-1911 observations showed that social polarization of the peasantry did not proceed as envisioned by Lenin. In actuality, he argued, peasants of average means represented the only expanding social group because out-migration and, to some extent, economic advancement affected the poor while heightened birth rate took its toll on the rich thus contributing to the promotion of the middle.[50] Comparing 1882-1891 data with the 1917 situation in selected districts, Chayanov concluded that the "picture that unfolds before us shows the disappearance of the most prosperous and the poorest strata in the village against the backdrop of its general impoverishment."[51] For Chayanov, real stratification would mean "chipping independent agricultural enterprises off the main body of working peasantry" rather than simple differentiation in terms of cropping area, implements, or even profit. "Some development of the capitalist mode of production," Chayanov goes on, "is surely in place. However, one ought to look for them not by grouping landholding sizes and the like but through determining whether households actually use hired labor for making profit and whether they lease land on crushing terms and practice usury."[52] According to Chayanov's data, the proportion of hired labor in different regions of rural Russia ranged between 3 and 10%. Combined with other factors this figure prompted his conclusion that the "social stratification of peasantry is yet in its embryonic stage."[53]

It appears that the social differentiation of peasantry that Marxists were fixated on did not unfold according to their vision. Communal arrangements worked against it forcefully, and no more than one quarter of Russian peasants opted for withdrawal from communes. Subsequently, the portion of land held by the gentry diminished steadily, thus adding to the overall levelling of the social playing field. In 1916, 89% of all the area under crops was in peasant landholding both communal and individual; peasants also held 94% of the country's cattle and 95% of its pigs. The only crop large landholders had a major share of (about half) was sugar beets.[54]

First Years Under Communist Rule

According to a view gaining ground in historical thinking, it was the social polarization in a few major urban centers vis-a-vis persistent tradition of communal egalitarianism that was mainly responsible for the October 1917 communist takeover.[55] A corollary point is that the so-called urban "proletarians" were yesterday's peasants raised in peasant communes. In fact the overwhelming majority of industrial workers on the eve of the communist takeover were either rural migrants themselves or second-generation urbanites. Culturally they were still communal peasants revolting against rampant social inequality in urban areas. The polarization in the countryside itself was hardly at issue, although the reluctance of the peasant masses to continue fighting on the fronts of the First World War was another factor contributing to the course of events.

Perhaps no less significant was the hostile sentiment of many peasants towards the Stolypin reform, as it definitely went against the grain of their communal instincts. It is not by accident that some authors labelled the first post-October land re-allotments as the "anti-Stolypin revolution."[56] These repartitions were attempted under the slogans "All are equal" and "To divide equally." Interestingly enough, neither Bolsheviks themselves nor other political organs temporarily in power during the ensuing Civil War of 1918-1920 participated in that conflict. Peasants themselves decided whom to rob and whom to spare.[57] This fact adds credence to some earlier hypotheses that the newly emerging Soviet power itself was essentially a reincarnation of the traditional communal arrangements of Russian peasantry, now extended over the entire Russian society under a smokescreen of Marxist code words and symbols.[58]

The first direct confrontation of the new regime with peasants took place during the years of the so-called War Communism, that is, in 1919-1921. The Red Army could confiscate all peasant belongings at will. Bolsheviks should also be credited in part for the Big Hunger of 1921, because on the eve of and during the drought in the Volga regions the new regime robbed peasants of their security grain stocks, including their seeds.

The first post-revolution years were marked by a new reinforcement of peasant communes, which now had control over 90% of the land.[59] The wealthiest peasants in large measure fell victim to the hostilities of the First World War and the Russian Civil War, and to the levelling confiscatory practices of War Communism. Overall, agriculture became even less commercial than before those wars, due to which agricultural exports reduced by one half.[60]

The New Economic Policy (NEP) of 1921-1928 was a major concession to the needs of the peasantry. It was an attempt to combine the socialist command economy with market forces. Confiscatory policies gave way to proportional taxation (per unit of land, per horse, per cow, etc.). The new Basic Law of Land Use adopted in 1922 included some provisions very reminiscent of the Stolypin reform, whereby peasant families opting for withdrawal from communes were once again given a chance to consolidate their strips of land in one holding either establishing a separate farmstead or retaining their residence in a nucleated village. Once again western and northwestern Russia led other regions in the formation of dispersed settlements, which were usually favored by wealthier peasant households desirous of getting away from communal vigilantes.

It was the peasants who benefited from the opportunity accorded by the 1922 land-tenure laws that facilitated a vigorous growth of agricultural output in the mid-1920s. Between 1924 and 1927, Russia's grain output increased 44% and its sugar beets output almost tripled.[61] Most peasant farms that contributed to that growth did not hire labor: even in the wealthiest category, the so-called *kulaks* (from the Russian word for fist), non-family labor accounted for no more than 20% of the total labor input.[62] It is thus important to point out that the economic advancement of selected farmers during a fairly short time, following wars and political shocks that had substantially levelled the playing field, should be credited to the farmers' own ingenuity, efforts, and skills. But it is no less important that as in the years of the Stolypin reform, the majority of peasants were still holding on to their communes.

Victor Danilov, the unsurpassed authority on Russian peasantry during the Soviet pre-collectivization period, using a 1927 sampled survey of peasant households, showed that only 3.7% of Russian peasants were "petty capitalists." The vast majority of peasants in all the regions (63.3% for Russia as a whole) were classed as "petty producers" (*melkii tovaroproizvoditel*), roughly equivalent to "peasants of average means." Needless to say, however, there were no crisp borderlines between the social groups. Between 1922 and 1926 the percentage share of the peasant "middle class" increased somewhat (from 60% to 63.6% in all of Russia) as did the share of the wealthy (from 3% to 6%). However, this levelling did not occur everywhere. Whereas in the Industrial Center and in the Northwest, wealthy peasants (having three or more head of cattle) accounted only 0.3-0.8% on the eve of collectivization, in the southern Volga they made up 5%, in the Northern Caucasus, 9-11%, and in Siberia, 25-27%.[63] In this way it is the regional cross-section of peasant stratification that shows to what extent communal shackles restricted the peasantry in the regions where colonization had occurred earliest.

In 1924 a book was published in Moscow with the peculiar title: *Village Under NEP: Who Is To Be Considered a Kulak and Who a Toiler? What Peasants Say On This.* It was put together using peasants' letters to a newspaper with the no less peculiar title *Bednota*. The title literally means "the poor" but its flavor cannot be understood outside the emerging political context. While in a Russian village it had never been considered a sin to be a loser, in the 1920s it was in fact deemed a virtue because it actually suggested loyalty to the regime, which, temporary concessions notwithstanding, continued to stake its political future on the poor. The general conclusion of the book is that while there was a considerable difference of opinion as to who is who in a Russian village, peasant correspondents were in unanimous agreement on one thing. All genuinely despised commerce and all forms of brokered transactions. For them both were equivalent to theft.[64]

Given these attitudes, the most progressive articles of the 1922 Land Use Law had little chance of consistent implementation, all the more because repartitional communes still reigned supreme in Russia. According to Danilov, "in 1927, just as in the early years of the NEP, repartitional communes were a dominant form of peasant land use. In the Russian Federation they occupied 95.5% of all land in agricultural use."[65] Small holdings and multi-strip-farming prevailed with the average size of one strip being about 0.2 ha in the non-Chernozem regions and between 0.3 and 0.9 ha in Chernozem regions.[66]

Cooperation and Collectivization

Although the first socialized farms emerged as early as 1918 as *kommuny,* *arteli,* and so-called *TOZy* (comradeships for collective working land), it was not until 1928 that collectivization began to be enforced fully.

Forms of cooperation in the production, procurement, and distribution of goods of many kinds had existed in Russia from the last quarter of the 19th century. According to Danilov, in 1915 up to one-third of Russia's population participated in some kind of cooperative arrangements.[67] In the 1920s, Russia seemed to be on the eve of the most consistent realization of a cooperative ideal. At that time Chayanov elaborated his theory of agricultural cooperation,[68] in which he distinguished between its vertical and horizontal varieties. He warned against the consequences of direct horizontal cooperation under which overall management is conducted on behalf of everybody involved. Such consequences included unjustified amalgamation of production units, uncontrollable bureaucracy, the severing of habitual organization links, etc. The program of vertical cooperation that Chayanov advocated would preserve many features of peasant farms and their specialization. Only production-related services, procurement, and sales were to be handed over to cooperatives, and peasants would decide themselves on the degree of their involvement in cooperative arrangements.

However, the concentration of power in Bolshevik hands, the creation of a powerful mechanism of coercion, and constant tensions in relationships with peasantry pushed in the other direction. The article "To lead the village, to implant collective and state farms" by Stalin[69] became the official reproof of Chayanov and other prominent figures in the organization-production school. Most of them lost their lives in the 1930s.

By 1927 the implementation of land-tenure regulations stipulated in the 1922 code of laws was suspended. This act signalled that the tug of war in the Bolshevik establishment was not going to be won by those who, like Bukharin, championed successful peasant farmers while labelling the policy of his opponents as "military-feudal exploitation of peasantry."[70] Politically doomed and purged were scholars like Kondratyev who openly stated that by "favoring the poor while restraining the growth of economically powerful strata in the village we do not only undermine the opportunities of industrialization but also deprive ourselves of the possibility of truly aiding the poor that we favor."[71]

The immediate spur to collectivization was the 1928 "bread" crisis, when grain sales were short of plan projections despite an expansion of the cropping area. This situation was partially due to the growth in the number of cattle: peasants were using more grain as feed. Retail prices on bread increased and with low farm-gate prices set in advance, peasants preferred to pay their taxes not in kind but in money. The government had a choice:

to increase purchasing prices; to acquire a certain amount of grain abroad and thus beat down domestic prices; or to forcibly expropriate grain. This third option became both the government's immediate choice and long-term policy. In 1928, reneging on the pre-set quotas of grain delivery by peasants became punishable by law. The penalty was confiscation of property with distribution of one-quarter of the confiscated property to the local needy.[72]

Officially, the tasks of the collectivisation campaign undertaken in 1929-1936 were to achieve technological advancement in agriculture quickly, reportedly only possible on large farms, and to eradicate sprouts of capitalism. However, an even more important goal for the Bolsheviks was to ensure speedy accumulation of means for their crash industrialization program. While peasants could not be relied on politically, only they could be counted on as the source of finances for the program. It was, therefore, important to subjugate peasants and to deprive them of political clout and of their freedom to dispose of their produce in their own ways. Doing all this would make it possible to regulate the commercial exchange of food for non-food items so that huge savings could be accumulated in government hands.

"Beginning in January 1933," writes Akhiazer, "collectivization allowed farm-gate grain and other food prices to be set at a level 10-12 times lower than the market would require. It became the most important mechanism for the financial resources being forcibly pumped out by the state. There is the view that to support industrialization it was sufficient to secure a 2% annual growth of grain output between 1928-1940. One may point out for comparison that during the entire period between 1801 and 1914, grain yields increased on average 0.5% annually. The question thus remains how could one achieve a 2% increase absent a market, that is, when a peasant is not interested in the growth of commercial farming and at the same time hates ferociously private property and private capital in the village."[73] The answer to this question was pursued along the lines of combining modern technology with the deeply ingrained communal custom.

It therefore no longer seems paradoxical that newly established collective farms, whose overt mission was to eradicate peasant communes and thus modernize agriculture, in fact drew on the peasants' patriarchal ethos of primitive egalitarianism. There is a clear affinity-based succession between a *kolkhoz* and the traditional Russian peasant commune, in that both alienated peasants from property and uprooted the sense of individual responsibility, replacing it with reliance upon the collective. It is unlikely that collectivisation would have ever succeeded had it not been for that sociocultural line of succession and continuity. A flip side to that syncretic egalitarianism was that poor peasants were eager to get hold of their wealthier neighbors' property. Those independent farmers who had withdrawn from their communes and been successful were subject to ferocious hatred. And the principal hate-mongers were those the Bolsheviks staked their future on.

The implantation of collective farms was, nevertheless, painful. Russian experience with various forms of collectivism did not include experience with the sweeping methods by which the new Soviet collectivism was being implemented. Communal "worlds" had been encroached upon on many occasions, but not with such indiscriminate and undisguised coercion which now repelled many people. Those unwilling to "voluntarily" sign up for kolkhoz were subject to dispossession: for household heads that meant being exiled or shot and having their property divided. No one has described the ins and outs of this campaign better than Robert Conquest,[74] and perhaps nobody has described a peasant's orgy of malicious rejoicing at the division of a *kulak*'s property (honey, fabrics, valenki, coats, and skirts from a dowry chest) better than the Soviet conformist Mikhail Sholokhov in his *Podniataya Tselina*,[75] mandatory reading for three generations of Soviets.

The campaign was essentially plunder and genocide of the most energetic, skillful, and dedicated part of the Russian peasantry. But with all its excesses there was something incredibly familiar about it all. "What is so striking about the terror that occurred," writes Akhiazer, "is not so much its scale as the overall complicity and at the same time passive attitude toward it. Stalin did not create people prone to conduct terror; millions of people themselves singled him out...and idolized him because he gave external sanction to those millions to follow their own values. Everybody bore responsibility for the terror. Naturally, nobody knew everything, but all, without exception, knew something. And the fact that this knowledge, however minuscule, did not nudge them to learn more, confirms that ignorance was not the reason, but the consequence."[76]

Who were all those village party members taking over real power in a village? They could have been a mounted patrolman, a bookkeeper, a chief cop (militiaman), a bailiff, a storeman, a chairman of a rural Soviet, or a cleaner. Of course, they were of peasant origin, raised as peasants, but did not work the land themselves. Most frequently they were recruited from among the rural poor who were unable to become successful farmers. Those who aspired to independence from locally imposed control fell into disgrace. The pervasive culture of envy of successful and independent people (branded *kulaks*) instilled by the communal way of life was what Bolsheviks put to use more than anything else. That *kulaks* were officially labelled ill-intentioned, perfidious people only helped to unleash pent-up passions.

The passions themselves, however, had long been there only waiting for a signal. In 1879 Engelgard commented on a rural witch hunt of an incomparably smaller size that had been instigated by the authorities: "A hunt for ill-intentioned people was to everybody's liking, so the authorities were not so much required to spur it as to restrain. Mouzhiks thought that the ill-intentioned are those who rebel against the tsar, for the tsar wants to bestow land on mouzhiks; landlords thought that ill-intentioned people want to take land away from landlords; priests, that they are those who insist on reducing the number of parishes, and on an accurate accounting of how candle money is spent...; railway officials, that after train collisions the ill-intentioned stir up people against them, take a close look at rotten ties... and, finally, solicit the elimination of the red service caps that had been conferred upon station chiefs. In summary, everybody went out of his way to lend a helping hand to authorities in their pursuit of the ill-intentioned."[77]

This time, however, this Manichean predisposition to unmask the were-wolves of evil combined with what Yuri Chernichenko defined as the "pent-up energy of looting" and a "union of the upper crust with lower classes against the uncontrollable middle,"[78] went farther than in the past. Under the collective farm umbrella not only were communal arrangements being restored but, in a way, pre-1861 bondage was as well. Because some peasants were fleeing collective farms, administrative restrictions were set up to stop them. Collective farmers, for example, were not allowed to bear internal passports, a condition which in Russia has always carried severe limitations to mobility.

The immediate economic consequences of sweeping coercion were so disastrous that only persistent faith in the patience, resilience, and vitality of peasantry, a faith which in turn led to unlimited license for expropriation, could force the authorities to stick to their policy. In reaction to those coercive policies, 25 million heads of cattle including 10 million cows, 17.7 million horses, over 10 million pigs, and 71 million sheep and goats were prematurely slaughtered by peasants in 1929-1933 to avoid surrendering them to collective farms.[79] The pre-collectivization level of livestock was restored only in the late 1950s. Yields of most crops plummeted; collective farms retrieved the 1913 level only by 1940, though the per capita production of grains was almost only half of what it had been earlier due to the population growth (see Table 2.1). Grain exports in the 1930s were one-fifth of the 1913 level.

Having lost in production, Bolsheviks managed to set up stiff controls over distribution. However, these controls did not keep the country from food crises. The Big Hunger of 1933[80] was the apotheosis; it was not caused by crop failures as were famines in the past but by state requisitions of grain that left peasants without seeds.

Villages on the Eve of the Second Half of the 20th Century

The preceding discussion describes the legacy that Russian agriculture carried toward the end of the first half of the 20th century. However, it had not yet endured the vicissitudes of the Great Patriotic War (the Russian name for the Second World War, reflecting the extent of Soviet participation in it). Although it was the Ukraine and Byelorussia that experienced the most damage, a substantial chunk of the Russian countryside was also under occupation and was the scene of devastating fighting such as that near Moscow in November-December 1941, the battles of Stalingrad in the winter of 1942 and of Kursk in summer of 1943, and the 1944 breakdown of the blockade of Leningrad.

According to the newest estimates, the immediate war-time losses of human lives sustained by the Soviets amounted to 25.3 million people.[81] Given that the 1939 rural population of Russia accounted for 38% of the whole Soviet population (within 1946-1991 borders), we may hypothesize that rural losses were about no less than 9 million people.[82] The war thus dealt a severe demographic blow to the countryside, already long affected by out-migration that proceeded despite tight migration control. A traditional way to sidestep this control in the immediate pre-war and especially in the post-war years was for young army recruits to remain in a city or town upon demobilization and succeed in getting a passport denied the rural population.

Yet the immediate losses of agricultural output of Russia due to the war were not as great as they might seem when assessed by the official statistics. In fact, they pale next to losses sustained during the collectivization. The pre-war production of milk, for example, had already been achieved in the early 1950s, as was the number of cattle. And the cropping area in Russia, after reaching its lowest level in 1942 (87,714,000 ha vs 92,065,000 ha in 1940), exceeded the pre-war level by 1943 because of new land under plow in the East.[83]

Table 2.4 shows that Russia began to recover from the combined cataclysms of the first half of the century only by the late 1950s. Despite a sizable expansion of the cropping area already by 1940, outputs of most products did not increase because of major dislocations of the rural population and diminishing yields per unit of land. In summary roughly forty years of would-be agricultural growth were squandered. There was no tangible progress in farming produce up until the 1950s. At the same time there were enormous structural changes. In 1940, collective and state farms contributed 30% of the country's meat and lard with the rest produced on tiny subsidiary plots belonging to collective and state farm members; by 1958 the proportion of the socialized sector increased to 58%. In crop harvesting socialization's lead was more sweeping. Whereas in 1928, 96% of the cropping area was in peasant households both communal and private, and in small-town households as well, in 1940 collective farms already held 86% and state farms 10% of the cropping area. By 1958, households could use only about 2% of the cropping area for their own needs.[84]

According to Nikonov, in 1950 peasants contributed 73% of their total working time to socialized farming, 10% to other state and cooperative insti-

TABLE 2.4 Selected Indicators of Russia's Agriculture (Within Current Borders of the Russian Federation).

Indicators	1913	1928	1940	1950	1960
Cropping area, million ha	69.8	No data	92.1	89.0	120.7
Total output of grain, million tons	50.5	50.0	55.6	46.8	76.2
Wheat yields, centners per ha	8.0	8.1	7.9	7.2	10.7
Total output of flax-fibre, thousand tons	314.0	298.0	239.0	172.0	240.0
Flax-fibre yields, centners/ha	3.2	2.3	1.6	1.2	2.4
Number of cattle (cows and bulls), million	33.0	No data	27.8	30.2	38.2

Sources: *Narodnoye Khoziaistvo RSFSR* 1958. Moscow, TSSU 1959; *Narodnoye Khoziaistvo RSFSR* 1970. Moscow, Statistika 1971.

tutions; only 17% of their working time was spent on personal auxiliary farming, that is, on tiny subsidiary plots allowed for personal use. However, only 20% of a household's income was earned on collective and state farms.[85] So while toiling for a *kolkhoz* peasants actually made their living on their own. It was in essence a throwback to corvee, only this time it was more cruel than prior to 1861.

Initially collective farms were small, about 500 hectares in 1940. Subsequently the number of collective farms was reduced from 167,000 in 1940 to 68,000 in 1950 and to 37,000 in 1958. The average landholding of a collective farm amounted to 4,000 hectares in 1958.[86] Power utilized in Russian agriculture increased fourfold in the late 1950s compared to 1913 because of collective and state farms' 0.5 million tractors and 300,000 grain harvesting combines. So Stalin's plans of ensuring technological renovation in agriculture were finally being fulfilled. However, even in the 1950s there was still no sizable growth of output.

Spatial shifts (Table 2.5) were also quite significant: in European Russia the relative roles of the Volga region and of the North Caucasus dwindled. A primary cause was the promotion of regional self-sufficiency and the growing expansion of arable land into the southern margins of West Siberia and the Urals.

This concludes our summary of the phenomena that we believe ought to stay firmly within one's field of vision when trying to comprehend the developments of today and to forecast the prospects for the future. While this chapter has dwelt on the historical background and on the formative

TABLE 2.5 Regional Percentage Shares in the Total Output of Wheat.

	1913	1940	1953
North & Northwest	0.7	0.8	1.0
Industrial Center	0.7	0.8	1.0
Chernozem Center	1.2	5.4	3.7
Volga	24.2	21.5	14.2
North Caucasus	34.7	23.9	23.6
Urals	14.7	14.4	19.5
West Siberia	16.6	16.4	21.9
East Siberia	1.3	6.4	6.7
Far East	1.5	1.8	1.0
Total	100.0	100.0	100.0

Source: *Narodnoye Khoziaistvo RSFSR* 1958. Moscow, Gosstatizdat 1959:192.

social experiences of the Russian peasantry on the eve of and during the major upheavals of this century, we now turn to the smoother, more evolutionary trend Russian agriculture experienced between the 1960s and the late 1980s, that is, before the new breach of continuity, market reforms of the 1990s, occurred. We will focus on the three aspects of this evolution: organization and management (Chapter 3); the geography of agricultural output and the production factors (Chapter 4).

Notes

1. Chernichenko, Yuri. "Eto Sluchilos v Rosii." *Izvestia* 17 February 1996, p. 5. The mind-boggling term "dekulakization" is actually abundantly used by Robert Conquest in his famous *Harvest of Sorrow*.

2. Thorner, D. "Peasant Economy as a Category in History." *Peasant and Peasant Societies. Selected Readings.* Edited by T. Shanin. London: Basil Blackwell 1987, pp. 62-66.

3. Lenin, V.I. "The Agrarian Question in Russia at the End of the Nineteenth Century." In V.I. Lenin. *Selected Works*, Vol. 1. Moscow: Cooperative Publishing Society of Foreign Workers in the USSR 1934, pp. 139-140.

4. Studensky, G.A. "Sostav i Rynochnost Selskokhozyaistvennogo Dokhoda v 50 Guberniyakh Rossii." *Ocherki Selskokhoziaystvennoi Ekonomiki.* Moscow: Novaya Derevnia 1925.

5. Kondratyev, N.K. "Export Selskokhozyaistvennykh Tovarov SSSR." In *Puti Selskogo Khozyaistva* 1927, No 10.

6. Mironov, B.N. *Istoria v Tsyfrakh.* Leningrad: Nauka 1991, p. 150.

7. *Plan Elektrifikatsii RSFSR. Doklad 8 Syezdu Sovetov.* Moscow: Gospolitizdat 1955, pp. 79-136.

8. Vishnevski, A.G. *Vosproizvodstvo Naselenia i Obshchestvo.* Moscow: Finansy i Statistika 1982, p. 183.

9. Maslov, Piotr. *Usloviya Razvitiia Selskogo Khoziaistva v Rossii.* Saint Petersburg: Izdaniye M.I.Vodovozovoi 1903, pp. 256, 257. The same source characterizes the consequences of land hunger in 1901 in the province of Voronezh in this a way: "Poverty *permanently* exists in such an acute form that the availability of bugs and cockroaches in a hut serves as an indication of relative prosperity" (p. 256).

10. Maslov, Piotr. *Uslovia,* p. 226.

11. As quoted in Maslov, Piotr. *Uslovia,* p. 227.

12. Chayanov, A.V. *Krestianskoie Khoziaistvo.* Moscow: Ekonomika 1989, pp. 232-282.

13. Mironov, B.N. *Istoria,* p. 152.

14. Maslov, Piotr. *Usloviya,* pp. 219-220.

15. Kartsev, E. *Ekonomicheskiye Etiudy.* Saint Petersburg: Nabliudatel 1882, No. 3, p. 255.

16. Hettner, Alfred. *Das Europaische Russland.* Leipzig, B.G.Teubner 1905, pp. 176, 189.

17. Chelintsev, A.N. *Russkoye Selskoie Khoziaistvo Pered Revoliutsiyei.* Moscow: Novyi Agronom 1928.

18. Rybnikov, A.A. *Perenaseleniye i Borba s Nim.* In: *Osnovnye Voprosy Plana Selskogo Khozyaistva.* Moscow: Novaya Derevnia 1928, pp. 57-99.

19. *Ekonomicheskaya Geografiya.* Edited by N.Baransky. Moscow: Sverdlov Communist University, 1929, Vol. 2, p. 105; Tikhonov, V. *Pereseleniya v Rosii vo Vtoroi Polovinie XIX vieka.* Moscow: Nauka 1978:107.

20. Robinson, G.T. *Rural Russia Under the Old Regime.* New York: Macmillan 1957:10.

21. Danilova, L.V. and Danilov, V.P. "Commune." In *Great Soviet Encyclopedia.* A Translation of the Third Edition. Vol.18, New York: Macmillan 1978, p. 47.

22. Robinson, Geroid Tanquary. *Rural Russia.* p. 125. About region-specific patterns of repartitioning land see: Pallot, J. "The Commune in the 1870s." In Pallot, J. and Shaw, D. J. B. *Landscape and Settlement in Romanov Russia.* Oxford: Clarendon Press 1990, pp. 136-163.

23. Ibid., p. 125.

24. As quoted in Maslov, Piotr. *Usloviya,* p. 361.

25. See the comparison of Russian *Obshchina* (commune) with the German *Mark* and *Allmende* in Danilova and Danilov (note 21).

26. Starikov U. "Obshchina: Ot Russkoy 'Marki' k Uravnitelnym Peredelam." *Znaniye-Sila,* 1994, No 3, p. 17.

27. Kliuchevsky, V.O. *Russkaia Istoria. Polny Kurs Lektsii.* Moscow: Mysl, Kniga Vtoraya, pp. 128-129.

28. Starikov U. *Obshchina: Ot Russkoy "Marki" k Uravnitelnym Peredelam.* Znaniye-Sila, 1994,3:20.

29. Riasanovsky, Nickolas V. *A History of Russia.* New York: Oxford University Press 1993, p. 431.

30. Akhiazer, A.S. *Rossiya: Kritika Istoricheskogo Opyta.* Moscow, FO SSSR, 3 volumes.

31. Riazanov, V. "Reformy i Tsykly Modernizatsii Rossiiskoi Ekonomiki." *Rossiiski Ekonomicheski Zhurnal* 1992, No 10, pp. 69-78.

32. See for example: Wcislo, F.W. *Reforming Rural Russia: State, Local Society, and National Politics, 1855-1914.* Prinston University Press 1990.

33. From this perspective the fact that today's Russian communists have expurgated "Proletarians of all nations, unite!", a formerly mandatory Marxist motto, from the front pages of their newspapers and openly embraced the nationalist "Russian idea" does not seem to be an aberration. Rather their erstwhile identification with Marx was a misnomer. This identification, however, was almost unanimously endorsed by American sovietologists. While it helped them as a political clan, it was, in our view, a major hindrance in their search for truth.

34. Akhiazer, A. "Dialectika Transformatsii." *Rubiezhi* 1995, No 6, p. 99.

35. Lurye, Svetlana. *Metamorphozy Traditsionnogo Soznaniya.* Saint Petersburg: Tipografia Kotliakova 1994, p. 99.

36. Lurye, Svetlana. *Metamorfozy,* p. 99.

37. One historical analogy is irresistible -- the affinity between the two cycles of attempts at reform associated with the names of the policy-makers. These are the Bunge-Stolypin and the Kosygin-Gorbachev cycles. In both cases the idea of reform initially (under Bunge and similarly under Kosygin) failed to succeed in its implementation due to the anti-reform stand of the upper echelons of power. Likewise in both cases it was already too late when twenty years down the road (under Stolypin and Gorbachev respectively) the reform met more consistent support.

38. As quoted by Robinson, G.T. *Rural Russia,* p. 194.

39. Maksudov S. *Neuslyshannye Golosa. Dokumenty Smolenskogo Arkhiva. Kniga 1. Kulaki i Partiitsy.* Ann Arbor: Ardis 1987, p. 21.

40. Dubrovski, S.M. *Stolypinskaya Zemelnaya Reforma.* Moscow: Academy of Sciences of the USSR, 1963, pp. 588-591.

41. Rybnikov, A.A. "Perenaseleniye i Borba s Nim." In: *Osnovnye Voprosy Plana Selskogo Khozyaistva.* Moscow: Novaya Derevnia 1928, p. 91.

42. Danilov, V.P. "Agrarnaya Reforma i Agrarnye Revolutsii v Rossii." In *Vielikii Nieznakomets: Krestyanie i Fermery v Sovremennom Mire.* Moscow, Progress-Akademia 1992, p. 317.

43. Danilov, V.P. "Agrarnaya Reforma," p. 317.

44. For an account of the respective discourse in today's Russia see: David, A.J. Macey "Stolypin Is Risen! The Ideology of Agrarian Reform in Contemporary Russia." In *The 'Farmer Threat': The Political Economy of Agrarian Reform in Post-Soviet Russia.* Boulder: Westview Press 1993, pp. 97-120.

45. Lenin, V.I. *Sochinieniya,* Second Edition, Volume 3, Leningrad: Gosizdat 1926, p. 132.

46. Ibid., pp. 96-101.

47. Lenin, V.I. "The Agrarian Question," p. 174.

48. Kritsman, L.N. "K Voprosu o Klassovom Rassloyenii Sovremennoi Derevni." *Na Agrarnom Frontie* 1925, No 7, pp. 11-17.

49. Niemchinov, V.S. "O Statisticheskom Izuchenii Klassovogo Rassloyenia Derevni." In Niemchinov, V.S. *Izbrannye Proizvedeniya* Moscow: Nauka 1967, pp. 44-62.

50. Chayanov, A.V. *Krestyanskoye Khoziaystvo.* Moscow: Ekonomika 1993, pp. 144-160.

51. Ibid., p. 425.

52. Ibid., p. 426.

53. Ibid., p. 428.

54. *Ekonomicheskaya Geografia. Tom 2 "SSSR."* Moscow: Sverdlov Communist University 1929, p. 365.

55. Akhiazer, Alexander. *Rossia: Kritika Istoricheskogo Opyta.* Moscow, FO SSSR: 1991, Vol. 2, pp. 8-49.

56. Maksudov, S. Nieuslyshannye Golosa. p. 23.

57. Ibid., p. 23.

58. This idea was first elaborated by Nikolai Berdiaev; later it was adopted by Nickolas Vakar (see Vakar, N. *The Taproot of Soviet Society.* New York: Harper & Brothers 1962.); still later it became integral to the cyclic theory of Russian history developed by Alexandr Akhiazer.

59. Danilov, V.P. *"Agrarnaya Reforma i Agrarnye Revolutsii."* p. 319.

60. Kondratyev, N.K. "Eksport Selskokhoziaistvennykh Tovarov SSSR." In *Puti Selskogo Khoziaistva* 1927, No 10.

61. *Itogi Desyatiletiya Sovetskoi Vlasti v Tsyfrakh, 1917-1927.* Moscow: TSSU SSSR 1928, pp. 118-119.

62. *Nieuslyshannye Golosa,* p. 25.

63. Danilov, V.P. *Sovietskaya Dokolkhoznaya Derevnia: Sotsialnaya Struktura i Sotsialnye Otnosheniya.* Moscow: Nauka 1979, p. 34.

64. See about this in: *Iz Istorii Reformatorstva v Rossii.* Moscow: Rossiiski Otkrytyi Universitet 1991, p. 90.

65. Danilov, V.P. *Rural Russia Under the New Regime.* London: Hutchinson 1988, p. 173.

66. Danilov, V.P. *Sovetskaya Dokolkhoznaya Derevnia: Naselenie, Zemlepolzovanie, Khoziaistvo.* Moscow: Nauka 1977, p. 130.

67. Danilov, V.P. *Sovetskaya Dokolkhoznaya Derevnia.* p. 210.

68. Chayanov, A.V. *Osnovnye Idei i Formy Organizatsii Selskokhoziaistvennoi Kooperatsii.* Moscow: Nauka, 1991.

69. Stalin, I.V. *Voprosy Leninizma.* Moscow: Partizdat 1939, pp. 233, 277-281.

70. *KPSS v Rezolutsyiakh i Resheniiakh Syezdov, Konferentsii i Plenumov.* Moscow: Politizdat 1970, Vol.4, pp. 189-190.

71. As quoted in: Makovsky, D., Moryganov, A. *Sotsialisticheskaya Perestroika Khutorskoi Derevni.* Smolensk, Zapgiz 1936, p. 60.

72. *Nieuslyshannye Golosa,* p. 31.

73. Akhiazer, A.S. *Rossiya: Kritika Istoricheskogo.* Vol.2, pp. 148-149.

74. Conquest, Robert. *Harvest of Sorrow.* New York: Oxford University Press 1986.

75. Sholokhov, Mikhail. *Harvest on the Don.* New York: Alfred A. Knopf 1961.

76. Akhiazer, A.S. *Rossiya Kritika Istoricheskogo.* Vol.2, p. 174.

77. Engelgard, A.N. *Iz Derevni. 12 Pisem 1872-1887.* Moscow: Mysl, 1987, p. 431. This is missing from the American edition (see endnote 4 to the Introduction); the following note replaces the deleted material: "Engelgard here describes the kind of petty surveillance he was subjected to and the interrogations of the peasants in his

area about his behavior and the activities on his estate."

78. Chernichenko, Y. "Eto Sluchilos v Rossii." *Izvestia* 17 February 1996, p. 5.

79. Ibid., p. 5.

80. Conquest, Robert. *Harvest of Sorrow.*

81. *Naseleniye Sovetskogo Soyuza 1922-1991.* Moscow: Nauka 1993, p. 75.

82. Our estimate is indeed very rough, but it may not be far off target. According to the *Demographic Yearbook of the Russian Federation 1993* (p. 19), the 1939 rural population was 72081 thousands. According to *Naseleniye Sovetskogo Soiuza 1922-1991* (p. 48), the overall Soviet population was 188794 thousands in 1946-1991 borders. The recruitment of able-bodied men 16 years old and older was universal except for the areas along the western frontier occupied within the very first days of the war. These and other western regions, however, stayed under occupation longer than most Russian areas and sustained additional losses. On the other hand, at the time the rural population was in general younger than urban population and contributed more recruits; fewer people claimed deferment in the countryside and the proportion of rank-and-files among recruits was higher than in urban areas. Given all these factors acting in opposite directions, it seems plausible that the proportion of the rural population of Russia is a reliable predictor of the proportion of human losses sustained during the war.

83. The estimates are based on *Narodnoye khoziaistvo SSSR v 1941-1945 gg.* Moscow: Goskomstat 1990, pp. 109-120; *Selskoye Khoziaistvo SSSR.* Moscow: Statistika 1971.

84. *Narodnoye Khoziaistvo RSFSR v 1959 Godu.* Moscow: Gosstatizdat 1960, p. 225.

85. Nikonov, A. *Spiral Mnogovekovoi Dramy: Agramaya Nauka i Politika Rossii (18-20 Veka).* Moscow: Nauka 1995.

86. *Narodnoye Khoziasitvo RSFSR v 1958 Godu.* Moscow: Gosstatizdat 1959, p. 225.

3

The Evolution of Russian Agriculture, 1960-1990: Organization and Management Priorities

Possibly, state socialism is the nearest modern concept, in a general way, with which we can compare the feudal state.

--Norman Scott Brien Gras[1]

The Soviet Union, of which Russia was a bulwark and integral part in 1922-1991, was a supercentralized and immensely ideology-laden state. The turning points of its economic development have been traditionally related to changes in leadership and the ensuing adjustments to the political course.

In agriculture the Communist Party apparatus not only determined the general strategy but through its local organs kept in check production units and collective and state farms. It would be safe to say that the Ministry of Agriculture, as an element of executive power, enjoyed less real authority than various industrial ministries did over their respective domains. In industry, after all, many issues were resolved by professionals, that is, by intellectual technocrats. In agriculture professional control was relaxed while Party control was more immediate and disruptive and was a combination of micromanagement and arbitrary rule. There were at least three inter-related causes of this situation. First, in the 1930s "agricultural science had been bludgeoned by the command-administrative methods more painfully"[2] than any other scholarly field in the former USSR; so all potential dissent, even on presumably ideologically neutral issues, had evaporated. Second, a persistent tradition of treating the countryside and its people as inferior has to be taken into account. Thirdly, rural out-migration (see Chapter 5) used to steadily drain agriculture of its most enterprising people.

Yet it has to be pointed out that, as it was for the entire country, the period following Stalin's death in 1953 was for rural areas incomparably less tumultuous and disruptive than the post-1861 agrarian history of Russia and especially of the Soviet Union prior to 1953. "It is very important," writes Georgy Gachev, one of the premier authorities on Russian culture, "to distinguish between the two periods in each society's being: the first which is usually coercion-laden and bloody, when the structure is in the making, and the second one, when it evolves by inertia and bears fruit peacefully."[3] Gachev believes that the "tolerable and ramified Soviet civilization of the 1960s to the 1980s, on which one could thrive and create" came as recompense for all the victims and pangs of the Sturm und Drang period of 1917-1953 and was much more placid than what commenced in the 1990s. By and large we share this nostalgic view. Even though developments on the agrarian scene were in fact less than serene, the coercive and cruel practices of the past were no longer pursued.

Khrushchev, Corn, and Virgin Lands

Nikita Khrushchev's years in power (1956-1964) are widely regarded as the Political Thaw. But in peasant memory it was rather an ambivalent period: of emancipation and at the same time of a continuing assault on the remnants of the private sector.

Born and raised in a Chernozem village in the province of Kursk, Khrushchev always remained in essence an ill-educated, commonsensical man, yet he was a self-proclaimed expert on agriculture who did not think twice when making a decision about it. Both the grandeur and absurdity of his agricultural decisions and their implementation were of truly legendary proportions.

The crash industrialization that had commenced in 1928[4] was conducted against the backdrop of grave underachievements in agriculture. The official propaganda used to attribute them to the aftermath of the Second World War and to isolated instances of political mistakes. In reality, however, the countryside was still recovering from dislocations caused by collectivization and the plunder that accompanied it. Whereas the 1913-1960 growth of industry was forty-fold, the total output of agriculture only doubled.[5]

Spatial shifts in farming were tangible indeed. As mentioned in the previous chapter, an attempt to push agricultural colonization eastward had already been undertaken under Stolypin. However, during the 1950s the scale of this strategy's accomplishments eclipsed all previous developments along these lines. Acquisition of additional arable land in the East had already taken place during the war when population and enterprises were moved to the Urals, Kazakhstan, and West Siberia. The colonization of virgin lands picked up appreciably following the March 1953 Plenum of the Communist Party's Central Committee. The economic rationale for the ambitious virgin lands program was that a sizable portion of the USSR's

TABLE 3.1 State Procurement of Grain Across Economic Regions.

	1949-1953, thousand tons	1954-1958, thousand tons	1954-1958 as a % of 1949-1953
North	121	35	29
North-West	140	51	36
Industrial Center	2173	1495	69
Chernozem Center	2772	3118	112
Volga	3195	4041	126
Northern Caucasus	3944	5005	127
West Siberia	3094	7539	244
East Siberia	1131	1820	161
Far East	180	210	117
Russia Total	19587	26436	135

Source: *Narodnoe Khoziaistvo RSFSR v 1958 godu*. Moscow, Gosstatizdat 1959.

European lands lay within the so-called risk-ridden farming zone. For example, the principal breadbasket regions of the Volga and the Northern Caucasus are subject to periodic droughts. Long-term observation of climate showed that droughts in European Russia tend to coincide with sufficient precipitation in Siberia, and vice versa. So under the appropriate plowing techniques and the policy of "melioration," which included such improvements as the introduction of woodland belts into the vast open spaces of virgin lands, the primordial steppes of the East could become a necessary supplement to the primary granaries west of the Urals.

Initially, the conversion of extensive sheep pastures into cropland resulted in substantial grain surpluses. Between 1953 and 1958, state procurement of grain on newly-emerged arable lands almost doubled (from 17.9 to 31.7 thousand tons[6]). Table 3.1 reflects the growth in state purchases across economic regions: the highest increment is in West and East Siberia. In 1954 former virgin lands contributed almost half of Russia's output of grain. However, beginning in 1956 the role of those lands had already begun sliding down. The accelerated, crash methods of colonization and the neglect of the ecological consequences of subjecting gigantic chunks of semi-arid lands to plowing had diminished the overall effect of the ambitious endeavor. Hot dry wind, droughts, and lack of sufficient amounts of fertilizer took a heavy toll on natural fertility. The enthusiasm which marked the first years of the campaign had also subsided. Whereas in the 1960s, the Southern Urals and western and eastern Siberia contributed 36% of Russia's grain, by the end of the 1970 their proportion dwindled to 27%.[7]

In animal husbandry a tangible growth in cattle went along with accelerated socialization. In 1941-1966 the amount of livestock in collective and state farms almost tripled while in personal auxiliary farming (PAF) it decreased somewhat. While we will focus on PAF's role and specific nature at the end of this chapter, we note here that in 1956 the allowances specifying the maximum number of privately owned cattle were dramatically reduced. Not only cattle but also pigs, geese, and chicken in excess of newly set thresholds were to be surrendered to collective farms. Aside from this, the government cut back on personal orchards and vegetable gardens and increased taxes on their produce. At the same time the administrative attachment of peasants to collective farms was loosened; even without internal passports they could now leave the countryside without fear of legal penalties. The combination of the above policy measures is astonishing in its short-sightedness, particularly considering the Soviets' view of their economic interests. It appears that Khrushchev not only threw the doors open but also kicked peasants in their buttocks by cutting back on subsidiary farming, a primary source of peasant earnings. In the 1960s, rural out-migration was given a decisive spur and became landslide in the 1970s when rural dwellers were granted internal passports.

The impact of what were later considered Khrushchev's mishaps was such that, according to some Russian experts, there actually were two collectivisation campaigns in the Soviet Union, not just one: the first one in the 1930s dealt a mortal blow to commercial private farming, while the second one unfolded in the 1960s depriving peasants of their subsidiary economy and thus converting them completely into agricultural hourly-wage workers.[8]

The growth of the amount of livestock on socialized farms and of the proportion of cattle kept in stalls created shortages of animal feed. Cattle plagues and premature slaughter due to hunger became rampant. As a result the Party began looking for methods of improving fodder crops. Since increasing the yields of existing crops had already been proven hard, the decision was made to restructure the cropping area in favor of crops that would ensure rich, low-cost feed. After a long period of alienation from the West, Western advice and experience were solicited. The results of the Soviet Agricultural Minister Matskievich's trip, published as a book,[9] produced a stunning impression. However, American experience was being studied in a very one-sided fashion. "What was being sought," according to Akhiazer, "was just its technological and technical aspects. The organizational side, the whole system of values pivotal for agricultural developments in the US, were vastly ignored. The conclusion that Khrushchev himself drew from American experience boiled down to the following: "There are no subtleties here that would be genuinely American. They just fertilize a lot."[10] Following Khrushchev's 1961 visit to the US, an almost ubiquitous introduction of corn was embarked on in Russia. Although given the Soviet Union's climatic conditions, corn for grain could only grow ripe in the southern Ukraine and on the plains of the northern Caucasus region, the

Party demanded it be sown virtually everywhere. Such an arbitrary approach resulted in Khrushchev's being popularly remembered as Nikita-*kukuruznik* (Nikita-corn-cob). It is now believed that his policy hindered the introduction of animal feed that fits Russian conditions like clover and other perennial leguminous plants, and that it shifted attention away from traditional grain fodder.

"Intensification" of Agriculture

The late 1950s - early 1960s saw the slowing down of agricultural growth. The seven-year economic plan of 1958-1965 was not fulfilled. The downturn prompted the Party, which after the Brezhnev takeover had a more collective style of leadership, to change its agrarian policy. In particular, it was acknowledged that further eastward expansion of arable land would be inexpedient. The March 1965 Plenum of the CPSU's Central Committee announced the start of the so-called intensification campaign. After that time "intensification" became a byword of the Soviet political and economic vocabulary for as long as the Soviet Union continued to exist. The term referred to making agriculture more capital intensive *per unit* of land (rather than expanding it indefinitely), a strategy which was expected to result in higher yields. In addition to insufficient funding, in explaining the downturn, the Plenum also pointed to insufficient material incentives in collective and state farms, to mistakes in planning and pricing, and to micromanagement of farms units from distant urban centers. So a timid attempt was also made at economic reform that would loosen the reins of central planning. The reform, which became known in Soviet memory as *Kosyginskaya* (after Brezhnev's Prime Minister, A.N. Kosygin), was just as abortive as the Bunge reform of the 1880s which, in hindsight, it appears to be distantly reminiscent of. Of all the farming-related decisions of the 1960s, the expansion of invested capital (agricultural fixed assets) turned out to be the only one that lent itself to implementation under the auspices of central planning.

In the decisions of the May 1966, October 1968, and October 1984 Plenums, and in the 1982-adopted Food Program, the same directives concerning agricultural development were reiterated: melioration, "chemicalization," and industrialization of farming. The proportion of agriculture in the overall budget allocation was driven as high as to 29% in selected years.[11] As a result, in 1965-1990, the monetary value of fixed assets of Russian agriculture increased sixfold. That included a fivefold growth in the overall power of the country's fleet of tractors and a sixfold growth in the application of fertilizers[12]. About half of all agricultural investment was directed to the construction of gigantic cattle-breeding farms and to land reclamation projects involving artificial drainage and irrigation. A special

Ministry of Melioration was created, a successor to a subdivision of the infamous *GULAG* responsible for building canals under Stalin. The newly-created Ministry soon became a substantial drain on the budget as it devoured a gargantuan portion of agricultural investment, effectively burying it in ditches being dug and bringing about inestimable ecological damage that dwarfed the related growth of output. Nevertheless, the Ministry proved to be an unassailable bureaucratic monster that even contemporary Russia has been unable to rid itself of.

The unprecedented 1965-1990 growth in invested capital resulted in only about a 50% growth in output,[13] exceeding population growth (35% during the same period) but not to the extent it was expected to. By the end of the 1980s it was already clear that such a ratio of investment to returns could not be further sustained, and, therefore, systemic reform might be the only remedy. Also the quality of technological supplies did not keep pace with increases in quantity. In the late 1980s, for example, the USSR produced six times(!) more tractors than the USA (in physical units),[14] but while 350,000 tractors were supplied to Russia's agriculture in 1989 alone, 300,000 were written off during the same year. Most were taken apart to make up for spare part shortages or because they were unrepairable. This despite the fact that about one million workers were employed in farming-related technological maintenance, that is, about the same number as in agricultural machine-building, implement production, and packaging combined.[15]

From Exporters to Importers

In the 1960s the Soviet Union began to import grain. Although the initial stimulus was several consecutive crop failures, the very ability of the Soviets to sustain and expand imports of food rather than spur domestic production and pursue systemic reforms hinged upon commercial exploitation of gigantic reserves of oil and gas in western Siberia. The glut of petrodollars did not favor a reform-minded spirit. Aside from grain the Soviets dramatically expanded imports of raw sugar, fruit, groats, etc. In the 1980s food accounted for almost one-third of Soviet imports and almost one half of the country's imports of consumer goods. Occasionally annual imports of grain exceeded 40 million tons. It is difficult to single out Russia's proportion of the imports total. The 1985-1991 annual grain imports accounted for about one quarter of all the foreign food imported to the USSR. Another quarter was sugar imports, and about 20% was livestock produce.[16]

In 1990 food shipments out of Russia amounted to 792 million roubles, that is, one-twelfth of the shipments into Russia, which amounted to 9207 million roubles and included 3357 million roubles from the other union republics and 5850 from abroad.[17]

In the early 1990s, Russia harvested about 100 million tons of grain whereas the demand was about 120 million tons. The deficit was covered by imports, which by 1992 actually amounted to 27 million tons. Of Russia's total need, 31 million tons were required for human consumption, 21 million tons were for seeds, and slightly over 60 million tons for animal feed, this latter figure representing more than one-half of all the grain harvested in Russia. The grain deficit, therefore, was linked not so much to insufficient production as to excessive consumption, including an outdated composition of animal feed. In 1986-1990, for example, the per capita production and consumption of grain in Russia amounted to 788 and 827 kg respectively, whereas in the EEC these figures were 512 and 447 kg. In the US per capita consumption was 852 kg, or more or less on a par with Russia; however, the US production was 1253 kg.[18] (Note that the category of "consumption" in this case includes human consumption, animal feed, and non-food derivatives.) However, in the US, corn, which is a very important animal feed, accounts for 67% of the grain produced, whereas wheat accounts only for 19%. In Western Europe, aside from corn there is a high proportion of barley: together corn and barley make up over half the total output of grain. In Russia, however, wheat accounts for 44% of the output with a substantial part of it consumed by cattle, and 12% is rye. The principal grain feeds in Russia are barley and oats, which together make up 35% of the grain output.[19] It is the overabundance of concentrated grain feed (and the shortage of green and succulent feeds) that lead to excessive use of grain. Aside from that, the USSR had a low livestock production per unit of land, which used to increase production costs per unit of livestock output. It is therefore little wonder that upon switching to a market economy, it became more profitable for Russians to import meat rather than grain. By 1994 Russia's per capita import of grain dropped from 194 kg to 25 kg, whereas the per capita import of meat almost doubled, from 3.9 to 7.6 kg.[20]

Changes in Collective and State Farms

Socialized farm production units established under collectivization were not carved in stone. They were intermittently amalgamated and broken up into smaller units; many changed their status, mostly from collective to state farms. Such changes were invariably caused by their lingering inefficiency as perceived by the Party organs holding onto the reins of their management. Party officials seemed to believe that once the "correct" administrative rearrangements were introduced, all problems would be resolved. For example, with the passage of time the proportion of state farms, vis-a-vis the overall number of production units, grew. Initially the difference between *kolkhoz* and *sovkhoz* was that the latter was a state-owned entity, whereas

TABLE 3.2 Regional Land in Collective and in State Farms as Percentage of Total Agricultural Land.

	1960 Collective	1989 Collective	1960 State	1989 State
North and Northwest	62	27	17	60
Industrial Center	64	40	31	53
Volgo-Viatka	83	55	13	38
Chernozem Center	70	66	26	28
Volga	54	42	42	53
Northern Caucasus	58	50	49	44
Urals	55	42	39	50
West Siberia	46	24	43	56
East Siberia	54	24	29	63
Far East	29	11	33	66
Russia Total	56	38	36	53

Sources: Zemelnyi Fond RSFSR. Prilozheniye k Statisticheskomu Bulletenuiu TSSU RSFSR 19 (171). Moscow, TSSU RSFSR 1961:46; Zemelnyi Fond RSFSR na 1 Noyabria 1989, Moscow, Goskomstat RSFSR, 1990:98.

the former held all its fixed assets (buildings, implements, livestock, perennial plants), its output, and its profit in communal (formally defined as "cooperative") property. The idea was to keep collective farms at a somewhat lower level of state control over property so that their dependence on public funds to make ends meet would be held in check. Land, however, in all cases remained public property, and in the case of collective farms was administratively attached to them free of taxation and for an unlimited time. The difference between collective and state farms, however, was purely nominal when it came to their relationships with the state concerning produce. No wonder that before long more subtle differences between the two types of farms virtually evaporated. Nevertheless, the transformation of collective farms into state farms continued to preoccupy the authorities (Table 3.2). To some extent this change was linked to a new fad in the 1970s: concentration of farming produce.

As Table 3.2 shows, the collective farms were dominant in the south, while in the north and east they accounted for only about a quarter of the agricultural land. This regional disparity was related to a combination of two factors: the specialization of a production unit and the unit's performance. Units that tended to be transformed into state farms included three categories: highly specialized livestock farms that were mostly profitable and located in exurbia; collective farms on whose land expensive reclamation projects were implemented; and the most chronically unprofitable farms. For the latter, transformation into a state farm was a locally publicized act of rescue from economic ruin, which, however, seldom resulted in recovery. For all these reasons, state farms became the most heterogeneous farm category.

Official Soviet statistics about collective and state farms' profitability were largely unreliable -- not because they were intentionally falsified but because in most cases the scale of subsidization and written-off debts remained unaccounted for. For example, according to the *Narodnoye Khoziaistvo RSFSR* data-books the number of unprofitable collective farms had been reduced from 74% in 1980 to 3% in 1990; for state farms the respective percentages are 67% and 3%.[21] The progress thus documented seems fantastic, until one discovers that the 1982 Food Program introduced the practice of farm-gate price mark-ups, which altered the statistics of profitability. Along with writing off debts, these mark-ups became the most important mechanism for allocating budget support for agriculture; a second "mechanism" was preferential treatment by selected chairmen and directors regarding technical supplies. Actually the prices at which collective and state farms used to sell produce to the state had been regionally differentiated even before 1982; however, the differentiation was at the macro-regional not local level and it was intended to make up for variable natural fertility. The new policy provided that in cases where the flat regional price for a product (milk or wheat, for example) did not cover its production costs, the price could be locally adjusted to ensure a certain profit margin. The mark-ups over the established regional prices ranged from a few to several hundred percentage points. Needless to say, such a policy created disincentives to economic recovery: why bother, if one's mismanagement will be bailed out anyway? Such a policy, of course, was far different from agricultural subsidies in the West where contract prices are negotiated in advance with the central government, which subsequently covers the difference between contract and actual market prices when the latter prove lower than the former. The crucial dissimilarity between the Soviet and Western practices is that in the West, contract prices, however favorable, are not a guarantee against bankruptcy. No wonder that Russia entered the 1990s with two thirds of its agricultural enterprises unable to be profitable without mark-ups.[22] It is important to understand that the flagrant economic irrationality of Soviet agricultural supports did not confuse their advocates. Their main argument was that the economic polarization of collective and state farms was social injustice. "Americans vindicated their technocratic ways through suffering," wrote Anatoly Salutsky, one of the most passionate preachers of the cradle-to-grave collective farm welfare that was supposed to create social equity in agricultural planning. "However, our ways are those of social justice, morality, and spiritual richness."[23] The type of social justice this was

all about was in fact very reminiscent of the joint responsibility in patriarchal peasant communes, only this time the responsibility was to be borne by
broader social entities. It is, therefore, important to point out that the basic
precepts of the Soviet agrarian policy can be viewed as reincarnations of the
traditional communal sentiment. In the name of "spiritual richness" policy-
makers were now imposing obligations on a district, province, republic, and,
ultimately, the whole nation.

An important aspect of this policy was the so-called *shefstvo*, a Party-led
system of patronage or sponsorship which emerged around 1965. At that
time information about rural out-migration made it into the limelight for
the first time. And many rural district Party bosses realized that as conspicuous as it was, this exodus offered them a kind of sinecure insofar as
they could attribute much of their underachievements to it. In fact it was
one of the first instances of a still currently powerful agrarian lobby at work,
a lobby composed of collective farm chairmen and district and provincial
power brokers. According to rules established in the 1960s and persisting
throughout the remainder of the Soviet Union's lifespan, non-agricultural
institutions (colleges and industrial factories primarily) were all supposed to
extend sponsorship over collective and state farms, especially during
harvests. That urban institutions, as it were, "adopted" rural farms meant
that they were required to render technical support and send personnel to
assist in harvesting. In either case such assistance would not be reflected in
the books. This sponsorship thus further obscured agricultural efficiency
accounting practices (already obscured by price markups), so much so that
some agrarian economists argued that "*shefstvo*-assistance [was] an economic
reality requiring regulation."[24] Scores of urbanites participated in harvesting
and in weeding. In the 1980s in the province of Moscow, for example, about
200,000 people from Moscow itself annually in the 1980s spent an average
of ten days on a farm[25] while they continued to be paid for their regular jobs
(students continued to receive their stipends and were supposed to cut back
on their classes without changing their curriculum). In the province of
Leningrad, 2.7 million person-days were invested by urban personnel on
farms; their labor input accounted for 10% of the total labor in agriculture
for that province, including 35% of the total labor input in crop harvesting.
Such an inflow of manual labor -- and it could be generated any time and
directed to any farm through a kind of emergency response planning
apparatus set up by provincial Party committees -- effectively discouraged
farms from using machines. In the province of Leningrad, which in the
1980s topped all the Non-Chernozem provinces in terms of agricultural
efficiency, only 25% of the vegetables and potatoes were harvested by
combines. In terms of acreage coverage, the proportion of mechanized labor
was even lower.[26]

However, all the levelling practices notwithstanding, the economic disparity of production units proceeded. The ensuing gap between the strong and the weak resulted in part from the personal qualities of their leaders. However, for most part, it was the result of more objective reasons considered in detail in the following chapters. Here we only note that even in similar natural environments the output per unit of land varied from 2-5 times. Consequently, by the end of the 1980s, 20% of all the collective and state farms produced 80% of the total output.[27]

One pervasive factor that reduces output has been huge losses of already grown produce. According to the World Bank, such losses can reach up to 40% of particular kinds of output.[28] These losses consist of immediate harvesting losses (for example, in 1993 about half of all flax was not harvested because of the shortage of labor and combines and the produce was subsequently destroyed by snow) and transportation, storage, and processing losses. According to some estimates, the overall Soviet losses of grain were equivalent to Soviet grain imports.[29]

Finally, the fact that collective and state farms were not just production units told upon their economic performance. Aside from agricultural production, collective and state farms built, owned, and maintained residences and schools and repaired local roads. By the end of the 1980s investment in the rural social infrastructure accounted for about 25% of the entire agricultural investment.

Personal Auxiliary Farming: The Backbone of Rural Survival

Subsidiary farming has always played a substantial role in Soviet agriculture. When collective farms were set up in the 1930s, their goal was to put the squeeze on peasantry to ensure reallocation of capital in favor of industry and to stabilize the supply of food to urban areas. The task of feeding rural laborers was left to laborers themselves. So for decades their family-based economy supported the bulk of the rural population and, through the system of so-called *kolkhoz*-markets, a part of the urbanite population as well.

Note that in the USSR land has never been privately owned. Peasant families were accorded tiny parcels of land that in most cases were attached to their residences. As these parcels did not legally belong to those who worked them, the term "personal" (*lichnoye*) farming rather than "private" is appropriate.

In 1940, 87% of the meat, 97% of the milk, and 97% of the eggs that Soviet collective farmers consumed were produced by personal auxiliary farming (PAF).[30] PAF had a very high proportion of the total produce sold to the state at that time: 54% of the total amount of potatoes, 55% of the meat, and 51% of the milk purchased by the state were PAF-produced. By 1958 these proportions decreased to 49%, 27%, and 16% respectively (in the USSR as a whole). By 1990, PAF was still contributing 26% of the total agricultural output (in Russia only, 24%) including two-thirds of the potatoes and one quarter of the meat and milk. These last statistics are all the more impressive considering that Soviet authorities had long been on the offensive against rural "proprietors." Size norms for subsidiary plots had been reduced in 1939 with the aim of fighting self-seekers and scroungers. In 1956, allowances for the number of cattle in PAF were reduced and new taxes on PAF output were imposed. A time series of cattle in PAF shows that in the post-war period the first major cattle reduction consequently took place between 1956 and 1958, that is, at a time when it could not yet be attributed to landslide rural depopulation. Correspondingly, the biggest reduction of arable land in PAF took place in 1958-60. At that time there were no single norms for cattle and land allowances; rather there was a general Party guideline to cut back on them. In response local bosses began to compete with each other at zealously following the Party line. But despite the inestimable losses it experienced, family economy proved to be resilient, but only up to the point when irretrievable demographic changes began to undermine it.

Surveys during the 1980s showed that 90% of the USSR's peasants conducted PAF because of insufficient food supplies in stores, while 20-60% (depending upon the age group) pointed to insufficient income.[31] In the 1970s, the net income earned from one cow in PAF was about 300-400 roubles a year, the equivalent of 3-5 months of collective farm wages. The proportion PAF contributed to rural family incomes, nevertheless, steadily declined due to a variety of reasons. In 1960 it was 43% and by 1985 only 24%. Nonetheless, PAF-based earnings contributed to levelling the difference between rural and urban incomes. In the 1980s, rural incomes were 83% of the latter.[32]

There are substantial regional disparities in PAF (Table 3.3 and Figure 3.1). They are caused by uneven regional supplies of arable land and by a spatially inequitable demographic situation. According to Table 3.3, in 1990, PAF involved only 1.8% of all the arable land in Russia. This percentage was highest in the North because of the low fertility there and widespread land abandonment by rural out-migration. In the Komi republic, for example, PAF's share of the arable land was 7.2%, in Archangelsk, Novgorod, and Pskov -- about 4%. Regional PAF proportions in arable land and in total produce do not correlate closely with each other (see Table 3.3). Research has shown that Siberia and the Northern Caucasus non-mountain provinces are the regions that stand out in terms of PAF efficiency. Both areas lead the rest in the average amount of cattle per PAF unit and in PAF's proportion of meat and milk output. In the 1990s these regions had

TABLE 3.3 Personal Auxiliary Farming in 1990.

	% in Arable Land	% in Number of Cattle	% in Total Output	% in Potato Output	% in Meat Output	% in Milk Output	% of PAF Units without Cattle	Number of Cattle per 100 Dwellers	PAF's Arable Land per one Dweller	Potato Yield Differential (PAF minus Socialized Sector) in centners/ha	% of Abandoned Residences
North	3.6	10	20	67	20	17	50	9	0.03	22	22.6
Northwest	4.1	10	21	56	16	18	54	13	0.07	14	23.5
Center	2.5	9	23	45	23	18	45	14	0.08	13	16.0
Volgo-Viatka	3.0	15	29	36	26	29	36	22	0.09	15	8.9
Chernozem Center	3.3	12	23	88	23	21	44	19	0.13	1	4.1
Volga	1.0	18	24	74	22	27	32	33	0.06	10	5.6
North Caucasus	2.1	19	22	66	29	24	40	18	0.04	-5	0.9
Urals	1.2	22	26	70	28	28	25	39	0.05	22	5.1
West Siberia	1.0	20	26	80	26	25	28	37	0.05	16	3.1
East Siberia	1.4	28	28	76	26	29	28	44	0.05	22	3.4
Far East	2.3	20	22	60	25	22	40	18	0.03	15	4.0
Russia Total	1.8	17	24	66	25	24	37	25	0.06	11	7.2

Source: Lichnye Podsobnye Khoziaistva Naseleniya RSFSR. Moscow, Goskomstat 1991:22, 85-120.

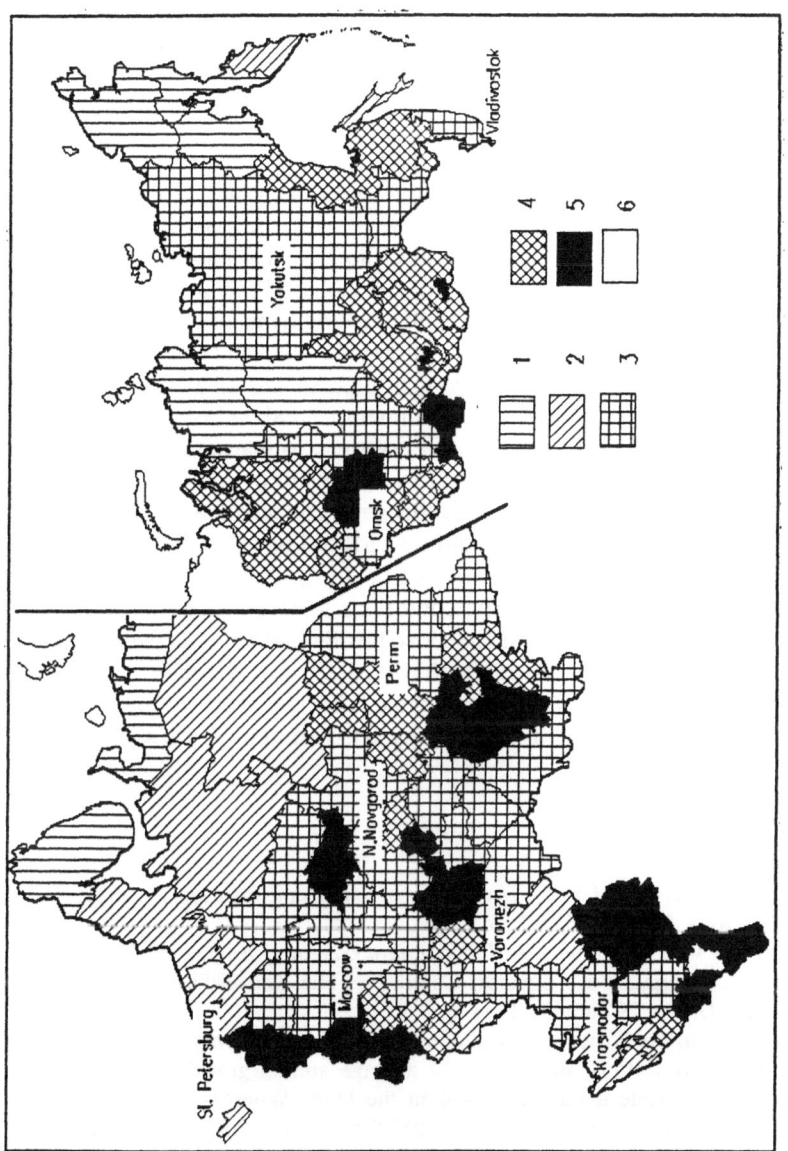

FIGURE 3.1 Percentage of Personal Auxiliary Farming (1990) in Total Agricultural Output. 1: 0-15%; 2: 15-20%; 3: 20-25%; 4: 25-30%; 5: 30-50%; 6: No data.

produced more private farmers than any other areas. The situation is more varied and complicated in the European non-Chernozem regions. In many of them increased out-migration and depopulation (see Chapter 5) undermined both socialized farming and PAF. In the Moscow and Leningrad provinces PAF took on a suburban character. However, family plots are small there and produce only 13-15% of the areas' total agricultural output versus 24% in Russia as a whole. This disparity is attributed to the predominantly recreational use of those parcels and with readily accessible urban food stores.

It is important to point out that collective and personal farming in Russia, though apart from each other in account books, have never been very far apart in real life. In fact one can say that with the passage of time a symbiosis between the two has evolved. Collective and state farms used to sell their laborers flour and animal feed, and assigned them tractors to plow their subsidiary plots. On the other hand, stealing from collective and state farms anything that one can physically carry away has long been too accepted to qualify as a misdemeanor, even though in the 1930s theft from a collective farm used to entail stiff punishment. Stealing became especially pervasive in outlying areas, in part as a response to inaccessible retail outlets, poor roads, and low incomes, that is, as a result of conditions as widespread and accepted as stealing itself. However, collective farms themselves resorted to PAF's help when poor milk yields did not allow them to fulfill the state order and mobilized retirees free of charge during hay-mowing and harvesting campaigns. In summary, conditions fostered a kind of informal social contract between PAF and collective farming, which effectively helped both eke out a living. By the 1980s they had effectively become mutually dependent to the point of being inseparable. This fact is exceedingly important to keep in mind when contemplating Russian agriculture's destiny.

Notes

1. Norman Scott Brien Gras. *A History of Agriculture in Europe and America*. New York: F.S. Crofts & Co 1925, p. 256.

2. Interview with Nikonov, A.A. *Literaturnaya Gazeta*, 5 August 1987, p. 10.

3. Gachev, Georgy. "Ya -- Sovetski Cheloviek i Nie Znayiu Drugogo Obraza." *Niezavisimaya Gazeta* 29 January 1994, p. 8.

4. According to the Soviet statistical data-book *Struktura Narodnogo Khoziaistva,* Moscow: Gostatizdat 1967, the 1918-1963 average annual growth rate of Soviet industry was 10% while it was only 3.4% in the USA. Whereas in 1913 Russian industry produced 12% of the US level, in 1965 the proportion had already climbed to 65%.

5. *Narodnoye Khoziaistvo RSFSR v 1987 godu.* Moscow: Goskomstat RSFSR 1988, p. 7.

6. *Narodnoye Khoziaistvo RSFSR v 1958 godu.* Moscow, Gostatizdat 1959, p. 210.

7. *Narodnoye Khoziaistvo RSFSR v 1987 godu.* Moscow: Statistika 1988, pp. 149-158.

8. Aleksandrov. Y.G. "Budushcheye Kolkhozno-Sovkhoznogo Stroya v Rossii." In *Krestianstvo i Industrialnaya Tsivilizatsiya.* Moscow: Nauka 1993, pp. 13-55; the actual unit of pay in collective farms was a *trudoden'* or a work-day, whose real value in money and in kind depended on the economic situation of a collective farm vis-a-vis that of the region and the country.

9. Matskevich, V.V. *Chto My Videli v SSHA i Kanadie.* Moscow: Politizdat 1956.

10. Akhiazer, A.S. *Rossiya: Kritika Istoricheskogo Opyta.* Moscow, FO SSSR, Vol. 2, pp. 209-210. "This viewpoint," comments Akhiazer, "could serve a classic example of how the issues vital for one culture cannot be comprehended by another which has not yet encountered the respective problems and layers of reality, as with organizational revolution, with market influences on economic growth, etc. All the wealth of [American] experience was, in Khrushchev's statement, reduced to its technological element, a clear testimony to the existence of a cultural ceiling peculiar to a given historical period of a given society."

11. Nikonov, A.A. *Spiral Mnogovekovoi Dramy: Agrarnaya Nauka i Politika Rossii (18-20 Veka).* Moscow: Nauka 1995.

12. *Narodnoye Khoziaistvo RSFSR v 1987 godu.* Moscow: Goskomstst 1988, pp. 149-158.

13. According to *Narkhoz RSFSR 1987,* pp. 166, 218-240, and *Narkhoz RSFSR 1990,* pp. 424-468, the 1986-90/1956-60 percentage ratio of farming produce (in physical weight) was as follows: grain total: 148%; vegetables: 147%; meat: 215%; milk: 157%; and potatoes: 78%.

14. *SSSR v Tsyfrakh v 1988 godu.* Moscow: Finansy i Statistika 1988, p. 290.

15. Alexandrov, Y.G. "Budushcheye," p. 36.

16. *Vnieshniya Torgovlia SSSR v 1990 gody.* Moscow: Finansy i Statistika 1991, p. 32.

17. *Narodnoye Khoziaistvo RSFSR v 1990 godu.* Moscow: Goskomstat 1991, p. 46.

18. *Finansovye Izvestia* 21-27 April 1994; American data were verified through *Agricultural Statistics 1993.* Washington, D.C.: US Government Printing Office 1994, pp. 1, 46 -- and appear to be correct.

19. *Finansovye Izvestia* 21-27 April 1994.

20. *Finansovye Izvestia* 20 July 1995.

21. *Narodnoye Khoziaistvo RSFSR v 1990 godu.* Moscow: Goskomstst 1991, pp. 403, 407.

22. Alexandrov, Y.G. "Budushcheye," p. 36.

23. Salutsky, Anatoly. "Slabye i Silnye." *Nash Sovremennik* 1987, No 9, p. 119. See also by the same author the book *Proroki i Poroki.* Moscow: Molodaya Gvardiya 1990. Mr. Salutsky is currently one of the chief columnists for the communist daily *Pravda.*

24. Zhikharevich, B.S.*Agro-Promyshlennyi Komplex Krupnogo Goroda.* Leningrad: Nauka 1989, p. 186.

25. *Moskovski Stolichnyi Region: Territorialnaya Struktura I Prirodnaya Sreda.* Moscow: Institute of Geography 1988, p. 118.

26. Zhikharevich, B.S. *Agro-Promyshlennyi,* p. 187.

27. Alexandrov, Y.G. "Budushcheye," p. 40.

28. *Food and Agricultural Policy Reforms in the Former USSR. An Agenda for Transition. Country Department 3. Europe and Central Asia Region.* Washington, D.C.: The World Bank 1992, p. 206.

29. *Argumenty i Facty* No. 30, 1990.

30. *Lichnoye Podsobnoye Khoziaistvo.* Moscow: Nauka, 1988.

31. Sidorova, M.I. *Stimulirovaniye Truda v Selskoi Sere APK.* Moscow: Nauka 1986, p. 78.

32. Ibid, p. 78.

4

Agricultural Output and Production Factors Prior to the 1990s

Historical heritage may influence the pattern and productivity of agriculture to a greater extent than the entire natural setting.

--Alexander Chelintsev[1]

In this chapter Russia is considered in the context of the former USSR, that is, in the context of the indivisible economic space in which the Russian economy formerly functioned and in large measure still does, an economic space that has proved more viable than the USSR itself.

Contrasts within the natural environments of Russia and the USSR are enormous. For example, the mean January temperature ranges between plus 4 degrees Celsius in the Transcaucasus along the Black and Caspian sea coasts to minus 50 degrees in the Yakut (Sakha) republic. The mean July temperature varies from plus 32 degrees Celsius in Central Asia and from 24-28 degrees along Russia's Caspian coast, to 0-8 degrees in northern Siberia and in the elevated areas. The growing season commences in February-March in the south and in late June in the north. Natural vegetation ranges from tundra to desert (Fig. 4.1).

Spatial contrasts within what Russians mean by the capacious, untranslatable term *osvoyennost'* (its meaning embraces "colonization," "settlement," "development of land," and "habitability of space") are even greater than those in the natural environment. We discussed this subject in Chapter 1 in general terms. One can talk of continuous settlement only in the triangle with corners in Saint Petersburg, Odessa, and Novosibirsk. Outside this triangle half the land does not have any permanent settlement at all; within the other half, human colonization is patchy. Needless to say, the picture of the rural population's pressure on land is a component of the more general picture of *osvoyennost'* (Fig. 4.2).

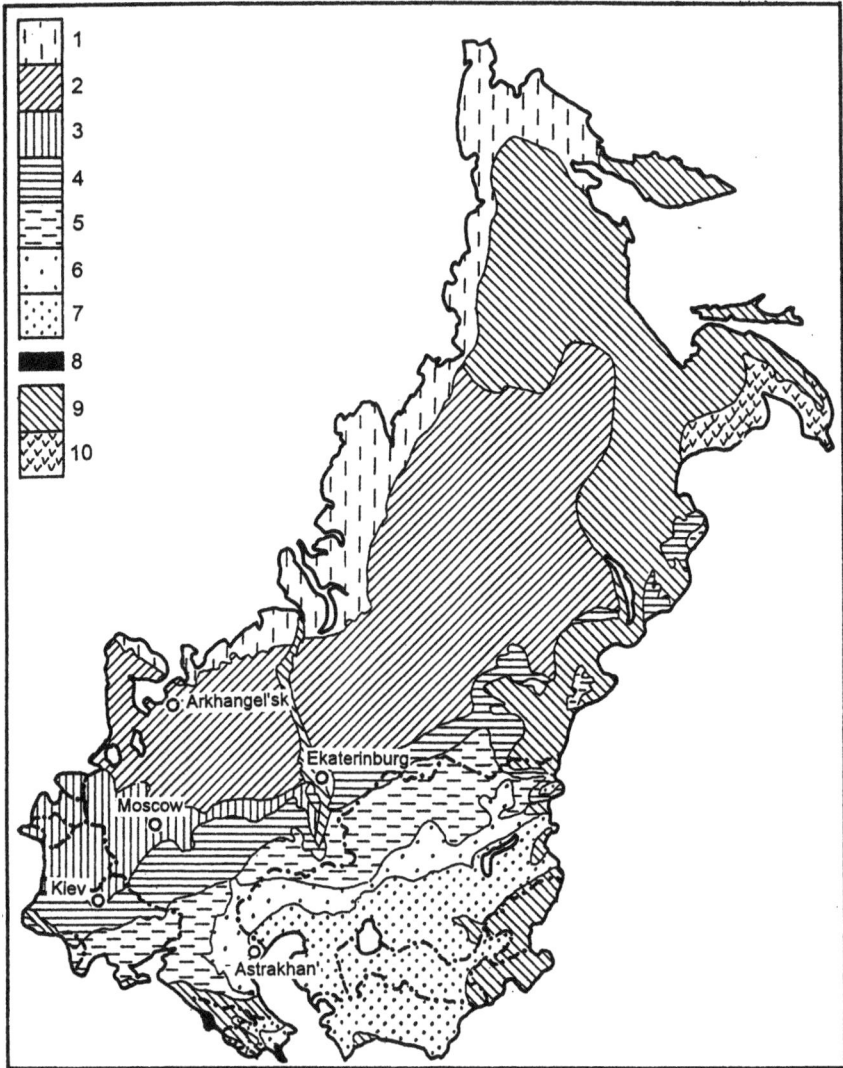

FIGURE 4.1 Natural Vegetation Regions of the Former USSR. 1: Tundra; 2: Taiga; 3: Mixed forests; 4: Forest steppe; 5: Steppe; 6: Arid steppe; 7: Semi-desert; 8: Humid subtropical forest; 9: Mountainous forests; 10: Mixed monsoon forest.

The centuries-old expansion of agricultural land, and of cropping area in particular, virtually came to a halt by 1960 (Table 4.1), although the crop-land continued to increase somewhat in a few regions as a result of emerging commercial exploitation of mineral riches (in East Siberia) and of land reclamation projects (in the Northern Caucasus and in the Volga region).

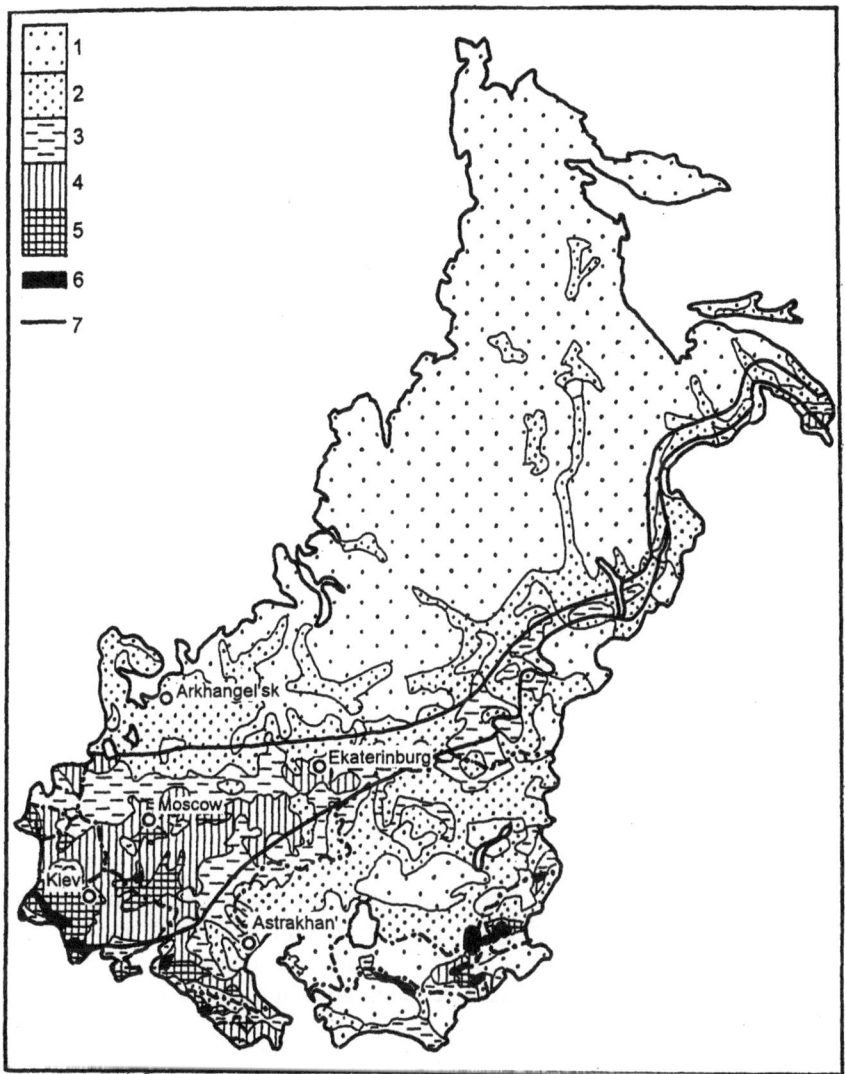

FIGURE 4.2 Rural Population Density in the USSR per Sq Kilometer. Adapted from Alexeev, A.I. *Mnogolikaya Derevnia*. Moscow, Mysl 1990:82-83. 1: Below 1; 2: 1-10; 3: 10-25; 4: 25-50; 5: 50-100; 6: Over 100; 7: Main settlement belt.

At the same time, however, substantial losses of agricultural land were incurred through rural depopulation in the north-west and in central Russia, where pastures and hayfields suffered the greatest losses. By the end of the 1980s, the Non-Chernozem zone (i.e., the Industrial Center, the Northwest, the North, the Volgo-Viatka, plus the Perm and Sverdlovsk provinces, and

TABLE 4.1 Regional Agricultural Land in 1989 as a Percentage of That in 1960 and Regional Percentage Share of Russia's Arable Land.

	Agricultural Land Total	Arable Land	Hayfields and Pastures	Regional % Share in Arable Land
North & North-West	63	87	54	3
Center	87	93	83	11
Volgo-Viatka	90	94	88	6
Chernozem Center	97	95	100	8
Volga	102	102	102	19
Northern Caucasus	93	101	83	12
Urals	97	100	89	17
West Siberia	101	97	106	15
East Siberia	101	111	101	7
Far East	85	134	74	2
Russia Total	95	99	92	100

Sources: Zemelnyi Fond RSFSR. Prilozheniye k Statisticheskomu Bulletieniu TSSU RSFSR 19(171). Moscow, TSSU RSFSR 1961:46; Zemelnyi Fond RSFSR na 1 Noyabria 1989, Moscow, Goskomstat RSFSR 1990:98.

the Udmurt republic of the Ural economic region) accounted for 19% of Russia's arable land, while the southern, mostly black-soil regions of European Russia made up 39%, and all the eastern regions, 42%. Overall the regional proportions of arable land are a close correlate of the spatial layout of vegetation (Fig. 4.1). While the percentage of arable land in Rus-sia's overall land area is 7.8%, it exceeds 60% in the Central-Chernozem region, and is over 50% in the Volga region and in the piedmont areas of the Northern Caucasus; in Central Russia and in some provinces of Siberia the proportion of arable land ranges between 20% and 40% and in the North it is as low as 1%.

Regional Specialization

In Russia as a whole there are 0.9 hectares of arable land per capita, which is similar to the level in the Industrial Center; in the Volga region this indicator is 2.2 ha, and in Siberia it ranges between 1 and 2 ha. One could, therefore, expect substantial regional variation in land use and production. By 1990 about 23% of Russia's total grain output was produced in the Northern Caucasus and 18% in the Volga region; the Central Chernozem, the Urals, and Western Siberia each accounted for 11-12%, while the

TABLE 4.2 Regional Percentage Shares in Areas Under Grain and Potatoes and in Grain and Potato Output in 1994.

	% of Russia's Grain-sown Area	% of Russia's Grain Output	% of Russia's Potato-sown Area	% of Russia's Potato Output
North	0.6	0.4	2.0	3.3
Northwest	0.7	0.5	3.3	4.0
Center	9.9	9.8	25.6	29.5
Volgo-Viatka	5.8	4.4	11.0	9.9
Chernozem Center	8.4	12.7	9.1	7.4
Volga	19.7	18.4	10.7	8.5
Northern Caucasus	12.0	23.0	5.6	5.4
Urals	18.3	11.0	13.0	12.8
West Siberia	15.8	11.9	10.0	9.5
East Siberia	6.6	6.0	5.3	5.4
Far East	1.5	1.4	3.9	3.7

Source: Narodnoye Khoziaistvo RSFSR v 1990 godu. Moscow, Goskomstat 1991: 420-435.

proportion from the Industrial Center was about 10%. In all but the Northern Caucasus, each region's proportion of arable land under grain exceeded its share of total grain output. In contrast, while possessing merely 12% of Russia's grain-sown area, the Northern Caucasus made up 23% of all of Russia's output, a clear indicator of low grain farming efficiency and commercial significance in most of the other regions (Table 4.2).

A largely latitudinal band of Russian provinces, from Leningrad and Smolensk in the west to the Urals and Western Siberia, stands out in terms of potato produce. However, in most of these provinces potatoes are produced mostly for internal consumption. Only in the Industrial Center does the proportion in potato output exceed the area's proportion of total cropping area because of consumer demand generated by Moscow.

Russia's south and exurban areas lead in terms of livestock produce.

But as Table 4.3 implies, all economic regions produce for their own needs as a result of a long tradition of diversifying regional farming output in order to avoid inter-regional exchange, the managing of which was never a successful element of Soviet central planning of agriculture.

It can be concluded from the above that Soviet agriculture was not nearly as variable as were the natural environment and the habitability of space. Of course, the northernmost areas, where crop harvesting is practically impossible, and, at the other extreme, the Northern Caucasus, let alone the oases of Central Asia, could not help but retain their crop distinctiveness.

TABLE 4.3 Comparison of Regional Percentage Shares in Total Population and in Gross Output of Agriculture in 1993.

	% of Gross Output	% of Total Population	Gross Output per One Rural Resident in Thousands of Roubles
North	3	4	403
Northwest	3	5	677
Center	14	20	616
Volgo-Viatka	5	6	471
Central Chernozem	8	5	628
Volga	13	11	658
N. Caucasus	16	12	415
Urals	14	14	586
West Siberia	12	10	590
East Siberia	6	6	555
Far East	6	7	698
Russia Total	100	100	561

Source: Economicheskoye Polozheniye Regionov Rossiiskoi Federatsii. Moscow, Goskomstat 1994:148; *Demographic Yearbook of the Russian Federation. Moscow, Goskomstat* 1994:19.

Still, in most other parts of the former USSR regional specialization on specific farming produce was not pronounced.

Figure 4.3 represents the major types of economy in the Russian countryside against the backdrop of the former USSR. This map is based on the *Atlas SSSR*[2] and books by Runova, Volkova, and Nefedova;[3] and by Alexeyev.[4] The characteristic features of the first type (which occurs in the northern territories) are patchy development, very low share (1%) of arable land, and a combination of forests (50% of the land) and natural meadows and pastures. Agriculture here specializes in reindeer breeding, fishery, and hunting fur-bearing animals. Settlements are very sparse and are mostly situated on the sea and lake coasts and river banks. Aside from this there is also a network of seasonally inhabited and movable settlements associated with deer pastures and trapping centers.

Type 2 covers the vast spaces of taiga which accounts for 50-60% of the land; arable land occupies 5-20%, and the rest is meadows and pastures. Habitable spaces tend to have a linear shape stretching along river valleys. The major activities here are forestry, game-shooting, and gathering. This is Russia's main felling area. The settlement network associated with forestry includes temporary settlements in active felling areas, small permanent settlements with services for logging teams, larger settlements in places

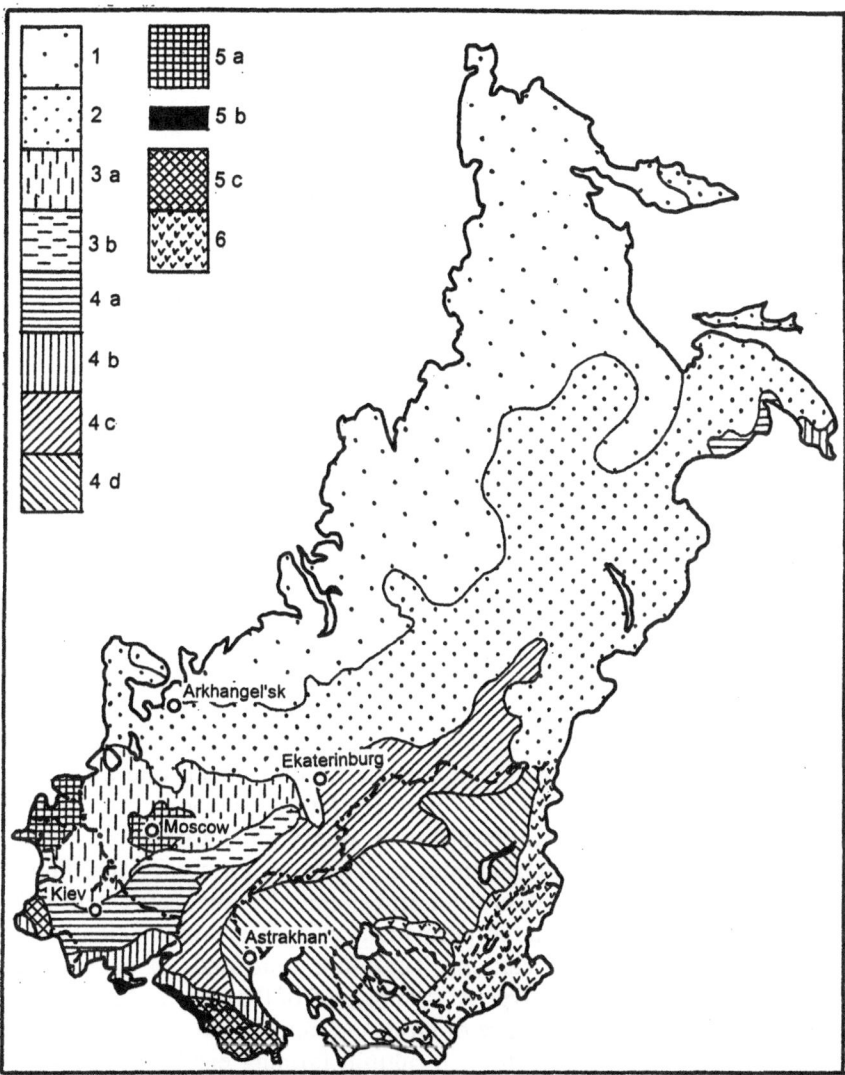

FIGURE 4.3 Types of Rural Economy (based on *Atlas SSSR*. Moscow, GUGK 1983). 1: Reindeer-breeding and hunting fur-bearing animals; 2: Forestry with small pockets of farming; 3: Agriculture and forestry, (a) both in equal proportions; agriculture: grains, potato, and flax; (b) plow-land exceeding forest area by 2-3 times; agriculture: grains, potatoes, and sugar beets; 4: Primarily agriculture: (a) general farming (grains, sugar beets, sunflower, and animal husbandry; plowland = 60%, forests = 10-20% of land area); (b) southern farming (grains, fruit, and vine; plow-land = 60%, forests = 10% of land area); 5: Agriculture, with forestry and recreation, (a) areas of moderate climate; (b) subtropical areas; and (c) southern mountainous areas; 6: Agriculture in arid and mountaneous areas, pockets of irrigated farming in river valleys and of animal husbandry in mountains.

where lumber shipments begin, and still larger settlements in lumber tran-shipment points with limited processing. Agricultural development in these areas is often very old, tracing back to the 15th and 16th centuries.

Type 3 is also associated with timber industry but mostly in the mixed-forest region. It is the core of Muscovy, settled by the Slavs in the 14th and 15th centuries. The area is a mosaic composed of plow-land, forests, and hayfields, each of which occupies about one-third of the land. Proceeding southward (sub-type 3b) the share of plow-land expands to 50% while the share of hayfields drops to 10%. The rural settlement network is spatially uneven: it is sparse in the north and inside large forest tracts but dense in plowed up spaces with more fertile soil (so-called *opolye*). Rural settlement is also more dense in the vicinities of large cities. The average size of a rural settlement is small (even prior to commencement of rural depopulation, not exceeding 130 people), which corresponds to historically small patches of plow-land (no more than 2-5 hectares in areas not subjected to land-recla-mation projects) carved out of forests. Both fields and settlements grow in size as one proceeds from the north-west to the south-east. Agriculture in this regional type emphasizes animal husbandry; but about one-half of the cropping area is under grains; potato farming predominates in exurbia; and some areas used to be prominent for flax.

Type 4 is mainly agricultural and is in forest-steppe and steppe regions. Arable land accounts for 60% of the total land area; forests, for 10-20% (in the south-east the share drops to 5%). Large parts of southern Russia and the Ukraine are located here. Fields under cultivation are large (up to 150-200 hectares). Large nucleated villages with populations ranging from 500 to 5000 people are widespread. In the Northern Caucasus one can come across rural settlements (so-called *stanitsa*) with populations in excess of 10,000. Settlements gravitate toward river valleys while watersheds are al-most entirely plowed up. Oftentimes such villages form continual linear agglomerations. In the south-eastern arid steppe the rural settlement net-work is more sparse, and large permanent villages are supplemented by seasonally settled pastoral stations. There are four kinds of agricultural specialization in type 4. Sub-type 4a is composed of the Central Chernozem provinces, the adjoining province of Orel, and most of the provinces of the Ukraine. It is characterized by poli-cultural farming emphasizing grains, sugar beets, and sunflower. To the south, in sub-type 4b, the share of for-ested land drops to 5-10%, and in the overall area under crops, orchards and vineyards replace root plants. To the east lies sub-type 4c, where wheat dominates the area, similar to the way it does in Kansas. This is the steppe part of the Volga region (*Povolzye*) and southern Siberia, the former virgin lands (*tselina*). Forested areas are minuscule while fields under cultivation are boundless. Finally, the near-Caspian regions of Russia and the major part of Kazakhstan (sub-type 4d) are predominantly cattle- and sheep-

TABLE 4.4 Dynamics of Russia's Agricultural Productivity.

	1940	*1960*	*1970*	*1976-80*	*1981-85*	*1986-90*
Yields per hectare:						
Grains, centners	7.9	10.7	14.6	13.8	13.0	15.3
Sugar beets, tons	9.6	15.8	17.5	15.8	16.6	22.5
Sunflower, centners	5.8	8.3	11.1	9.9	10.0	12.7
Flax, centners	1.6	2.4	3.4	2.5	2.9	2.5
Potatoes, tons	8.9	9.1	12.3	10.6	10.4	10.8
Milk yield per cow in tons	1.2	2.0	2.3	2.1	2.3	2.8

Sources: Narkhoz RSFSR. 1962, 1988, 1991.

breeding areas with arable land accounting for 20-50% of the total land area, along with a high share of pasture-land.

Type 5 is a combination of agricultural and recreational land-use and a bit of forestry. It occurs in the area in the immediate vicinity of Moscow and in the Baltic states. The spatial structure of land-use and settlement is similar to Type 3, but population density and especially infrastructure-related amenities are higher.

Type 6 is not characteristic of most of Russia.

Grain Yields as an Indicator of Success

Table 4.4 reflects changes in agricultural output over a fifty-year period. The following discussion of the regional variance of these changes begins with grains. As mentioned above, in Russia grains are virtually ubiquitous, accounting for about one-half of most provinces' cropland. According to Rakitnikov, the most respected Russian agrarian geographer, in the Soviet Union "grain yields accurately reflect the achieved level of arable lands' power of producing, regardless of whether grains dominate the region or are of secondary importance."[5]

In the 1950s and early 1960s, that is, before the commencement of the "intensification" campaign (see Chapter 3), the geography of grain yields was closely correlated with favorability of soil and climatic conditions. Thus, given average grain yields for the USSR during that period of 8-10 centners/hectare, the highest yields (15 centners/hectare) were in the forest-steppe of the Ukraine and the piedmont areas of the Northern Caucasus. Yields consisted of 10-15 centners/hectare on other Chernozem soils, but for

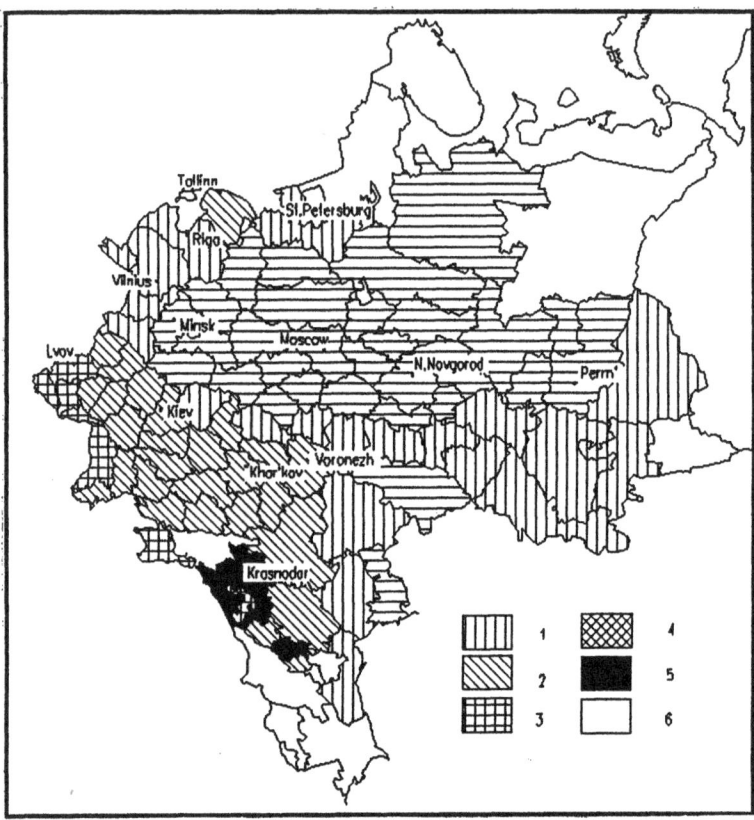

FIGURE 4.4 Average 1961-1965 Grain Yields (in European USSR) in Centners per Hectare. 1: 4-9; 2: 9-13; 3: 13-17; 4: 17-21; 5: 21-25; 6: Data unavailable.

the Non-Chernozem Zone, Byelorussia, and the Baltic republics, yields were no higher than 6-10 centners/hectare (Fig. 4.4).

By the mid-1980s the geography of grain yields had assumed a rather predictable meridional-zonal character (Fig. 4.5), yields reaching 20-30 centners/hectare in the entire western-south-western region of the country encompassing Karelia, the province of Leningrad, the Baltics, western Byelorussia, western and central Ukraine, and Northern Caucasus (without the province of Rostov). Yields were 10-20 centners/hectare in the bulk of the Non-Chernozem Zone and in the Central Chernozem region, and did not exceed 10 centners/hectare in the steppe areas of the Volga and Urals regions where spring wheat prevails. Interestingly enough, the major drop-off in grain yields (by a factor of 2 to 3) occurred at the western border of the Russian Federation, despite the fact that the natural conditions, say, in

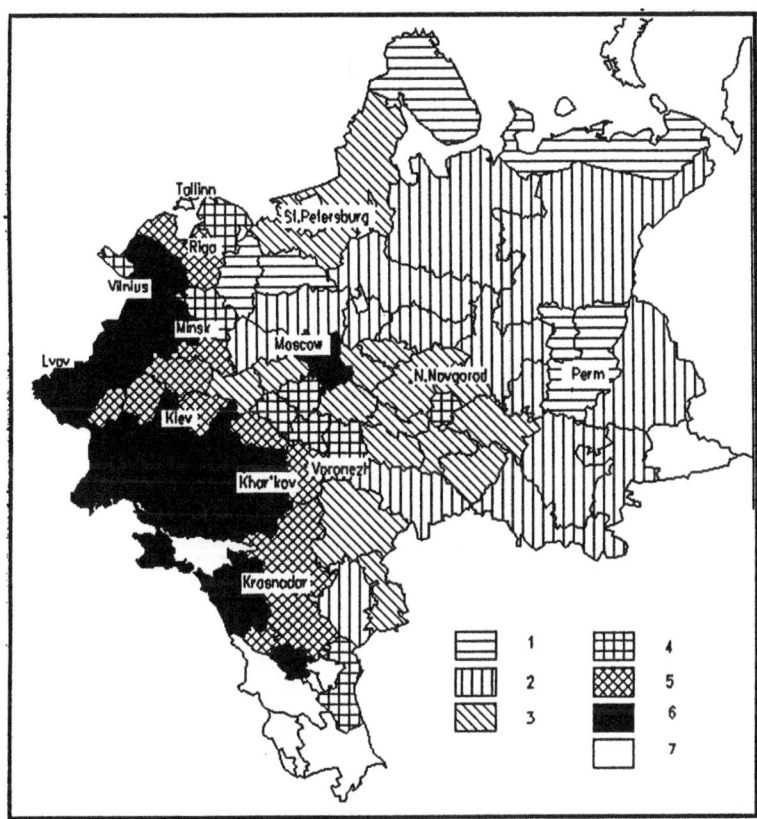

FIGURE 4.5 Average 1986-1990 Grain Yields (in European USSR) in Centners per Hectare. 1: 5-9; 2: 9-13; 3: 13-17; 4: 17-21; 5: 21-25; 6: 25-47; 7: Data unavailable.

the province of Pskov, are virtually identical to those in Estonia; by the same token, the Byelorussian province of Mogilev is no different from the Russian province of Briansk. Outside the westernmost part of the USSR, only some segments of the Central Chernozem region and the province of Moscow stood out for their heightened yields.

The yields of perennial herbs and the productivity of natural hayfields have spatial patterns similar to that of grain yields: both generally decline as one moves eastward. The same holds true for sugar beets and sunflower, and for milk yields per cow (Fig. 4.6).

The above testifies to the fact that the *east-west gradient in agricultural productivity* discussed in Chapter 2 as characteristic of Russia prior to the collectivization of agriculture *had once again emerged in the 1980s.* This re-emergence may be indicative of some major, if latent, factors behind the

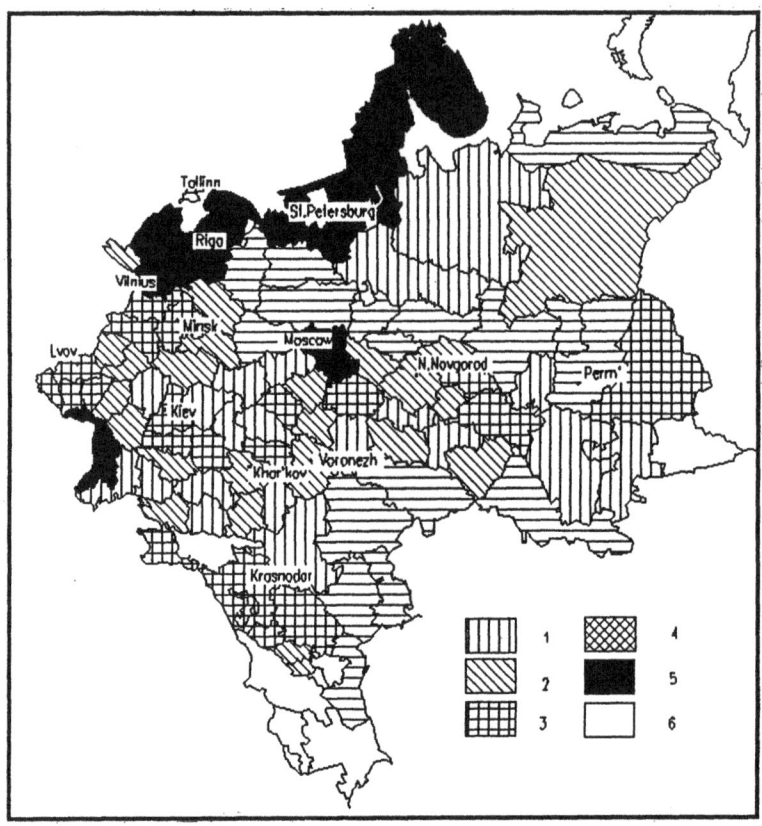

FIGURE 4.6 Annual Milk Yields per Cow in Kilograms (1990). 1: 1800-2500; 2: 2500-2800; 3: 2800-3000; 4: 3000-3500; 5: 3500-5000; 6: Data unavailable.

spatial variance of farming productivity in the European ex-USSR. Becauseof this and in view of the fact that the above gradient does not promise to be short-lived (in fact, it is evident in the geography of agricultural pro-ductivity of the 1990s), an historical overview is once again in order.

Historical Changes in Regions' Rank Order

Available time series of grain yields in the civil divisions of European Russia (gubernias, later oblasts[6]) reveal three distinct periods during which the macro-geographical proportions in output per unit of land shaped up differently:

1. Between the 1860s and the 1930s (prior to collectivization) grain yields in non-Chernozem areas had been at least on a par with those in Chernozem regions (Table 4.5 captures part of this period);
2. From the 1930s through the 1960s, following collectivization, the non-Chernozem regions slumped, and the overall geography of agricultural output per unit of land came to match soil fertility (this situation is captured in Figure 4.4).
3. Beginning in the early 1970s a new split into leaders and laggards began to emerge (Figure 4.5 captures its mature stage). As discussed above, output no longer matched soil fertility (directly or in a reversed fashion); rather an east/west or Russian/non-Russian split occurred: provinces of the Russian Federation, both Chernozem and non, were fading into the background, while the western fringe of the USSR, including both Chernozem (Ukraine) and non-Chernozem (Byelorussia and the Baltics) areas already assumed leadership. In fact, the Baltics had done so already in the 1960s; Byelorussia joined the leaders only later.

The macro-geographical proportions peculiar for the *first* period (predating the collectivization of agriculture) reflected in Table 4.5 find their explanation in the enhanced industrial and urban growth experienced by the north-central areas of European Russia as compared to the south-central and southern areas. Correspondingly, the burden of agrarian overpopulation

TABLE 4.5 Average Macro-Regional[a] Grain Yields and Their Differentials in the European USSR in Centners per Hectare.

	1913	1940	1955	1965	1981-85
North	9.0	6.8	5.3	8.7	16.0
Center	7.9	8.2	8.2	10.6	17.2
South	8.8	10.2	12.2	16.5	21.8
North-South Gradient	-0.2	3.4	6.9	7.8	5.8
West	9.3	10.8	10.8	18.7	26.0
Center	8.5	9.9	12.6	11.6	18.5
East	7.7	7.8	6.5	7.7	12.6
East-West Gradient	1.6	3.0	5.9	11.0	13.4

[a]*The North-South Arrangement*--South: Ukraine, Central Chernozem, Volga, and Northern Caucasus; Center: Latvia, Lithuania, Byelorussia, Industrial Center, Volgo-Viatka, and Urals; North: economic regions of North and Northwest and Estonia. *The West-East Arrangement*--West: Estonia, Latvia, Lithuania, Byelorussia, and western Ukraine; Center: North, Industrial Center, Central Chernozem, and central and eastern Ukraine; East: Volga and Urals.
Source: Compiled after *Narkhoz SSSR* data-books.

was heavier in the south while the center of gravity of consumer demand shifted to the north, spurred by the accelerated growth of Moscow, Saint Petersburg, Yaroslavl, Ivanovo, Vladimir, Tula, and other industrial centers.

Prior to collectivization the non-Chernozem regions of Russia have almost never taken a back seat to the more fertile Chernozem regions in terms of output per unit of land. The only region that the Non-Chernozem Zone of Russia proper has never been superior to is the Baltic. Moreover, in 1928-30, when the Russian south had already fallen victim to the collectivization campaign while the Center and the north had not, grain yields in many non-Chernozem Russian regions were on a par with, and in some years even higher than, those in North America.[7]

As was pointed out in Chapter 2, substantial variance in socio-economic and property relations existed in Russia's countryside prior to the collectivization campaign. This variation manifested itself during the Stolypin reform of 1905-11 when peasants in different regions showed different propensities to withdraw from their communes and embark on independent farming. When the cataclysms of 1914–1921 had levelled this differentiation, and the peasant commune once again became the dominant type of land use in Russia's countryside, the various regions again began revealing different degrees of attachment to traditional ways and different preferences for private farming.

But Russia has long traditions of centralism and of relentless imposition of uniformity upon all spheres of life including, obviously, economic activities. If any one area stands out in this regard, agriculture has had the dubious distinction of being the leader. As was discussed in Chapter 3, in agriculture two similar types of production units, collective and state farms, both micromanaged from a provincial center that subsequently ceased to notice the obscure property-related distinctions of those types, were universal norms by the mid-1930s. This homogeneity occurred in the face of quite a variance in both natural and social environments. Even Lenin, already in power, underscored in his later speeches what he defined as the "actual inevitability of a difference" in agriculture from place to place. "Crop harvesting in the gubernia of Kaluga," preached Lenin, "is not at all a fair replica of that in the gubernia of Kazan."[8] While this commonsensical comment looks like passing a guilty verdict on the theory and practice of central planning and its inherent drive to stifling uniformity, it might well be considered a sensible guideline given the ultra-revolutionary vigor of the time. But the drive proceeded inexorably.

In the wake of the pioneering work by Robert Conquest, the collectivization of agriculture under the Soviet system has mostly been described as the Great Terror. Another abundantly publicized aspect of a centrally planned agriculture has been its role in undermining incentives for hard, efficient work. For our purposes it is important that we underscore still one more

TABLE 4.6 Grain Yields in Selected Regions of the European USSR.

	1956-60	1961-65	1966-70	1971-75	1976-80	1981-85	1986-90
Non-Chernozem Zone	6.7	8.1	11.8	13.8	11.4	11.5	13.8
Central Chernozem Region	13.6	16.2	17.9	18.2	16.6	14.5	22.4
Volga	7.3	10.8	13.0	11.8	13.0	11.4	14.0
Byelorussia	8.4	13.1	13.1	21.3	21.3	18.6	25.2
Baltics	8.3	11.3	19.8	22.8	23.6	19.6	23.2

Sources: Narkhoz RSFSR. 1966, 1971, 1991. Narkhoz SSSR... 1973, 1991.

aspect: collectivization was an endeavor to *eradicate variance,* to install over-arching uniformity onto a kind of activity which by its very nature is doomed to be site-specific.

The lands on which not only soil but also social conditions were variable and arranged in a mostly mosaic pattern, were those of western and north-central Russia. It is then little wonder that these non-Chernozem regions were adversely affected by the encroaching impersonality and uniformity of collectivist farming much more than were the naturally monotonous regions of the Russian steppe, a land of enormous fields and few growing towns.

Because the capital intensity of agriculture began to vary greatly from one region to another only much later, in the 1960s, *soil fertility became the only production factor* in the land-labor-capital triad whose differentiating impact upon agricultural output could not possibly be obliterated. We can see, therefore, why the geography of yields had come to match the geography of soil fertility. In fact, this match marks the highest point ever in the course of Russian agriculture's painful adjustment to the extreme version of central planning that came to rule the countryside.

By 1940 the north-south gradient was somewhat steeper than the east-west gradient, and this was still the case shortly after the war (Tables 4.5 and 4.6).

The situation reversed only by the mid-1960s. Not that sameness of agricultural operations had diminished by that time. Rather, spontaneous factors of a social nature began to exert their influence. Two stand out: the spatially variable intensity of rural out-migration (see Chapter 5); and the development of urban-rural integration in exurbia, where agricultural operations in large measure fell out of centralized control (that is, were less affected by the vertical chain of command in the so-called Agro-Industrial Complex) due to the emergence of direct (horizontal) ties and links between farms and industrial enterprises. Both factors worked to the disadvantage of

TABLE 4.7 Milk Yields per Cow in the Non-Chernozem Regions of the European USSR, in kilograms.

	1940	*1960*	*1970*	*1980*	*1990*
Non-Chernozem Zone of Russia	1105	2090	2522	2180	2902
Byelorussia	886	1818	2304	2222	3220
Baltics	1863	2471	2998	3102	3658

Sources: Narkhoz RSFSR 1991; Narkhoz SSSR 1991.

the eastern margin of the vast Russian plain, both Chernozem and non, because of its vast inter-urban spaces. Correspondingly the Non-Chernozem Zone of European Russia and the Volga region became the most backward agricultural regions in the European USSR. The changing dynamics of grain yields are reflected in Table 4.6, which enables us to clearly identify some turning points in the parallel evolution of regional farming systems. Table 4.7 includes supportive evidence to the same effect.

Inside Russia, region-specific trajectories of grain yields and other components of output began to diverge significantly in the 1970s. Returning to Table 4.4 one can see that in many kinds of produce, productivity grew steadily up until the 1970s. While this was probably the case, some caution needs to be exercised in drawing such a conclusion, in this case because beginning in 1976 Soviet statistics switched from reporting yields of standing crops to reporting yields of already harvested crops, with the result that recorded yields dropped on average by 10%. However, there was no sizable growth in actual productivity after 1970 including the regions heavily invested in, that is, the Non-Chernozem Zone of European Russia and the regions east of the Urals. The upward grain yield trend in the Central region shown by Figure 4.7 is almost entirely due to the contribution of the province of Moscow, while in the remaining ten provinces of the region the trend rather resembled that depicted by the same figure for the province of Yaroslavl. Only the more fertile southern regions of Russia, long Cinderellas in terms of invest-ment priorities, showed higher returns when central planners finally turned to them after they had lost their faith in their prior investment decisions. Needless to say, a graph for Russia as a whole (Fig. 4.7) blends the upward and the downward regional trends into a somewhat rugged plateau indicative of stagnation.

Three Explanations of Failure

In Russia both Soviet and post-Soviet, low agricultural outputs are frequently explained by the overall *harsh natural conditions*. The idea that

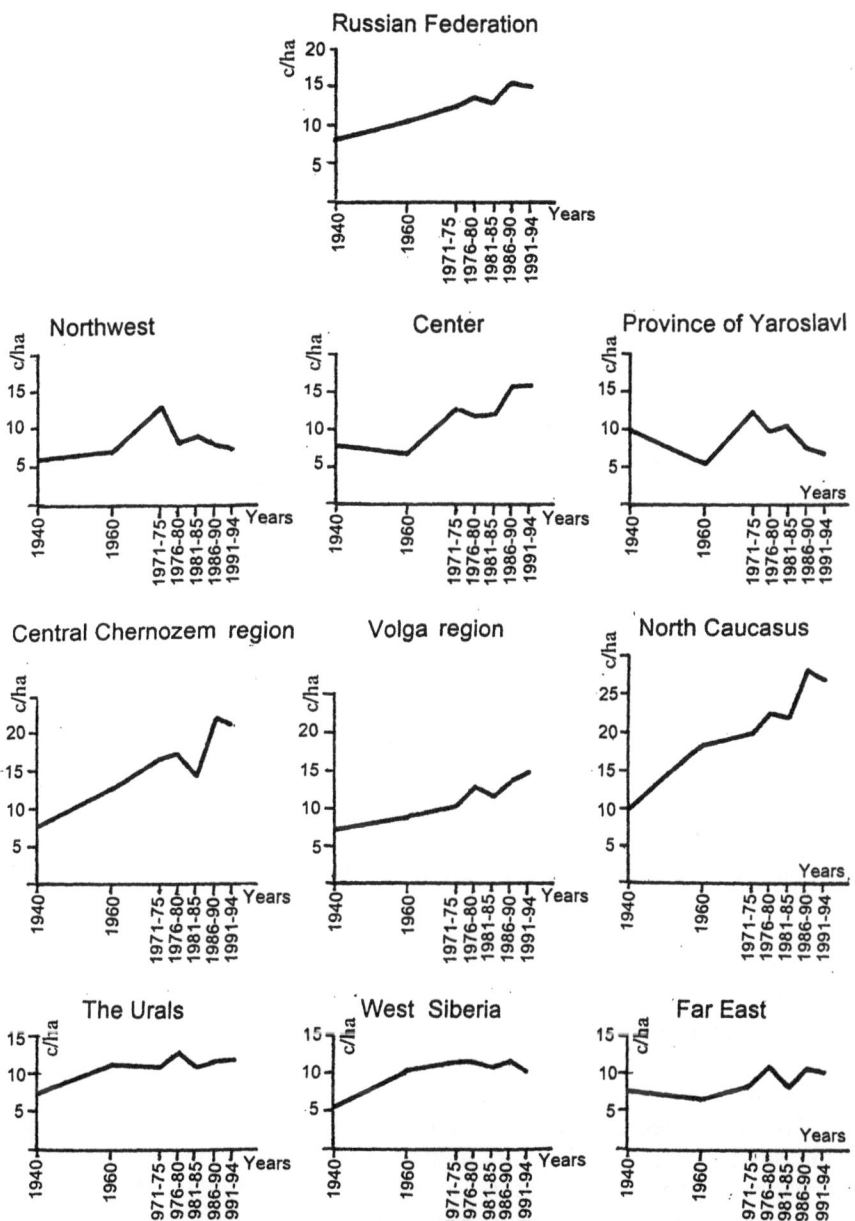

FIGURE 4.7 Regional Dynamics of Grain Yields in Centners per Hectare (1940-1994).

Russia is a country of a climatically risky or marginal farming (riskovannoye zemledeliye) is exceedingly popular. Typical Soviet reports from the so-called agricultural front looked like summaries of military operations and routinely commenced with the phrase, "despite unfavorable weather conditions."

Specific techniques have long been in use in Russia enabling one to assess the favorability of natural conditions for agriculture. They are based upon long-term records of yields on specially designated, regionally representative parcels of land not using irrigation or any other sophisticated cultivation method, that is, under natural conditions of soil type, heat, and moisture.[9] The estimates thus obtained are expressed in centners/hectare yields of respective crops. The highest grains-related estimate is that of the Northern Caucasus and of the western margin of the Central Chernozem region (about and over 30 centners/hectare). The favorability estimate is fairly high (20-25 centners/hectare) also in the middle belt from the province of Smolensk in the west to Bashkortostan in the east. The north, central Siberia, and the south of the Volga region fare the worst. Figure 4.8 shows the 1986-1990 differentials between actual yields and "normal" ones expressed as favorability estimates. The point is that in 80% of Russian provinces actual yields are short of "normal." These provinces are located in all climatic regions of Russia, but in the Non-Chernozem Zone, outside the province of Moscow, the shortages range from seven to 15 centners per hectare(!). Although this result is reported for the late 1980s, its relevance is not confined to this period only, as the 1970s situation in this regard was quite similar. Also note that Figure 4.8 shows a pronounced western bias reflecting a west-east gradient of productivity.

It would be logical to assume that if and when the actual productivity of land is systematically short of the environmentally prescribed norm (whose very idea is to eliminate the influence of the environment-specific variance upon the output per unit of land), the climatic explanation of agricultural failures can be easily called into question. The fact that the gap between the output and the above norm is especially wide not in the climatically extreme but in quite hospitable areas only makes the standard explanation more doubtful.

Two other versions of these failures circulating among journalists, the general public, and some experts deal with the *deficit of investment* and with *rural exodus.*

Putting a squeeze on the countryside was indeed one of the principal ways the Soviets tried to make crash industrialization succeed. To procure food cheaply was the major reason behind the collectivization of agriculture. However, as we showed in Chapter 3, beginning in the 1960s the financial flow was reversed and money began to flow back to villages in the form of investment, subsidies, written-off debts, etc. It is safe to say that whereas in the 1930s to 1950s the risky nature of agricultural activity was dramatically

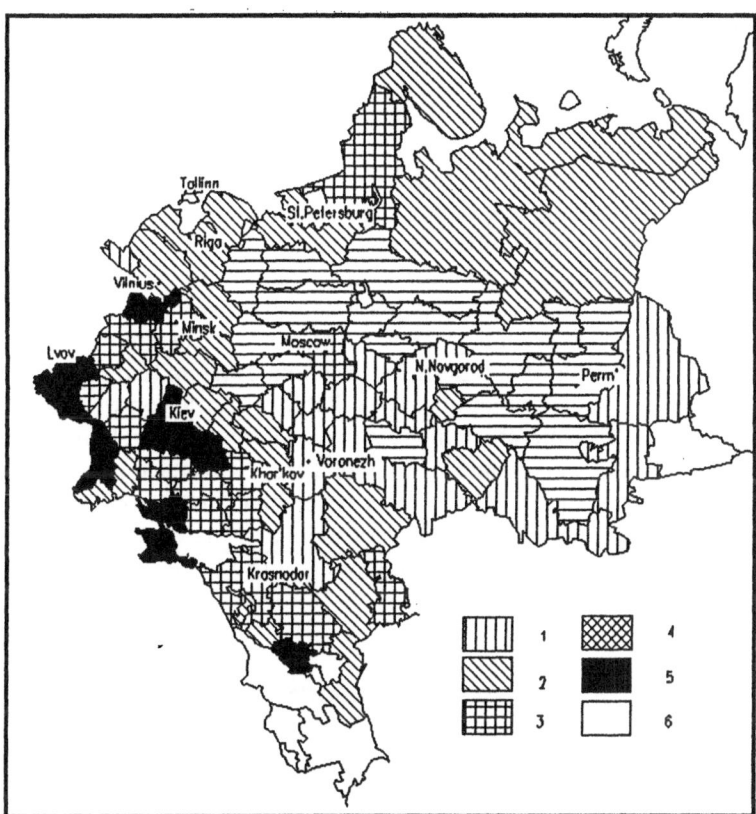

FIGURE 4.8 Actual (1986-1990) vs. Normative Grain Yields in Centners per Hectare. 1: Actual yields 6 to 11 below normative; 2: Actual yields 2 to 6 below normative; 3: Actual yields 2 below to 2 above normative; 4: Actual yields 2 to 6 above normative; 5: Actual yields 6 to 14 above normative; 6: Data unavailable.

exacerbated by outright robbery by the state, during the time period that followed, it was the state itself that lost money by pumping it into inefficient production units. By 1995, the percentage share of agricultural fixed assets (in the overall value of invested capital) had come to match that of farm employees (in the overall employment); therefore, the capital/labor ratio in agriculture had approached the national average.

Rural out-migration and a contracting labor pool in the countryside are very frequently used as an explanation of agriculture's failures. Russia's rural population had indeed halved and farming labor was reduced by a factor of three in 1926-1990, while the total population increased 1.5 times. The percentage share of retirees in non-Chernozem villages is 1.5 times higher than that of children. But, on the other hand, Russia's agriculture still accounts for 13% of the nation's labor force, that is, a proportion much higher than in economically advanced nations. So the true cause is not the number of people but their productivity, plus the lack of the economic system's successful adjustment to a shrinking pool of rural labor, a shrinkage that was by no means extraordinary and that befell all industrialized nations. Some Russian economists have long stated that a correlation between the demographic situation in the countryside and the labor deficit in agriculture is spurious[10] and derives from emotion rather than from rigorous analysis.

While each of the above three explanations looks reasonable, all are heavily laden with popular mythology. The root cause of Russian agriculture's poor yields is the result of an inefficient economic system that makes poor use of whatever production inputs are available, a conclusion we will try to make geographically concrete in the remainder of this chapter, based on the last period of relative economic stability of 1986–1990 which preceded the removal of government control over prices. Regional proportions of the subsequent period (beginning in 1991) do not seem to deviate much from those in the previous period, yet they are more difficult to monitor, and not only because of fluctuating prices.

Investment Priorities

The increase in the application of *fertilizers* used to be one of the most important aspects of the so-called intensification of agriculture. On average by the end of the 1980s, the dose of mineral fertilizers applied to one hectare of arable land was 120 kg (in terms of active ingredients). In Russia this figure was only 106 kg, including 150-400 kg in various non-Chernozem prov-inces; in the Baltic republics, 290-320 kg; and in Byelorussia, 340 kg. In fact, there was a steep north-west-to-south-east gradient in the fertilizers' application.

Comparing this information with data on optimal doses of the amount of fertilizer required for sound crop harvesting and subsequent compensation of soil nutrients[11] shows that the optimal requirements in fertilizers were met only in the Baltics, in Byelorussia, and in the provinces of Moscow and Leningrad. In many non-Chernozem provinces the efficacy of fertilizers was greatly reduced by their incorrect composition and by negligent, wasteful, and ecologically damaging application that included leaving paper and damaged plastic bags with plant food in furrows.

Another major aspect of the agricultural "intensification" was *land melioration and reclamation,* which in various Russian provinces consisted either of artificial drainage (in the Non-Chernozem Zone) or of irrigation (in the southeast). The enormous investment in melioration far outweighed its positive effects. In the moisture-sufficient regions of Russia melioration benefits were mostly confined to the provinces of Moscow and Leningrad and to the republic of Karelia. In many other provinces reclamation was in large measure beyond the control of soil scientists and hydrologists and occurred in response to the large-scale accumulation of labor and fixed assets in the hands of land reclamation management, assets that demanded justification of their existence. In the province of Yaroslav, for example, out of 180,000 artificially drained hectares handed over by land reclamation projects to the collective and state farms, about 40,000 hectares per year used to get waterlogged once again the late 1980s. That melioration was in large measure simply a drain on the budget and caused extensive ecological damage had already become common knowledge in the late 1970s. The ecological damage caused by land reclamation projects is well documented.[12]

The *mechanization of agriculture* was yet another priority of the Soviets. According to published records, by 1989 the share of mechanized operations at planting vegetables had reached 80%, at harvesting vegetables 25%. Half of potato harvesting operations are mechanized. In animal husbandry mechanization is at a 65-85% level.[13] However, these data concern only agriculture in the public sector. Given the high share of the almost entirely manual personal auxiliary farming (PAF) in the total output of potatoes, vegetables, and milk (see Chapter 3), assessment of actual mechanization would be much lower.

One peculiar feature of Soviet planners' agricultural mechanization is associated with their weakness for things gigantic. For example, most tractors and grain-harvesting combines weigh between 12 and 15 tons. As a result they compress the soil, which reduces yields by 20% on clayish and loamy soils.

The overall accumulation of prior investment in agriculture is best captured by the indicator of fixed assets, which includes monetary value of machines, structures, installations, and cattle. Figure 4.9 shows that in the geographical distribution of fixed assets there is a steep east-west gradient, which is a natural derivative of the gradient in the settlement network. By the late 1980s the lowest region had only one-ninth the value of fixed assets per unit of agricultural land compared to the highest. Fixed asset values were highest in the Baltics and lowest in the steppe areas of the Urals and West Siberia. This gradient had substantially steepened during the last 20 years of the USSR's existence.

The geography of agricultural labor (Fig. 4.10) looks like a correlate of the bio-climatic variance (Fig. 4.1) in the former USSR. The closeness of this relationship, which seems natural in the context of the spatially continual human colonization of a gigantic oikumene (whose natural environment lacks steep gradients), has been further accentuated by weak agricultural specialization and by the vital significance of subsidiary plots for

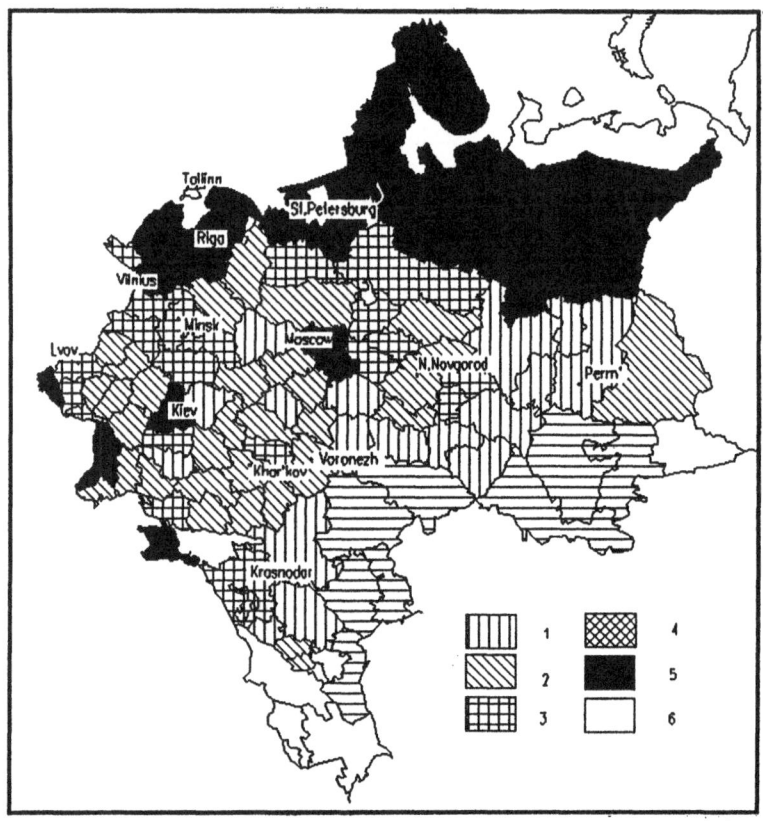

FIGURE 4.9 Fixed Assets of Agriculture in Roubles per 100 Hectares of Agricultural Land (1990). 1: 170-800; 2: 800-1200; 3: 1200-1600: 4: 1600-2000; 5: 2000-
5500; 6: Data unavailable.

rural and small-town populations. In the late 1980s, as in earlier periods, the
highest number of people engaged in agriculture were in the western parts
of the Ukraine and Byelorussia. Compared with these areas, most of the
Russian countryside looks like a social desert with solitary "oases" in the
piedmont areas of the Northern Caucasus, in the Chuvash republic, and
around Moscow and Saint Petersburg. These latter areas have a well-
balanced combination of all three major factors of agricultural productivity:
land, labor, and capital.

In the late 1980s we made an attempt to assess the spatial variability of
the influence that this combination and its components exert upon the
public sector's agricultural output of the European USSR. We wanted to
base our analysis on the non-monetary estimate of the *resource or input*

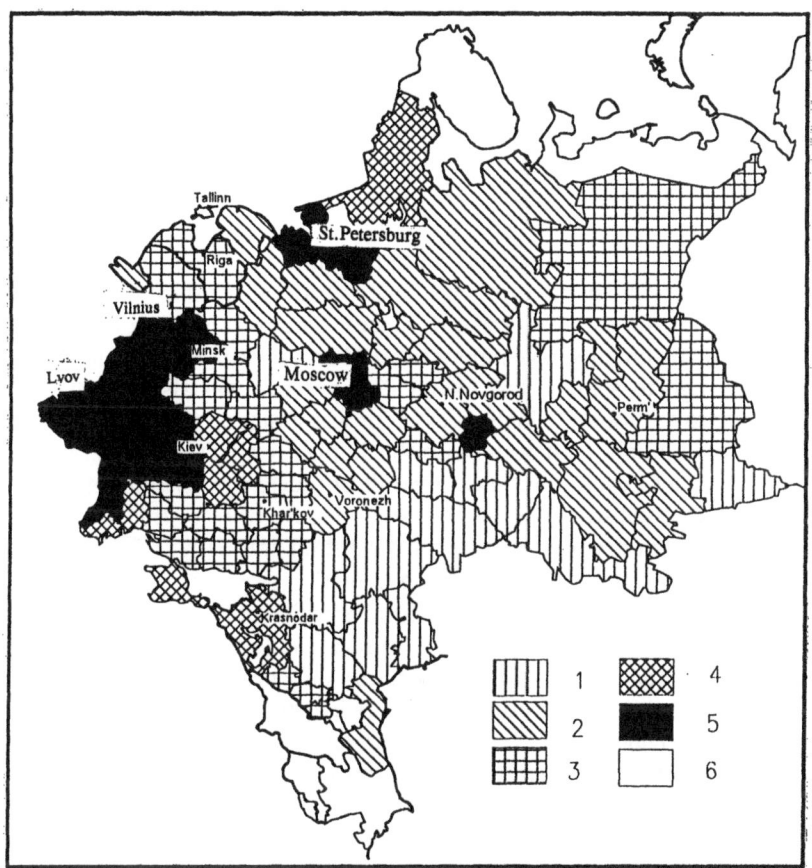

FIGURE 4.10 Number of People Engaged in Agriculture per 1000 Hectares of Agricultural Land (1990). 1: Below 50; 2: 50-75; 3: 75-100; 4: 100-125; 5: Over 125; 6: Data unavailable.

potential, and on the non-monetary estimate of *output.* That, we believed, would eliminate the distorting effect of preset prices which automatically ensure higher profit margins for certain types of produce (e.g., poultry, eggs, pork, etc.) making them look more "expensive" than others (e.g., milk, wheat, etc.) and thus distorting the overall picture of output distribution. We included monetary estimates only when they concerned commensurable entities (e.g., fixed assets). Each of 85 territorial units (*oblast,* autonomous republic, or a small union republic without *oblast*-rank sub-divisions) was thus assigned two sums of statistically standardized estimates: that of resources (inputs) and that of output. The former sum consisted of a "normative" grain yield (an approximation of the favorability of bio-climatic conditions); a monetary value of fixed assets per 100 hectares; and the number of people employed on the collective and state farms. The output estimate consisted of average actual grain yields, the output of meat in physical weight per 100 hectares, and average milk yields per cow.

The analysis[14] focused on the complementarity of specific inputs, and on the univariate (one-input) production function. Of our many conclusions the following three are important to this narrative.

- Though the actual practice of agricultural investment was not guided by clear-cut spatial priorities, to a considerable extent it was aimed at evening out the resource differences. Thus, given substantial variation in environments, investment favored areas with lower natural fertility. Though to a much less extent, this practice was also in effect when disproportionately high investment was assigned to areas with a poor supply of labor. In both cases this implicitly egalitarian (i.e., levelling) approach resulted in lower overall productivity: investment generally failed to substitute for deficient inputs. Because of this the center of gravity of the country's agricultural output gradually shifted toward the areas with balanced resource potential.
- The spatial variation in agricultural productivity is tightly linked to that in input or resource potential. In fact, the latter accounts for 81% of the variance of the former, which can only be the case if and when the spatial variance in returns (i.e., efficiency) is relatively small (of Russia's provinces, only Leningrad ranked well above average in this regard). Thus the overall "mass" of resources involved is almost the only thing that matters.
- Elimination of the land component of input from the calculations, i.e., assuming that the output is produced by capital and labor only, magnifies the spatial variance in agricultural efficiency. Table 4.8 shows the spatially generalized results of integral efficiency (returns to labor *and* capital) calculations. It appeared that of Russian regions only Northern Caucasus fared moderately well; all the other highly efficient agricultural areas of the European USSR were located outside Russia.

Ecological Damage Inflicted by Agriculture

Accelerated erosion has become a scourge of farming in areas with extensive plow-land and deficient woodland belts. The situation is worst in the Northern Caucasus and in the Central Chernozem region where land in which 10 tons of soil per hectare are annually washed out by splash erosion accounts for 12% of the total arable land and land with 5-10 tons washout accounts for another 12%.[15] Given that the rate at which soils are being

TABLE 4.8 Regional Efficiency of Socialized Agriculture in the European USSR in 1981-1985.

	Output per Worker, in 1000 Roubles	Output per 100 Roubles of Invested Capital, in Roubles	Capital/labor Ratio, in 1000 Roubles per Worker	Integral Efficiency $(I)^a$
Non-Chernozem Zone (of Russia)	4.60	30	15.2	- 10.0
Volga	4.62	21	21.6	- 10.3
Central Chernozem	4.29	25	17.5	- 7.1
Northern Caucasus	5.25	26	19.9	0.9
Byelorussia	4.42	30	14.9	20.9
Ukraine	4.29	34	12.5	4.3
Baltics	6.74	22	29.9	81.3

$^a I = \{(Y_i - Y_j \times l_i/l_j) - E(K_i - K_j \times l_i/l_j)\}$, where:

   i, j - provinces (oblasts, autonomous and union republics without oblast-rank divisions); for example, the Non-Chernozem Zone of Russia consists of 29 provinces;

   n - number of provinces in a region

   Y - agricultural output per 100 hectares of agricultural land;

   K - fixed assets (invested capital) in agriculture per 100 hectares of agricultural land;

   l - average annual employment;

   E - return to capital in roubles of output per 100 roubles - value of fixed assets.

Source: Ioffe, G.V. *Selskoye Khozyaystvo Niechernozemya...* Moscow, Nauka, 1990:24-25.

naturally replenished does not exceed 1.2-3 tons a year in various regions, such intensive erosion leads to a noticeable diminishing of the topsoil.

In the Non-Chernozem Zone the severest splash erosion is the result of melioration projects which created vast fields more common to natural steppes. Switching from a historical pattern of patchy land cultivation, in which small fragments of plow-land were interspersed with forest tracts, to the system of open fields is especially fraught with ecological damage to the long, gentle slopes that abound in the Non-Chernozem Zone. Here the arable land in the province of Perm and most provinces of the Volgo-Viatka region are most erosion-ridden. Overall, the eroded land accounts for 10% of the arable land in the subregion of southern taiga and for 60% of that in forest steppe. Yields are seriously impacted by this erosion. However, this effect oftentimes lags behind the practices which create it, which in part explains the insufficient attention attached to the prevention of accelerated splash erosion.

Ecological damage caused by farming in arid regions is usually more immediate. In Russia one can distinguish two regions with sizable effects of deflation (erosion by the wind): the European south-east and the southern area of West Siberia, which is in the same physiographic region as northern Kazakhstan. The overall area subject to dust storms is 900,000 sq km in European Russia and 700,000 sq km in Siberia.

The deflation processes in European Russia are potentially more catastrophic. They mostly afflict such areas as the *Salsk* steppe in the province of Rostov, the republic of Kalmykia, the Volga (*Povolzhye*) region, the plains along the low Don River, and the Stavropol plateau of the Northern Caucasus piedmont. On these areas no less than 10-15 dust storms occur annually. Strong winds can remove 2-3 centimeters of soil within 15-20 days. In Kalmykia and in the south of the Volgograd and Astrakhan provinces deflation combines with other human-accentuated ecological damage, especially with secondary salinization of soils, and leads to extensive desertification. About half of the total land area of Kalmykia is affected by this process. It is mostly caused by over-grazing which has always been difficult to keep in check because grazing norms used to be set only for public sector agriculture, whereas in fact every shepherd took care of many extra PAF sheep. As a result the biomass removed through grazing exceeds the naturally replenished mass by a factor of four. Extensive soil degradation in European Russia's south-east was first noticed in the 1950s when year-round utilization of pasture-lands commenced and sandy soils were plowed up. Degradation increased because of an attempt to introduce irrigated rice-growing. This undertaking was a disaster: acceptable yields occurred for 2-3 years only; yields then plummeted and soils quickly became salinized.

All in all, under current agricultural technology the south-eastern fringe of European Russia has to be characterized as a zone of very high ecological risk (Fig. 4.11).

In West Siberia dust storms are shorter (lasting a few hours) but more frequent and can lead to substantial losses of soil especially in the steppes of Kulunda and Ishim located on both sides of Kazakh-Russian border.[16]

A special group of adverse ecological effects is associated with pesticides and weed-killers. For that and many other aspects of the environmental pollution in the Soviet Union readers should consult *Ecocide in the USSR*, a well-documented book by Murray Feshbach and Alfred Friendly[17] (also translated into Russian), to the report by Alexander Yablokov at the conference Ecology-89 in Sweden,[18] and to *Environmental Resources and Constraints in the Former Soviet Republics* recently published by Westview Press.

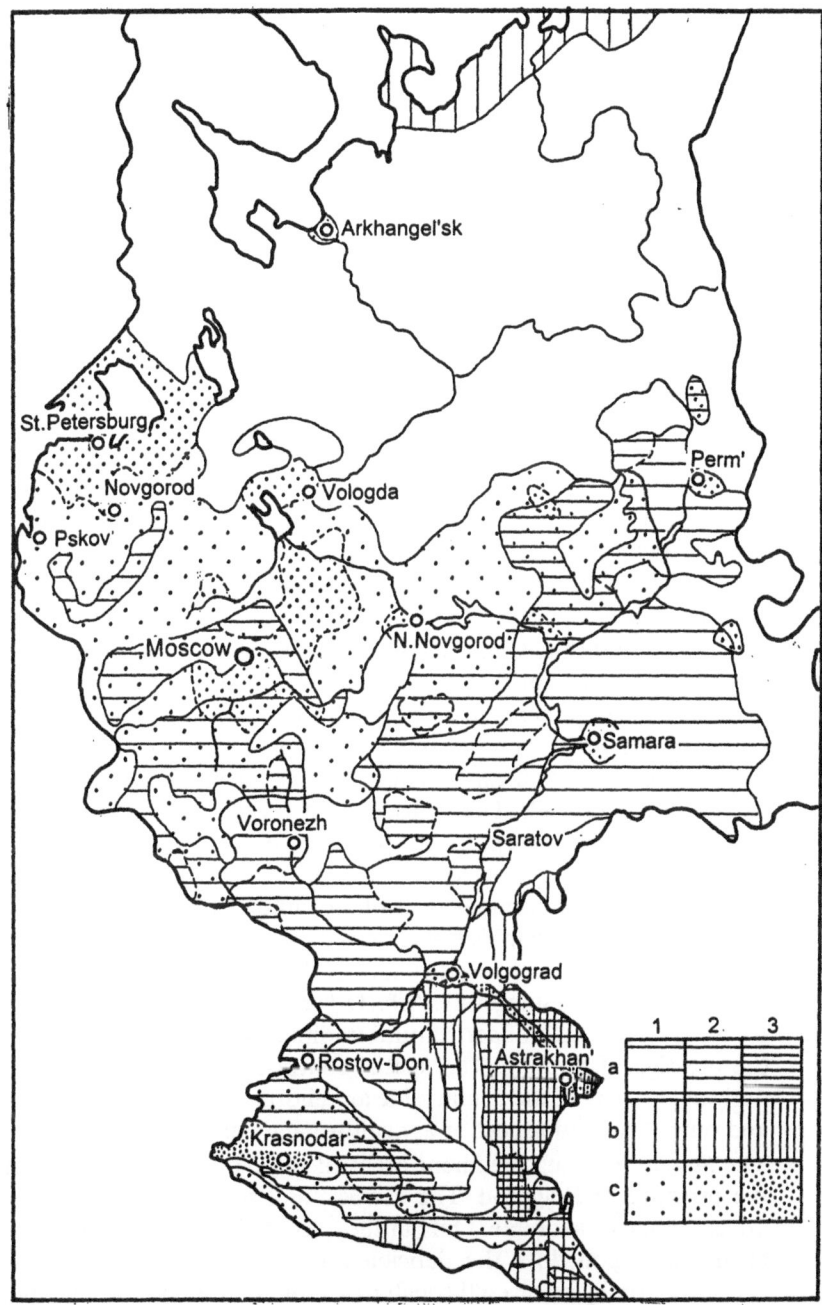

FIGURE 4.11 Ecological Problems of Agricultural Land Use (1990). Environment's condition: 1=on verge of crisis; 2=crisis; 3=catastrophe. Man-caused/accentuated process: a=arable land degradation; b=pasture deterioration; c=soil pollution.

Summary of Regional Variances in Russian Agriculture

It follows from the above that there is a substantial regional variance in Russian agriculture. We distinguish between several macroregions (covering almost all of Russia), each of which can be thought of as relatively homogenous in terms of output, factors affecting it, and the consequences they cause.

- The *North*. Here, natural, bio-climatically prescribed fertility is the lowest. A relatively high level of output per unit of land is created exclusively by high investment. But though the capital/labor ratio is one of the highest in Russia, labor productivity is low as is return on investment because of both deficient fertility and deficient labor. Reclamation practices in the region have been aimed less at the improvement of agricultural land long in use (20-30% of which has been abandoned in the last 20 years) than at draining swamps, clearing bushes, and amalgamating arable land. Local populations have not been prepared for working these enlarged tracts of land or for the ecological damage that has ensued.
- The *Center* and the *Northwest*. By and large these are the most ailing agricultural regions of Russia. They are characterized by the most severe underachievements in crop harvesting -- as determined by comparing actual yields with environmentally prescribed norms -- and by the low productivity of cattle. These regions were afflicted by crisis as early as the mid-1960s. They subsequently received lavish investments which did not help, except for the provinces of Moscow and Leningrad where investments were successful in large measure due to the stable -- and sometimes increasing -- pool of agricultural labor. Outside these provinces the demographic losses caused by the sucking effect of urban growth and the ensuing negative rate of the rural population's natural increase have been staggering. Affected regions were subsequently impacted by the breach of agricultural traditions and by a drastic drop in the prestige of farming. The spatial contrasts between exurban and outlying districts of the countryside are enormous here in every input- and output-related aspect of agriculture and rural infrastructure.
- The *South*, including piedmont areas of the Northern Caucasus and the province of Belgorod. The climatic and soil conditions of these areas are most favorable. Only in the east of the Northern Caucasus economic region does the deficiency of moisture pose a problem, which is partially resolved through irrigation. Labor and capital inputs per unit of land are average; they are above average in *Kuban'* (*Krasnodarski kray*). Rural out-migration affected most areas in the south but nowhere to the extent it affected the previous macroregional

types. Center-periphery contrasts in this area pale next to those of the previous type as well. The south has appeared to be the most responsive to investment. The output per unit of land reached one of the highest levels in the USSR (along with the Baltics and the provinces of Moscow and Leningrad) by the late 1980s. While land use intensity drops eastward, as in previous types, the monotony of landscape and the share of plowland grows eastward as well, exacerbating adverse ecological effects of improper farming techniques. Reduction of humus content, erosion, and overcompression of soil are among the most pervasive of those effects.

- The *Central Chernozem* region (except the province of Belgorod) and *Central Volga*. This area is a natural forest-steppe and it is clearly transitional between the previous two macroregions in terms of inputs, output, and related factors. While natural fertility and the types of ecological problems encountered bring this region closer to the most favorable type, socio-demographic problems make it closer to the Center and the Northwest. The output per unit of land is slightly above average.
- The *East*, including the southern part of the Volga region and southern Siberia. This is a semi-arid macroregion of low labor and capital-intensive grain farming. Crop harvesting has a low average productivity. Plowland dominates the landscape. Soils have been extensively damaged by deflation exacerbated by the deficiency of woodland belts. Current cultivation practices here are associated with an irretrievable loss of humus.

By and large the above variation is very persistent. It took shape despite overt policy targets and campaigns, all of which had a substantial levelling intention. In fact levelling has succeeded but not to the point of eliminating inter-regional variance. The introduction in the 1990s of something markedly new, that is, of market elements into the rural economy, only accentuated this variance, as we will later see.

Notes

1. Chelintsev, A.N. *Selskoye Khoziaistvo Sovremennoi Rossii kak Stadii Selskokhoziaistviennoi Evolutsii i Kulturnyi Uroven Selskogo Khoziaistva v Nikh.* Saint Petersburg 1910, p. 10.

2. *Atlas SSSR.* Moscow: GUGK 1983, pp. 130, 152, 154, 174.

3. Runova, T.G., Volkova, I.N., and Nefedova, T.G. *Territorialnaya Organizatsiya Prirodopolzovaniya.* Moscow: Nauka 1993, pp. 78-81.

4. *Alexeyev, A.I. Geografiya Selskoi Mestnosti.* Moscow: Znaniye 1989, pp. 16-20.

5. Rakitnikov, A.N. *Geografia Selskogo Khoziaistva.* Moscow: Mysl 1970, p. 210.

6. Yanson Y., *Sravnitelnaya Statistika Rossii i Zapadnoyevropeiskikh Gosudarstv.* Saint Petersburg 1878-1880; *Selskoye Khozyaystvo Rossii v XX Viekie.* Moscow 1923; Obukhov, V. M. *Urozhainost i Meteorologicheskye Factory.* Moscow: Gosplanizdat 1949; Cherevanin, F. A. "Vlyianye Kolebanyi Urozhayev na Selskoye Khozyaystvo v Techenye 40 Let, 1883-1923." In: *Vlyianye Nieurozhayev na Narodnoye Khozyaystvo Rossii.* Moscow: RANION 1927; etc.

7. Ioffe, G.V. *Selskoye Khoziaistvo Niechernozemya: Territorialnye Problemy.* Moscow: Nauka 1990, p. 28.

8. Translated by the authors from the fifth Soviet edition of Vladimir I. Lenin's Collected Works, Vol. 54, p. 198.

9. *Prirodno-Selskokhoziastvennoye Raionirovaniye Zemelnogo Fonda SSSR.* Moscow: Kolos 1983.

10. See, for example, Churakov, V. "Problemy Trudodefitsitnosti Khoziaistva i Zaniatosti Trudovykh Resursov Sela Rossii." In *Kadry APK v Sistemie Upravlenia.* Barnaul: SO AN SSSR 1987, pp. 7-8.

11. See, for example: Yurkin, O.N. *Povysheniye Effectivnosti Udobrenii v Intensivnom Zemledelii.* Moscow: Rosselkhozgiz 1979, p. 57.

12. See, for example: Zaidelman, F.R. *Melioratsiya Zabolochennykh Pochv Niechernozemnoi Zony RSFSR.* Moscow: Kolos 1981; *Prirodnaya Srieda Yevropeiskoi Chasti SSSR.* Moscow: IGAN SSSR 1989, pp. 129-139; etc.

13. *Razvitiye Agropromyshlennogo Kompleksa Rossiiskoi Federatsii.* Moscow: Goskomstat 1990, pp. 61-68.

14. Accounted in: Ioffe, G.V., Nefedova, T.G., and Pines, D.G. "Intensifikatsiya Selskogo Khoziaistva na Yevropeiskoi Territorii Strany." Part 1. *Izvestiya AN SSSR, Seriya Geograficheskaya* 1988, No 1, pp. 64-77; Ioffe, G.V., Nefedova, T.G., and Runova, T.G. "Intensifikatsiya Selskogo Khoziaistva na Yevropeiskoi Territorii Strany." Part 2. *Izvestiya AN SSSR, Seriya Geograficheskaya* 1988, No 5, pp. 1-15. The second article, methodologically less important, subsequently showed up in English in *Soviet Geography,* Vol. 30, No 1, pp. 49-64.x G0 R   w

15. *Erozionnaya Opasnost Selskokhoziaistvennykh Zemel Europeiskoi Chasti SSSR.* Map and Explanatory Text. Moscow: MGU 1987, p. 26.

16. Sazhin, A.N. "Regionalnoye Raspredeleniye Pylnykh Bur v Stepnykh Raionakh Vostochno-Yevropeiskoi i Zapadno-Sibirskoi Ravnin." In: *Geografiya i Prirodnyie Resursy* 1988, No 2, pp. 108-115.

17. Feshbach, Murray and Friendly, Albert, Jr. *Ecocide in the USSR.* New York: Basic Books 1992.

18. Yablokov, A. "The State of Environment in the Soviet Union." *Environmental Policy Review. The Soviet Union and Eastern Europe.* 1990, Vol 4, No 1.

5

Rural Population Change in 1959-1989 and Rural Infrastructure

The dates in the title of this chapter refer to two of six Soviet censuses (1926, 1937, 1959, 1970, 1979, and 1989). The period in between is characterized by unidirectional demographic trends, especially those with a pronounced spatial dimension.

While the description of production factors may be significant in and of itself, one has to keep in mind that the ability to influence any economic system rests upon the behavior of large groups of people. That they are not just a "resource of labor," but a principal actor on the economic scene used to be forgotten in the Soviet Union; a rule reigned supreme, according to which investment "produces" a required number of people no matter what. The rule rang true under agrarian over-population, and it seemed to be true under various administrative restrictions of mobility; but when migration had nevertheless effectively undermined the procreative potential of the rural population, it ultimately proved to be wrong.

After 1989 the long-lasting trend of rural population decline in Russia was reversed. This trend was once the focus of extensive research, to which we ourselves contributed. Does the recent reversal signify a true about-face in rural population dynamics, making earlier research obsolete? While it may seem so at first glance, a closer look at on-going processes that will be presented in Chapter 7 does not confirm this conclusion. Thus the issue of rural depopulation of several recent decades has not lost a bit of relevance.

Rural settlement change in Russia has been thoroughly studied since the late 1960s, when a crisis in agriculture concerned the Soviet political establishment for the first time. The research results can be divided into two parts: normative[1] and empirical.[2] Whereas the former came to be a response to the question "what has to be done?" in terms of required investment and its spatial distribution, the latter derived from the monitoring of actually unfolding processes, which used to deviate persistently from the

TABLE 5.1 1979 Rural Settlement and 1959-1979 Population Change in the District of Uglich.

| | Public Transportation Accessiblity of District Center, in minutes | | | |
	0 - 60	60 - 120	>120	District Total
% of the District's Land Area	11.7	26.9	61.4	100
% of the District's Population	31.4	28.1	40.5	100
% of the District's Settlement	22.3	28.7	49.0	100
Population Density per sq km	48.5	8.7	5.5	8.3
Average Population per Settlement	55	38	32	39
1979 Population as a % of That in 1959	84	56	36	53

Source: Ioffe, G.V. *Selskoye Khoziaistvo Nechernozemya.* Moscow, Nauka 1990:101. We monitored rural population change in the Uglich district up till 1985. In that year the population in the outer zone of accessibility of the district was only 25% of its 1959 level, whereas in the central zone it was still 84%.

TABLE 5.2 Population in the Provinces of Novgorod and Kursk in 1986/87 as a Percentage of That in 1959.

| | | Urban Total | | |
Province	Total	Province Center	Other Towns	Country
Novgorod	102	186		49
Kursk	90	211		50
Province Without Urban Agglomeration				
Novgorod	74	135		46
Kursk	68	228		47
With Urban Agglomeration				
Novgorod	265	367	143	115
Kursk	174	207	146	91
Between 0 and 0.5 Hours Accessibility Range				
Novgorod	387	367		163
Kursk	200	207		135
Between 0.5 and 1 Hour Accessibility Range				
Novgorod	100	110		83
Kursk	96	162		89
Between 1 and 1.5 Hours Accessibility Range				
Novgorod	83	96		66
Kursk	95	145		70

Source: Adapted from Zayonchkovskaya, Z.A. *Demografisheskaya Situatsiya i Rasseleniye.* Moscow, Nauka, 1991: 113.

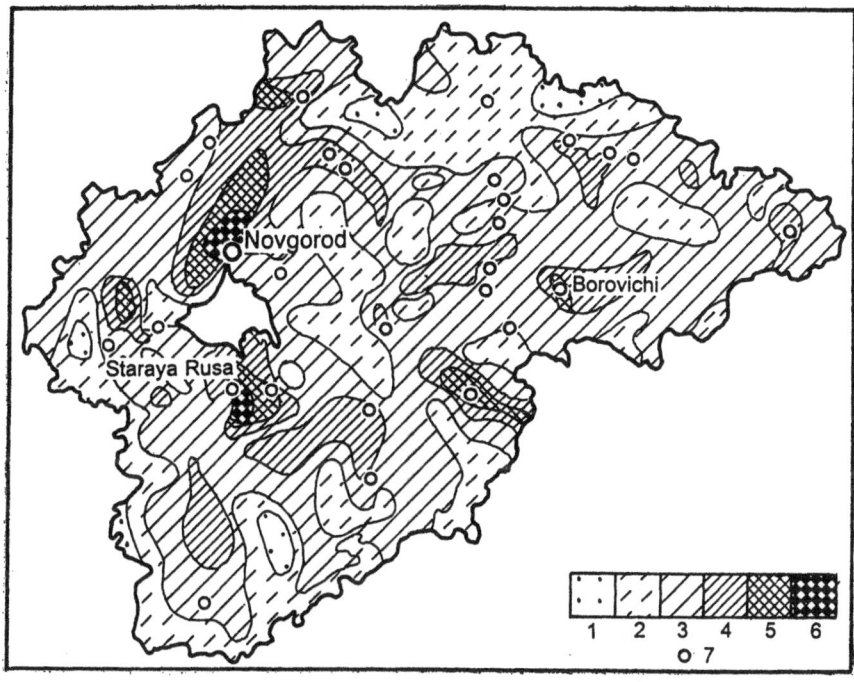

FIGURE 5.1   Rural Population in the Province of Novgorod in 1986 as a Percentage of that in 1959.   1: Less than 20%; 2: 20-40%; 3: 40-60%; 4: 60-80%; 5: 80-100%; 6: Over 100%; 7: Urban Settlements. *Source: Evolutsiya Rasseleniya v SSSR.* Moscow, IGAN SSSR 1989:124.

prescribed "norm." The following is a brief summary of the results of empirical research, derived from some of the sources mentioned in endnote 2, and from a few other sources as well.

Rural Polarization

Evidence overwhelmingly suggests that shifts in the rural population of the provinces of European Russia and in most of their minor civil divisions depend on proximity to cities and towns: rural settlements in outlying areas shrivel up while only those close to urban areas and/or major highways survive. Those "voiceless but evocative ruins of rural Russia today," to use Priscilla Roosevelt's language,[3] include myriads of abandoned wooden huts with planked windows and millions of formerly arable hectares of spontaneously forested land. This appeared to be typical for portions of hundreds of minor civil divisions (rayons) and dozens of provinces (oblasts) at large. Figures 5.1 and 5.2, and Tables 5.1 and 5.2 provide important illustrations.

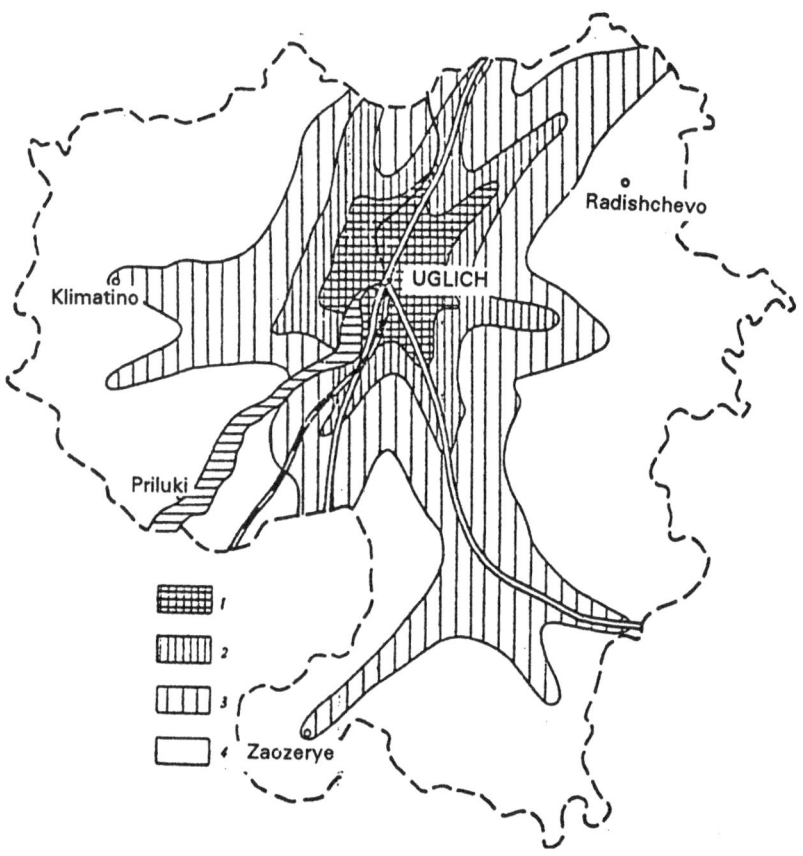

FIGURE 5.2 Delimitation of Travel-Time Zones from a Rayon Seat. Travel time to rayon seat (including walking to bus stop and waiting): 1 = less than 30 minutes; 2 = 30 to 60 minutes; 3 = 60 to 120 minutes; 4 = over 120 minutes. *Source:* Igudina, A.I and Ioffe, G.V. "Sdvigi v Razmechchenii Selskogo Naseleniya..." *Izvestia AN SSSR, Seriya Geograficheskaya,* 1985, No 5, p. 73.

It is essential to point out that the spatial polarization of the populated area, with its distinctive core-periphery dichotomy revealed by the above figures, was superimposed upon space that was sparsely settled. The presence of large and super-large population centers and centripetal population re-distribution trends only adds more prominence to the above point, as those centers appear as isolated islands in a rural vastness. For example, Moscow and Saint Petersburg -- two principal population centers 651 km apart (the same distance that separates Boston and Washington,

D.C., i.e., the northern and southern tips of the US Megalopolis) -- have not witnessed any sizable population growth between them. Rather than having become a megalopolis, the area is the most depopulated area in Russia. In 1927-38 the rural population between Moscow and Leningrad had already shrunk by 15-20%, in 1939-58, by 50%.[4] The process of rural population decline initially involved all the areas located in the Moscow-Leningrad inter-urban gap, including the Moscow and Leningrad provinces themselves; however, it subsequently stopped at their borders, and by the 1960s the rural population of these "capital" provinces had already begun to stabilize. In contrast, in the Tver, Pskov, and Novgorod provinces population decline peaked in 1959-1989 (Fig. 5.3). In the outlying districts of these provinces, the proportion of elderly people in the local population now ranges between one-third and one-half.

Such spatially uneven rural population dynamics produced sharp contrasts in rural population density. For example, the ratio of the province of Moscow to its southern neighbors in this regard is three to one; to its northern neighbors, it is four or five to one. Along with the province of Moscow, a rural population density of over 15 persons per sq km has been retained only in the Central Chernozem region (see Chapter 8), in the republics of the Volga region, and in piedmont provinces of the Northern Caucasus. Population density contrasts between the provincial center's (and sometimes the second ranked center's) exurbia and the outlying districts *inside* many provinces had increased even more than *between* them. This trend typified the Non-Chernozem Zone especially (see Chapters 10, 11, and 12 for a more detailed treatment) rendering Russian *prigorod* (exurbia) even more dissimilar to Western, especially American, suburbia[5] than ever before. Whereas the latter develop from the center outward, large groups of people steadily "ejecting" themselves from the corporate limits of a town, a Russian *prigorod* largely develops inward: on the one hand, the rural settlement network contracts, on the other, people are willing to move closer to a town to take advantage of its broader range and higher level of opportunities and quality of services. A restrictive policy of governing residence permits in large cities and a lasting absence of real estate market have created a peculiar barrier effect at the city line. *Prigorod* has thus become a stepping stone *to* a city or town, not *out* of it as in the West. The fact that Russian cities used to expand outward does not change the nature of the situation: typically they just extend a city line, and suburbs do not take shape. Only most recently, around the largest cities of Russia, has the situation has begun to change somewhat (see Chapter 13).

Another factor contributing to the *prigorod*-suburb dissimilarity is that in Russia cities exert their pull on the agricultural labor force even more than on the rural population as a whole. As a result, agricultural population tends to concentrate in more urbanized areas. This latter trend is clearly evident

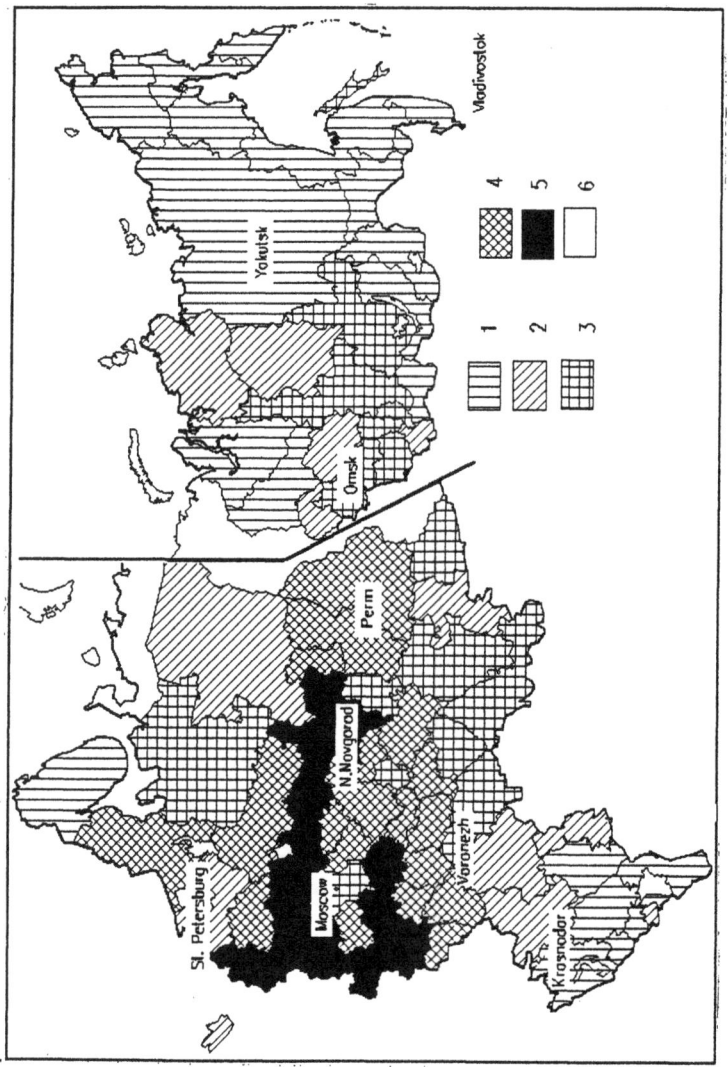

FIGURE 5.3 Rural Population in 1989 as a Percentage of That in 1959. 1: 200-230%; 2: 81-100%; 3: 66-80%; 4: 51-65%; 5: Less than 50%; 6: Data unavailable.

TABLE 5.3 Statistical Explanation of the 1959-1986 Dynamics of Rural Population in the Non-Chernozem Zone of Russia.[a]

Regression Equation

$\ln Y = -0.653 + 0.413 \ln X_1 + 0.077 \ln X_2$

Y: Rural Population in 1986 as % of That in 1959

X_1: Rural Population in 1959 as % of That in 1926

X_2: Percentage of Total Population Living in Towns >100,000 Residents

Validity: $F = 13.883$; $R^2 = 0.526$; $\delta r_1{}^b = 0.481$; $\delta r_2 = 0.045$

[a]Russia's Non-Chernozem Zone consists of 29 provinces (oblasts and republics) including all the provinces of the North, Northwest, Central, and Volgo-Viatka macroeconomic regions plus the Sverdlovsk (now Ekaterinburg) and Perm oblasts and the Udmart republic of the Ural region, and the free-standing Kaliningrad province. Murmansk was exluded from the sample.

[b]δr - partial determination coefficient, i.e., share of Y's variance accounted for by each of independent variables; $\delta r_1 + \delta r_2 = R^2$.

Source: Ioffe, G.V. *Selskoye Khoziastvo Niechernozemya.* Moscow, Nauka 1990: 101.

TABLE 5.4 Statistical Explanation of the 1960-1986 Dynamics of Agricultural Employment in the Non-Chernozem Zone of Russia.[a]

Regression Equation

$Y = -1.264 + 0.354 \ln X_1 + 0.285 \ln X_2$

Y: Engaged in Agriculture in 1986 as % of Those in 1959

X_1: Rural Population in 1959 as % of That in 1926

X_2: Percentage of Total Population Living in Towns >100,000 Residents

Validity: $F = 15.370$; $R^2 = 0.552$; $\delta r_1 = 0.277$, $\delta r_2 = 0.275$

[a]See notes for preceding table.

Source: Ioffe, G.V. *Selskoye Khoziastvo Niechernozemya.* Moscow, Nauka 1990: 101.

in a comparison of the partial determination in Tables 5.3 and 5.4. Its cause seems to be a fairly large and growing group of elderly people; it is especially large in less urbanized provinces: most of those elderly are no longer acting employees of collective and state farms, but they account for over one-fifth of the recorded rural population.

It should, however, be pointed out that the polarization process of the kind shown in Figure 5.1 affected the most fertile lowland provinces of the Northern Caucasus (especially the Kuban', also known as Krasnodarsky Kray) to a much smaller degree than it affected the rest of European Russia. But it impacted the Russian heartland causing outcries by self-proclaimed defenders of national roots. The grounds for their reaction were subsequently reinforced by the fact that the agricultural output itself, both in monetary and in natural terms, followed a core-periphery pattern as well. Not only are there easily distinguishable rings of outwardly descending output per unit of land but gradients get steeper as the spatial concentration of farm production grows.

It therefore looked like there was a cause-and-effect relationship between the spatial pattern of rural depopulation and the performance of agriculture. Because of this, the attempts to curb or even reverse the process of agricultural decay in outlying areas by virtue of agricultural investment typically raised the capital/labor ratio to a level too high for the existing technology to cope with, and, therefore, led to a lower level of returns.

Societal Response

The societal response to village decay and to the whole chain of issues stemming from rural depopulation has not been homogeneous in Russia. One can distinguish between two intellectual traditions that had re-emerged by the mid-1980s: the Slavophilic and the Westernizing traditions. Each has its own viewpoint on rural depopulation. Whereas neo-Slavophiles beg the question of the relationship between rural decay and the poor performance of agriculture (Fig. 5.4), Westernizers regard the economic fabric of a society as a driving force of change. In particular they point out that rural communities had effectively shrunk in all advanced nations without doing any harm to agriculture (Fig. 5.5).

The latter point is worth arguing for, if only because the spatial contraction of the rural settlement network is by no means an inherently Russian phenomenon. A similar process had been documented for Western Europe prior to the 1960s,[6] when terms like the German *Sozialbrache* (social fallow) in relation to land abandonment were coined.[7]

However, there are at least two meaningful distinctions regarding Russian spatial contraction. Although in Western Europe rural depopulation had been decried in terms quite similar to those which were later taken up by the neo-Slavophiles, the breakdown of a traditional village was not accompanied by food shortages (indeed, there was a conspicuous absence of the latter). Also, agricultural subsidies in Western Europe, linked (as they still are) to extra-economic, mostly socio-cultural factors, have exerted a tranquilizing influence upon the populist message.

Indeed, the downsizing of agricultural employment in the West (to 6% of the labor force in Germany and almost 2% in Britain, while it remains as much as 13% in the Russian Federation) was preceded and then accompanied by economic progress dwarfing everything that has so far occurred in Russia. For example, growth in farm labor productivity in countries like

RURAL EXODUS IS THE REASON FOR THE SHORTAGE OF LABOR IN AGRCULTURE AND THE RESULTING FOOD PROBLEM

FAILURE OF CENTRAL PLANNING TO STOP RURAL EXODUS IS A RESULT OF:

- •CENTROMANIA: UNDERESTIMATING THE SIGNIFICANCE OF RURAL INVESTMENT

- •ANTI-PEASANT ACTIVITY INHERENT IN MARXIST IDEOLOGY

- •JEWISH CONSPIRACY STRIVING TO ERADICATE A TAPROOT OF RUSSIAN TRADITION

OPERATING ASSUMPTIONS:

- •A VILLAGE OUGHT TO BE POPULOUS

- •UNPROFITABLE FARMS SHOULD BENEFIT FROM PUBLIC FUNDS

- •THE COMMUNAL SENSE OF MUTUAL ASSISTANCE HAS TO BE RESTORED

- •WESTERN EXPERIENCE IS INAPPLICABLE TO THE RUSSIAN SOIL

FIGURE 5.4 Why Russian Villages Die: A Neo-Slavophilic View.

West Germany, France, the UK, the Netherlands, etc. in 1953-1967 was either about or over twice the growth in industry.[8] In the West one farmer feeds 40-80 people, while in Russia he feeds merely 12. No wonder that in Russia the process of rural depopulation, essentially similar to that earlier experienced in the West, has been viewed differently. A village is looked at by many as the prey of rapacious, soulless urbanization; and long-term food shortages have boosted that image.

Village Consolidation

Some of today's staunch Russian nationalists, like writer Vasily Belov, first emerged on the public scene as advocates of rural revival back in the 1970s during the campaign against the officially endorsed rural consolidation plans (so-called *sseleniye*). The essence of rural consolidation lay in designating

RURAL EXODUS IS INEVITABLE BECAUSE OF INDUSTRIALIZA-
TION AND URBANIZATION; THE INHERENT INABILITY OF CEN-
TRAL PLANNING TO FORESEE AND ADJUST TO THIS
INEVITABILITY RESULTS IN FOOD PROBLEMS

POLICY ASSUMPTIONS:

- •SHORTAGE OF LABOR IN AGRICULTURE HAS NO IMMEDIATE
  CONNECTION WITH RURAL DEPOPULATION

- •UNDERINVESTMENT OF SOCIAL INFRASTRUCTURE IS A
  SCOURGE OF A NATION AS A WHOLE, NOT ONLY ITS
  COUNTRYSIDE

- •UNDERDEVELOPED PERIPHERY IS TO A LARGER EXTENT
  CAUSED BY AN UNDERDEVELOPED CORE THAN BY
  UNEQUITABLE DISGRIBUTION OF PUBLIC GOOD

- •A VILLAGE WILL NEVER BE AS POPULOUS AS IN THE PAST:
  QUALITY, NOT QUANTITY, OF HUMAN CAPITAL IS CRITICAL

- •ALTERNATIVE LAND-USE IS URGENT IN PERIPHERAL AREAS;
  A RELIABLE SOCIAL SAFETY NET FOR RURAL RETIREES IS
  URGENT AS WELL

FIGURE 5.5 Why Russian Villages Die: A Westernizer's View.

growth poles in the countryside in order to concentrate investment into residential construction and social services.

The precursor to this idea dates back to 1928-29 when hamlets (*khutora*)
became labelled legacies of the accursed past while large "socialist" villages
were called the seeds of a shining future. At the time of the dramatic shift
to collectivization, dispersed settlement in the countryside was viewed as a
breeding ground of peasant insurgency and a hotbed of bourgeois individu-
alism. From the vantage point of the Soviet establishment, large nucleated
villages always looked better because they lent themselves to easier control
and micromanagement than shadowy hamlets did. Only later did the notion
of economy of scale come to reinforce what initially had been an ideologically charged concept.

Needless to say, the consolidation of settlements was enforced mostly, if
not exclusively, in areas where villages were small, that is, in the non-
Chernozem regions (Table 5.5). As is typical for the rural realm in general,
the size of a settlement has been historically a function of the acreage of
arable land within walking distance of a village. An important mediator of

TABLE 5.5 Average Number of Residents per One Rural Settlement in the Regions of the European USSR.

Non-Chernozem Zone		*Volga*	*Chernozem Center*	*N. Caucasus*	*Byelorussia*	*Ukraine*	*Baltics*
Year							
1897	135	447	391	1494	110	331	No data
1979	122	389	355	894	165	655	26

Source: Ioffe, G.V. *Selskoye Khoziaistvo Niechernozemya.* Moscow, Nauka 1990:32.

this cause-and-effect relationship has been the spatial pattern of plow-based land use. Such use could be *selective,* in the sense that only part of the land is used, as in a natural forest, or *all-around,* as in a steppe. In compliance with the layout of vegetation regions, the character of rural settlement in the European part of the former USSR experienced a gradual spatial change from the northwest to the southeast. A dense network (up to 30-32 settlement units or clusters per 100 sq km) of small villages existed in the northwest, while a sparse network (4-5 units per 100 square km) of large villages dominated the southeast. Aside from the Baltics, the tiniest villages were most numerous in Byelorussia, and the Novgorod, Pskov, and Smolensk provinces of Russia. In many other forest-based provinces of Russia, like Yaroslavl, Vologda, Archangelsk, and Kostroma, tiny villages (below 10 people) used to be the largest single size category.[9]

Already by the beginning of 1941, 689,245 peasant households were moved from the smallest villages to large ones in Byelorussia, the Northern Ukraine, and 16 provinces of European Russia.[10] However, the far-reaching program of village consolidation was thwarted by the war, and was subsequently assigned to the back-burner during the years of post-war recovery.

It re-emerged only in the early 1960s, in a much more relaxed political atmosphere, a product of Khrushchev's Thaw. This time the economy of scale (i.e., increasing returns through enforced spatial concentration) was more the concern than anything else. (After many years of Stalinist repression nobody feared peasant insurgency any more.) All rural settlements were divided into two categories: a) those to be retained and subject to investment (so-called "*perspectivnye*"), and b) those not to be retained whose inhabitants were subject to resettlement (so-called "*nieperspectivnye*"). These categories quickly became household names in the countryside and beyond. Later, in the 1970s, following the official acknowledgement of poor agricultural performance in non-Chernozem regions, this approach became radicalized in that the number of designated growth poles was significantly reduced. For example, when in 1974, as a Moscow University graduate, one of the authors was assigned to participate in the physical planning of the

Uglich district of the Yaroslavl province, only 33 of the 550 villages in the district were designated as *perspectivnye*.

A consequence of this approach was a ban on all sorts of new construction in the majority of villages, i.e., in those destined to be abandoned. This ban spurred the spatial concentration of rural services (however squalid they might be), a process that had been unfolding spontaneously anyway due to rural depopulation. But it also triggered many local conflicts concerning construction permits in demographically viable villages but not in the promoted category.

The quintessential irony is that as has usually been the case in Russia -- pre-Soviet, Soviet, and post-Soviet -- prohibitive measures were enforced more vigorously and consistently than positive incentives. Bans and restrictions became the mainstays of a policy, while permits and allowances sat on the sidelines. Changes in political climate do not seem to affect this pattern; what they do affect is the people's ability to work around the policy-makers. Even in the 1930s it was possible for people to escape arrest by abruptly changing their place of residence and thus falling out of a targeted pool. But what happened only on occasion in the '30s was a matter of course during the 1960s and '70s: at that time a rural family consigned to another village was unlikely to follow the decree. If anything, such a family would rather abandon the countryside entirely and move to a nearby or distant town.

The heated public debate that later emerged over the policy and strategies of village consolidation overlooked this discrepancy between policy and action. On the one hand, plans for consolidation requiring that rural households be moved to larger villages were systematically frustrated and in most cases not undertaken, and faked reports were concocted and filed; on the other hand, bans on new construction actually did work, thus undermining life in many villages fairly accessible from a city simply because they did not happen to be large enough to be promoted.

The public debate, however, focused on the facade of the policy, i.e., the resettlement guideline (which in post-war time never amounted to forced resettlement), and whether large villages are indeed preferable to small. Such debates over phantom issues are not uncommon in Russia and elsewhere, while real, actually unfolding processes stay undetected by the public eye.[11] In reality it was the spontaneous interplay of migration and births and deaths that promoted some villages while subjecting others to decay. As a rule, in the vicinity of a growing town the majority of rural settlements were viable, no matter how large or small they were; but in the outlying areas, the majority of villages were fading away, also regardless of their size.

Such substantial urban-directed shifts in the rural population's spatial distribution have been largely spontaneous rather than pre-planned. It is a little wonder that the policy of village consolidation was first abandoned in the Moscow and Leningrad provinces and only later elsewhere.

TABLE 5.6 Regional Percentage of Rural Homes with Specific Amenities (1994).

Region	Plumbing	Flush-Toilet	Central Heating	Bath-tub	Natural Gas	Hot Water
Russia Total	33	22	22	19	72	11
North	20	16	20	14	48	11
Northwest	34	31	31	28	70	24
Province of Leningrad	51	50	52	47	67	42
Center	39	31	29	28	79	19
Province of Moscow	57	51	53	48	70	44
Volgo-Viatka	32	17	20	14	76	6
Central-Chernozem	28	22	21	19	82	9
Volga	30	18	24	15	82	9
Northern Caucasus	43	24	15	22	74	11
Ural	31	17	23	14	73	7
West Siberia	36	22	22	15	69	7
East Siberia	13	9	13	8	39	6
Far East	26	24	30	22	35	11

Source: Sravnitelnye Pokazateli Sotsialno-Ekonomicheskogo Polozheniya Naseleniya Regionov Rossiiskoi Federatsii. Moscow, Goskomstat 1995:263-326.

Rural Infrastructure

The condition of the rural infrastructure exacerbated the core-periphery pattern of rural population change. On a highly aggregated, province-to-province level, the outstanding positions of the Moscow province and the Leningrad province in European Russia could be demonstrated with respect to virtually all kinds of social facilities. Figure 5.6 singles out plumbing. Central heating (oftentimes the only alternative to the traditional firewood and/or peat burning stove, it is the urban solution of having a centrally located "boiler-house" that serves all or most dwellings in a village) is characterized by the same spatial distribution. The latter is all the more important since in Russia's countryside the spread of electric heating devices is fairly limited.

The only kind of amenity, the availability of which has brought Russia's countryside close to its urban areas, is *natural gas* for cooking. It is mostly supplied in gas-bags unless a locality happens to have the advantage of access to a nearby pipeline. According to Table 5.6, about two-thirds of all

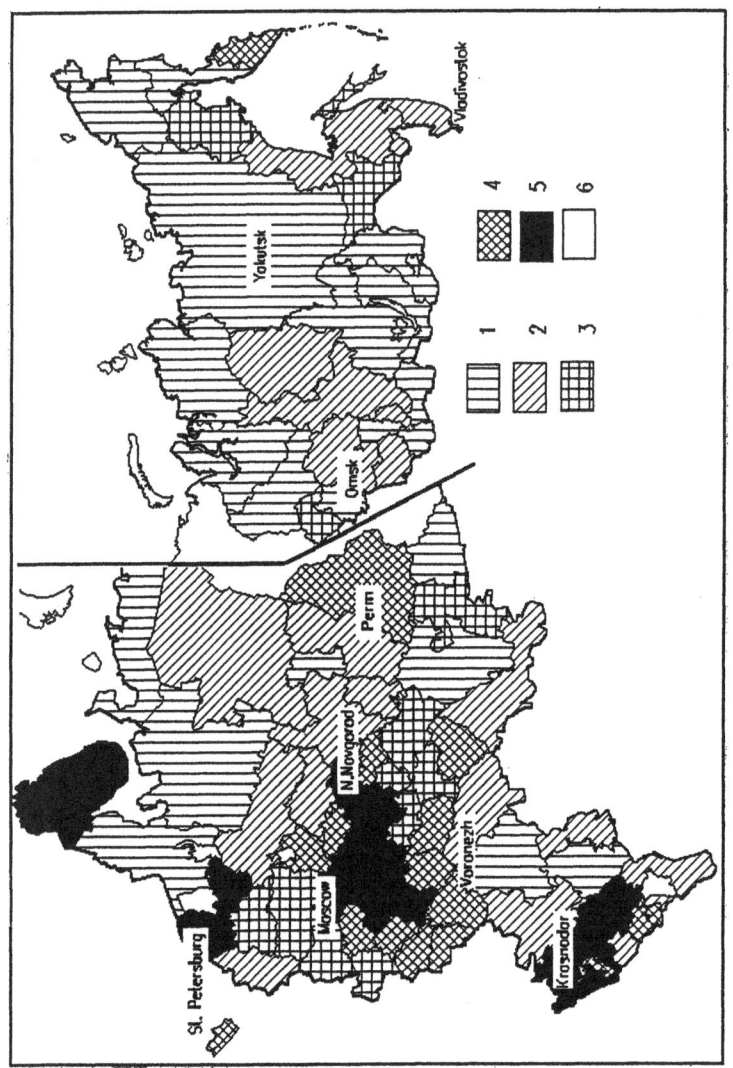

FIGURE 5.6 Percentage of Rural Houses (Other Than *Dachas*) with Plumbing. 1: 1-20%; 2: 20-30%; 3: 30-40%; 4: 40-50%; 5: 50-90%; 6: Data unavailable.

rural dwellings in Russia were served by gas in the late 1980s-early 1990s. Since natural gas is in most cases supplied for cooking only, running hot water for bathing and cleaning purposes is ubiquitously unavailable in country houses, even those served by central heating. In the majority of villages located outside immediate urban hinterlands and beyond Moscow and Leningrad provinces at large, *electricity* arrived only between the late 1960s and the early 1970s. In the Non-Chernozem Zone there are many provinces (Archangelsk, Novgorod, Pskov, Vologda, Chuvash and Mari republics, to name just a few), in which no less than one-half of all rural settlements were lacking electricity as recently as 1966.[12]

It would be appropriate to qualify the rural infrastructure in most of Russia's countryside as *rudimentary.* The following brief description of a residential setting mostly applies to *outlying districts,* as exurban ones enjoy upgraded standards.

The *residential setting* is represented by dwelling structures of three types given in descending order of number of residents:

- *Individual family log houses* with or without wooden siding assembled from trunks of coniferous trees; a foundation is optional. These are one or two bedroom houses, single-, rarely two-storied, with relatively high-pitched roofs to offset huge snow accumulation, and with fretted, sometimes quite fancy exterior window framing. In most cases a barn and a cattle-shed are attached to such dwellings. A hand-cranked well is located on a personal plot (usually 0.6 ha); in many cases construction of the well predated the house itself. A wooden outhouse stands beside the dwelling structure; as a rule, no indoor plumbing is available. This type of dwelling still dominates rural areas even though its proportion in new construction has been dwindling: in 1961-65 it was as high as 81%,[13] and now it is around two-thirds.

- *Serial pre-fabricated* one- and two-storied *brick and panel homes.* These buildings designed for rural areas have been in use since the late 1960s. However, they do not have a good reputation and have been typically assigned to newcomers, especially those unwilling to take root in the area. While such houses are served by piped water and have indoor toilets (not flush toilets, though), they are usually more difficult to heat than wooden houses, and dampness is their scourge, as well as bad ventilation. A "house that does not breath" is how country people would describe such a structure. Of no less importance is that, contrary to tradition, outbuildings for cattle and forage are not attached to these pre-fabricated structures and are usually at a distance from them, which makes it difficult to take care of cattle early in the morning during the cold season.

- *High-rise apartment buildings,* usually five-storied. These structures are an urban variety and are usually deemed least appropriate for the countryside despite the set of urban amenities that come with them. A fair number, however, are located centrally in collective and state farms and in district centers, many of which have a rural status. Major criticisms of high-rise apartment buildings replicate those aimed at serial prefabricated houses. To these must be added a lack of sufficient in-door storage space and cleaning facilities, which are especially important in rural areas when living space is confined to a small urban-style apartment. In a log house one would have at least a spacious passage to use as a hall or a corridor where dirty clothes can be shed. Usually only people not engaged in agriculture and lacking cattle of their own would occupy such a dwelling in the Russian countryside. Even so, it would not be considered attractive because it contradicts the rural way of life.

There is a huge difference between rural and urban housing stock in Russia in terms of ownership. Even before the privatization campaign of the 1990s, private dwellings dominated the countryside. In 1994, when that campaign was already in full swing in the cities, 39% of urban residences were in private ownership, whereas in the country, 81% were.[14] A privately-owned residence in the country is usually maintained by no one but its owner, as the needed support services are virtually non-existent, not in the least because of the low buying power of rural consumers.

Until recently there has been no dwelling structure in Russia that combines both amenities *and* accommodations designed for the rural way of life. One would normally expect *either* the former *or* the latter, so that in public consciousness "village" (*derevensky; derevnia*) is identified with backwardness and unsophisticated taste. An ever-increasing number of exceptions do exist in exurban areas, but they largely confirm the rule.

Other aspects of the rural infrastructure deserve brief characterization.

Most rural *retail facilities* are outlets with both food and non-food inventories, so-called "items of daily need." Only the largest villages and district centers have department and general stores. Currently retail outlets offer a wider variety of goods than ever before, but many items are unaffordable for the vast majority of the populace. Prior to 1992 when food prices were heavily subsidized, a typical selection of food in a rural retail outlet would include: bread, vodka and/or low-grade alcohol (popularly called portwine), macaroni, groats (pearl-barley, fine-ground barley, oatmeal, and, rarely, much favored buckwheat); canned fish, low-grade smoked and salted fish; canned pickles, and a few varieties of candy and cookies. Bread and vodka are essentially the main, if not the only, food items regularly purchased by country people. The vodka is not just for one's own consumption. In fact, half liters of vodka have long been a kind of hard currency in Russian villages. If, for example, a pensioner needed his/her (usually her, as old women dominate the village) land plowed or some repairs made, he or she would typically pay with vodka. The only non-food item of equal importance in rural retail outlets is matches.

It should be noted that a substantial number of villages in outlying districts are actually inaccessible during at least two "seasons of bad roads," the autumn and the spring. If a summer happens to be rainy, they are isolated for a third season. Only winter provides reliable accessibility. So in many cases rye and wheat flour are delivered in winter and people are supposed to bake bread themselves in an oven.

Heavy drinking, theft, and bad roads have always been a scourge in Russia. Of those, *roads* constitute the area to which major efforts of improvement have been applied since the late 1960s, but paved roads arrived in too many places all too late, after *bezdorozhye* (a term capturing both the absence of roads and the impassability of those that are available) had already drained them of population. Also the quality of pavement leaves much to be desired, the roads are too susceptible to temperature change, and every spring costly repairs are required.

The Russian *bezdorozhye* is especially vivid against the backdrop of West Europe (see Chapter 1). In 1994 Russia's total length of paved roads per 1000 sq km was 27 km, up from 19 km in 1980 (Figure 5.7). In the most densely-packed provinces of European Russia it is substantially higher; for example, in the Industrial Center and in the Central Chernozem region there are 163 km of paved roadways per 1000 km, while in the two leading provinces of these regions, Moscow and Belgorod, paved roadway density is 308 and 218 km respectively.[15]

Health care facilities in the countryside are mainly represented by so-called feldscher's practices; the respective practitioners are qualified to perform only routine tasks, so that any surgeries or chronic conditions require long-range trips to a district or even to a provincial center. In order to call for an ambulance one has to first cover the distance to the nearest phone. *Phones* are still a relative rarity in the country, especially home phones. Usually only collective farms' offices or rural Soviets have phones; a few very distinguished public personae may have telephones in their homes. But even when solicited, an ambulance may decline to come. The reason may be lack of gas, desperate roads, a broken engine, carelessness, or, most importantly, the attitude that rural people are inferior. It is not uncommon that a caller who has already covered 2-5 km on foot to the nearest phone is required by a hospital telephone operator to solicit a feldscher's confirmation of actual urgency. That may mean covering extra distance to another village. While on geographical travel in the 1970s and the '80s, we witnessed denials of urgent medical services in situations in

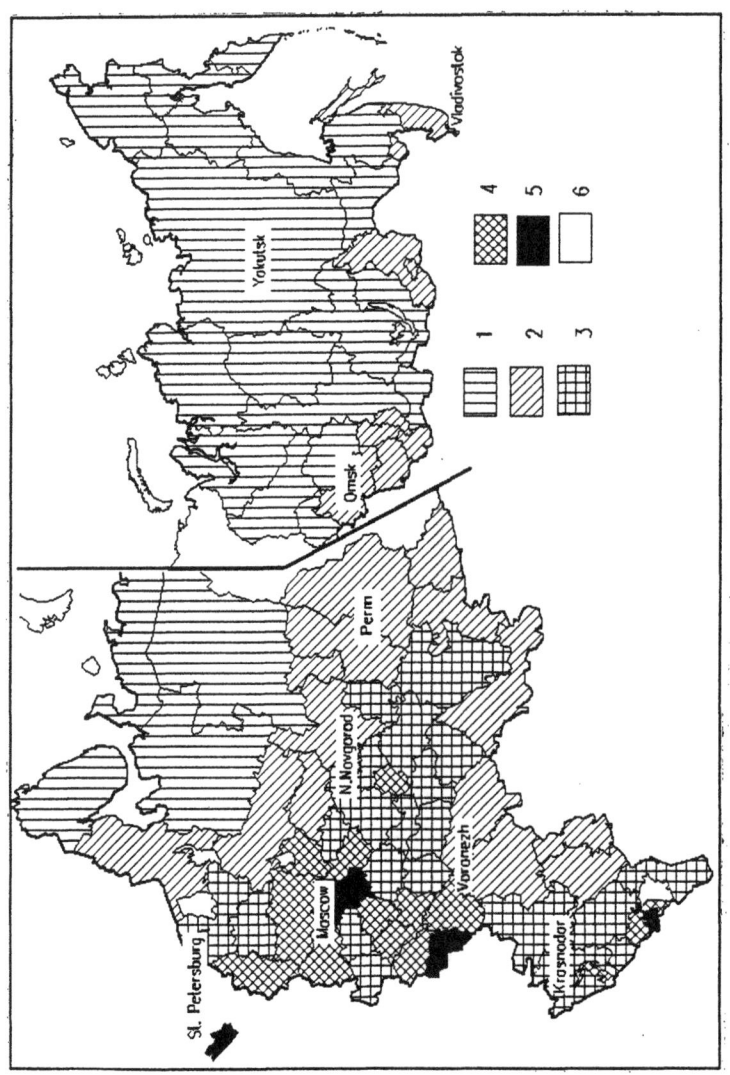

FIGURE 5.7 Length of Paved Roads in Kilometers per 1000 Square Kilometers (1994). Based on: *Sravnitelnye Pokazateli Sotsialno-Ekonomicheskogo Polozheniya Naseleniya Regionov Rossiiskoi Federatsii.* Moscow, Goskomstst 1995:250. 1: 1-20; 2: 21-100; 3: 101-150; 4: 151-200; 5: 201-310; 6: Data unavailable.

which urbanites would be guaranteed prompt treatment. In the late 1980s the rural/urban ratio in per capita ambulance services was 1:8.[16]

Rural schools, as in many other countries with sparsely settled regions, are caught between maintaining a reasonably high quality of facilities, which is feasible where there are large enrollments, and improving accessibility, which would bring about many tiny schools.

In rural Russia with its atrocious roads, concentration of schooling services was a preferred policy in the 1960s and the 1970s; it was pursued as a collateral to village consolidation. In practice it implied setting up boarding schools, whose districts would include dozens of rural Soviets. But because many parents were reluctant to send their children to such schools, the unavailability of a normal, day-time school nearby sentenced many villages to abandonment. Currently the policy is to maintain as many elementary schools as possible, for as few as two students, but schools with higher grade levels still tend to locate in a few larger settlements.

Day care facilities are also available only in the largest villages.

As noted earlier, the above picture is especially characteristic of agricultural settlements in outlying areas, from which it would take two or more hours to reach the nearest city with a population in excess of 100,000 by public transportation. Myriads of abandoned villages dot such areas. Many of them are heart-wrenching spectacles: empty, decaying houses, their windows nailed up with lumber, orchards run wild, nearby fields given over to forests. A few old women still languishing in the neighborhood congregate in one of the houses to wait out winter, so that at least one of them might somehow signal for help if needed. Some are visited by relatives in summer who help with home repairs and dig soil for a vegetable garden. Some are never visited; some simply do not have family anymore.

The closer an area is to a big city the better the quality of its rural infrastructure is. Some rural geographers were puzzled when they tested the regression models in which rural population change at a low spatial level (a district or a rural Soviet) served as a dependent variable. It usually turned out that the quality of rural services did not work as a reliable predictor of rural population dynamics. One researcher hypothesized that "either the set of independent variables is imperfect or the quality level of rural services is spatially homogenous."[17] Subsequently this problem was dealt with in-depth in a doctoral dissertation by Alla Igudina.[18] A major result boiled down to the following: local variance in availability and quality of social infrastructure is quite real; however, it is dwarfed by the variance in the accessibility of urban centers. Only by virtue of commuting to towns can country people make up for what they are deprived of at home.

Beginning in the late 1970s, many depopulated and abandoned villages have been experiencing a peculiar revival: urbanites from the largest cities pursuing second dwellings. People confined to their apartments all year always entertain the idea of getting a country house if only for seasonal, mostly summer, use. We will elaborate on this in Chapter 13.

Notes

1. Examples include: Belenky, V.R. *Problemy Agroindustrialnykh Poseleniy.* Moscow: Mysl 1979; Markov, E.M., Butuzova, V.P., and Taratynov, L.P. *Gradostroitelnye Osnovy Razvitiya Malykh Naselionnykh Miest Niechernozemnoy Zony RSFSR.* Moscow: Stroiizdat 1984; etc.

2. Examples include: Alexeyev, A.I. *Mnogolikaya Drevnia.* Moscow: Mysl 1990; Ioffe, G.V. *Selskoye Khozyaystvo Niechernozemya: Territorialnyye Problemy.* Moscow: Nauka 1990; Zayonchkovskaya, Z.A. *Demograficheskaya Situatsia i Rasselenyye.* Moscow: Nauka 1991; etc.

3. Roosevelt Priscilla. *Life on the Russian Country Estate.* Yale University Press 1995: p. xii.

4. Zayonchkovskaya, Z.A. *Demograficheskaya Situatsia i Rasseleniye.* Moscow: Nauka 1991, p. 41.

5. This dissimilarity was first discussed in the Introduction.

6. See for example: Saville, J. *Rural Depopulation in England and Wales, 1851-1951.* London: Routledge 1957 and scores of other sources.

7. Ruppert, K. "Sozialbrache in Westdeutschland." *Agrarwirtschaft* 1959, Vol. 8, No 3, pp. 69-77.

8. *Peasant and Peasant Societies. Selected Readings.* Edited by T. Shanin. London - New York: Basil Blackwell 1987, p. 126.

9. For the most detailed treatment of this issue see: Kovalev, S. A. *Geograficheskoye Izucheniye Selskogo Rasseleniya.* Moscow: MGU 1960. According to *Selskiye Naselionnye Punkty RSFSR.* Moscow: Goskomstat 1990, pp. 12-29, the average size of rural settlements across economic regions of European Russia in 1989 was as follows: 117 residents in the North, 76 in the Northwest, 107 in the Industrial Center, 188 in VolgoViatka, 321 in the Central Chernozem region, 404 in the Volga region, and 965 in the North Caucasus.

10. Ioffe, G.V. *Selskoye Khoziaistvo Niechernozemya: Territorialnye Problemy.* Moscow: Nauka 1990, p. 111.

11. This phenomenon was reflected in the following joke circulating among urban intellectuals in the 1970s Russia. An old woman approaches a rabbi and complains that hens are dying in her hen coop one by one. "Take a twig and draw a triangle in front of your hen coop," advises the rabbi. The old woman complied but came by again in a couple of days. "Rebe, hens are still dying," said she. "Take a twig and draw a circle in front of your hen coop," was rabbi's advice this time. Upon her third visit Rebe advised her to draw a trapezium. Soon afterwards the woman came yet again in tears. "That's it," she said, "my hens have all died." "Oh, what a pity," replied the rabbi, "I've still got so many ideas."

12. *Problemy Razvitiya Selskogo Khozyaistva v Niechernozemnoy Zonie.* Moscow: Kolos 1967, p. 181.

13. Alexeyev, A.I. *Mnogolikaya Derevnia.* Moscow: Mysl 1990, p. 116.

14. *Sravnitelnye Pokazateli Sotsialno-Ekonomicheskogo Polozheniya Naseleniya Regionov Rossiiskoi Federatsii.* Mocsow: Goskomstat 1995, pp. 291-299.

15. Ibid., p. 250.

16. Alexeyev, A.I. *Mogolikaya,* p. 122.

17. *Voprosy Sotsialno-Ekonomicheskoi Geografii Vierkhnievolzhya.* Kalinin: KGU 1983, p. 78.

18. Igudina, Alla. *Sotsialno-Geograficheskiye Factory Dinamiki Selskogo Naselenira.* Unpublished Doctoral Thesis. Moscow: IGAN SSSR 1982.

6

Crisis and Reform in the 1990s: The Economic Aspect

Present State of Knowledge

As recently as the late 1980s few Western analysts focused on the Russian countryside, on the regional patterns of agriculture, or on the processes unfolding in the realm of Russian villages. The Soviet Union's potential demand for grain imports and the inhumanity and other adverse implications of collectivization were topics commanding more attention.

Early in the 1990s the situation changed. A sudden and unprecedented Russian openness in regard to outsiders looking for statistics and interested in field trips attracted scores of commentators as did the pleasurable anticipation that something "hot" that was going to unfold in the wake of reform. As a result, in barely four to five years an extensive body of Western information about Russian agriculture grew up almost from scratch, and some true leaders in this area emerged.

The efforts of Don Van Atta were the first to stand out. He was both contributor to and the editor of the pioneering 1993 volume on the agrarian transformations in Russia. *The 'Farmer Threat'* is an informative account of pre-reform agricultural arrangements in the USSR and of reform itself up till 1992, including changes in the ownership of land, and in the management of large farm production units, and including the emergence of private farming and the obstacles in its way.[1]

The most impressive chapter of the book is by Stephen Wegren. Unlike the work of other authors, his approach is not merely descriptive (what happened?) and prescriptive (what must happen so that reform can succeed?) but also truly contextual and exploratory. Apparently inspired by his field experience in rural Russia, Wegren poses incisive questions about what people are becoming peasant farmers and whether truly independent peasant farms are being created. Commenting on these issues Wegren shows that actual developments in the countryside are confusing and appear to have little to do with the actual reforms. In particular, Wegren observes that the rural conservatism and demographic characteristics of rural Russia are hardly conducive to the development of private farms.[2]

While *The 'Farmer Threat'* has greatly contributed to the West's understanding the Russian agrarian scene, it also manifests shortcomings peculiar to Western interpretations of that scene. Some of them have to be attributed to the nature of the subject itself, Russian agriculture, as it goes through transition-triggered turmoil that generates a lot of contradictory information. The issue of land ownership exemplifies these shortcomings most clearly. This is not surprising, since even Russian-based commentators find it difficult to put events in perspective and extract inferences.

However, the shortcomings result, at least in part, from Western researchers' timing, inflated expectations, and insufficient historical research. Even in those cases where history is nominally factored into the train of an author's thought, agricultural reform in Russia is nevertheless studied as if it were unfolding in a vacuum or on an empty canvas. What is largely ignored are the implications from the communal experience of Russian peasantry. Ironically, some journalists whose period of accreditation in Russia normally exceeds that of most PhDs, had moved ahead of the latter in noting the importance of rural Russians' attraction to collective farming.[3] This attraction, in fact, casts additional light on why the notorious "farmer threat" has not yet materialized and is not about to do so. It also suggests that had more attention been paid to the numerous antecedents of current processes and to the scholarship addressing them, Western researchers' results might have been more meaningful and lasting.

To be sure, the ability to comprehend different cultures on their own terms has never been widespread. It is in fact a rare quality, a talent, which is why Stephen Wegren's insightful analyses stand out: they go beyond what we are used to expect from a person without extensive life experience in Russia. In his works published after *'The Farmer Threat,'* Wegren calls into question Western infatuation with private farming as the mainstay of Russian agrarian reform. Instead he shows persuasively that tenacious cultural tradition affects current developments even more than does the ongoing agrarian reform. Commitment to collective forms of economic activity and to egalitarian norms of profit distribution appear too powerful to be undermined quickly, if at all. Wegren addresses "the basic incongruity between reform goals and reform behavior and institutions"[4] in Russia and the lack of selective incentives that would elicit political support for those goals from at least some country people. During several field trips to Russian villages Wegren identified several political, economic, and cultural obstacles to reform as they apply to localities.[5]

However, despite all his insight, Wegren still acknowledges that "the lack of spontaneous decollectivization by collective and state farm workers themselves" was "one the most striking features of land reform in Russia."[6] This comment confirms to us that Russia's *cultural continuity* in action confuses Western commentators of Russian agrarian reform even more than anybody who has long been exposed to the writings of Nikolay Berdiaev, Alexander Akhiazer, etc. and/or to life experience in Russia might imagine. Ironically the West's own scholarship on Russian agrarian history is not at fault at all. Suffice it to note Robinson's superb book *Rural Russia Under the Old Regime*, which sustained nine editions. Also *Rural Russia Under the New Regime* by Victor Danilov has been available in English. But as Tucker pointed out, historical knowledge is not assimilated by those writing on current affairs,[7] while historians as well seem to be confined to their own business. This situation is best captured by the Russian word "*vedomstvennost,*" which defies direct translation. It refers to the separation of the functions of different departments responsible for the same entity, such as having two agencies in a pants factory, one to preside over the production of left trouser-legs, the other for the right ones.

Just as history is shortchanged in a typical Western analysis of today's agrarian scene in Russia, so too is protracted rural out-migration. We are rightly told, for example, that out-migration has just been reversed and that earlier it had drained the countryside of much of its people, especially younger people and women.[8] But neither the more far-reaching qualitative effects on human capital of this exodus, nor its polarizing influence are taken up. If they were considered, they would likely further cool some of the excitement in Western academic circles in regard to rural reform in Russia.

The flurry of Western academic writing about the Russian countryside will one day likely subside in much the same way that the surge of private farming in Russia already has, but to date this body of scholarship continues to grow.

Pre-Reform Condition

Current reform has been ripening for quite some time. Although the situation in agriculture was short of an outright economic catastrophe in the 1980s, sustained moderate growth in production and consumption was occurring at a disproportionately high cost. Keeping retail prices low at the expense of ever-increasing subsidies became a huge drain on the budget. Low efficiency and speedy reduction of farm labor exacerbated the problem. Still, it was the mid-1980s fall in the international oil and gas prices that seems to have prodded reform more than anything else, as it cut back on the ability to sustain huge imports of grain. That some kind of change unleashing initiative was imperative began to be recognized by the highest echelons of power in the late 1980s. It resulted in an attempt to promote internal cost-accounting in collective and state farms. The underlying idea was to switch from operation-based compensation for laborers to one based on results. In order to do that the whole cultivation cycle of a certain produce was to be assigned to a group of laborers whose payment schedule was to be set in advance. However commendable, this idea proved to be impossible to realize consistently in a collective farm whose general economic well-being did not depend solely on its performance and which was unable to rid itself of incompetent or lazy workers. Also the rural population did not demonstrate readiness for even a modest reform, so the idea began to degenerate into one more top-down formal campaign. Some innovations, though with difficulties, did make it to the rural scene. Among them was leasing land from socialized farms: usually ethnically non-Russian teams of vegetable-growers (e.g., Koreans from Central Asia) did that, as did construction teams from the Transcaucasus (mostly ethnic Armenians) and the autonomous republics of the Caucasus' northern slopes. Also the first semi-private, actually tenant, farmers appeared in some provinces and occasionally enjoyed the support of provincial administrations (as in the province of Orel). However, the attitude of the general public to all those pioneers was apprehensive, the dominant opinion being that if an independent farm is a success, it is because the farmer has special access to the administration and/or because he steals from a socialized farm. The fact that many of the pioneers were not locals exacerbated their alienation.

Along with internal cost-accounting on socialized farms, some steps were made to expand their economic rights. Thus in 1986, they were allowed to sell up to 30% of their produce to whomever they wanted at negotiated prices. However, with a market infrastructure being absent and farm administrators unwilling to take initiative, only a few, mostly exurban farms took the opportunity to sell vegetables and potatoes at urban-based peasant markets.

In the meantime the percentage share of subsidies was increasing; by 1989 it had exceeded 70% of the retail price of meat, 60% of that of milk, and 20% of that of bread. During the 1980s the overall sum of state agricultural subsidies increased from 25 to 100 billion roubles with three-quarters of the subsidies directed to animal husbandry.[9] This increase and growing food import were among the main causes of the national budget deficit. Already in 1990, the government was unable to pay in full for agricultural produce it was supposed to buy at official prices. The state-run distribution system thus began to crumble. Collective and state farms began to ignore state production plans. The government decided to alleviate the problem in April 1991 by setting higher food prices and promising to increase imports of grain fodder. However, the growth of salaries and wages got ahead of price hikes. As a result, all the signs of hidden inflation were already present in 1990: a short supply of many food items, long lines, and official limits on one-time purchases of deficient foods. Scared by looming food shortages and possessing enough money, the populace was sweeping everything it could off the food stores and putting it by for a rainy day. In response to this mass hysteria, provincial and municipal powers began to restrict food shipments across their corporate limits and introduced personal identification controls in local food stores to eliminate "aliens." Crop failure in 1991 and the reduction of food imports by 18% due to delayed payments on earlier purchases made the situation even worse. By November-December 1991 Russians spent hours in lines for bread, milk, butter, eggs, and other necessities. Lines could be avoided only at peasant markets but for exorbitant, to a rank-and-file citizen, prices.

In this way, a deep crisis of the whole food procurement system preceded the radical reform measures of 1992 and had resulted from previous economic arrangements governing Soviet agriculture.

Reform and Agriculture

The breakup of the USSR precipitated radical economic reform. The removal of state control over retail and wholesale prices became the first and the most publicized aspect of reform. As a result of this step, many food prices in Russia approached or even exceeded their Western counterparts. For example, in the summer of 1995 the average price of one kg of sugar in Russia was 94% of what it was in the US, beef was 75%, smoked sausage, 111%.[10] Between 1992 and 1995, the cost of a consumer basket composed of 19 standard types of food had grown from 11% to 50% of its cost in the USA.[11] Note that in 1994-95 the average monthly salary in Russia was $100-$140. True, on average it accounted for no more than 25% of a Russian's real earnings, most of which stayed unaccounted for and were derived from so-called side-employment; but even so, for many Russians scores of food items they had been used to became unaffordable. Shifts in per capita food consumption (Table 6.1) are characteristic in this regard: whereas the consumption of relatively cheap bread and potatoes, already quite high, had increased even more, expensive products were cut back on drastically.

Sharp retail price disparities entered the post-Soviet scene for the first time since the early 1930s. Table 6.2 featuring these disparities is not of interest for historians only: a much less significant growth of prices *after* 1994 (than that shown in the table for 1993-94) has left inter- and intraregional differentials intact. On a region-to-region basis, the biggest price hikes occurred in the north and east of Russia. By the end of 1991 the monthly price of a standard consumer basket did not vary much across Russian regions: virtually everywhere it cost between 150 and 200 roubles,

TABLE 6.1 Per Capita Food Consumption in Russia, in Kilograms.

	1989	*1993*
Bread and other Flour-Based Products	98	109
Potatoes	94	118
Vegetables, Melons, and Gourds	91	74
Meat and Meat-Based Products	75	58
Fish and Fish-Based Products	16	12
Dairy Products in Milk Equivalent	397	298

Source: Izvestia. 24 June 1994.

TABLE 6.2 Food Retail Prices (in Roubles) in Russian Cities in December 1993 and in December 1994.

	Potatoes		*Beef*		*Butter*	
	1993	*1994*	*1993*	*1994*	*1993*	*1994*
Archangelsk	700	1600	4000	8500	No Data	No Data
Vologda	220	750	2500	8000	No Data	No Data
St. Petersburg	228	905	4875	10650	11500	30000
Pskov	190	600	2200	10000	2300	10000
Moscow	379	1583	7227	15417	8000	32500
Yaroslavl	165	622	3100	8916	2710	7900
Orel	160	600	2300	6000	No Data	No Data
N. Novgorod	174	980	2988	8625	2900	16500
Cheboksary	90	800	2300	6000	2250	13000
Belgorod	83	990	1800	6000	2500	12000
Voronezh	130	1300	2400	9500	2500	13000
Ulyanovsk	120	1000	2200	5500	3000	8000
Samara	567	1100	2467	5683	3167	14000
Krasnodar	200	1500	3000	7000	3000	No Data
Nalchik	350	1000	1800	4000	2700	18000
Ekaterinburg	500	1300	3000	5500	No Data	No Data
Izhevsk	167	800	2500	5000	2200	14000
Omsk	400	1057	1896	4857	3860	14125
Krasnoyarsk	350	1200	3500	6000	No Data	No Data
Vladivostok	770	2500	4800	12500	No Data	No Data
Magadan	1000	2600	No Data	18000	No Data	No Data

Source: Statisticheski Bulletin 1 (APK) "Osnovnye Pokazateli Funktsionirovaniya APK v 1994 godu. Moscow, Goskomstat 1995 (computer printout).

accounting for about one-fifth of average personal incomes. But already one year later the Far-Eastern consumer basket was twice as expensive as that in the Volga region. It accounted for from three-quarters of the reported average income in the "poor" Northern Caucasus to two-fifths of that in "rich" West Siberia. In the fall of 1994, the same basket cost about 40,000 roubles in the province of Ulyanovsk, 200,000 roubles on Sakhalin island, and 72,000 in Russia on average. The joint analysis of income and price dynamics shows a high degree of correlation between the two: thus, the actual regional disparity in cost of living did not increase much because this indicator increased about the same, 3-5 times, in most areas. Still, there are provinces where no more than half of average real income has been lost; these provinces form two bands stretching north and south of Moscow. *Within* Russian regions, however, the highest losses in real incomes have been experienced in the largest cities which had long enjoyed a privileged status in terms of availability of food at low, state-controlled prices.

Aside from decontrolling prices, the government also shed its monopoly on foreign trade. Imported foods began to gush into Russia, first to the largest cities and then to smaller ones. The sagging buying power of consumers combined with an unrestrained flow of imports has quickly filled food stores' shelves. However, the reaction of people, long used to purchasing everything offered by the state-owned retail system and in huge amounts, has been ambivalent. The ensuing necessity to choose what to buy to fit one's financial means and the economic stratification of buyers has caused grumbling, especially among the needy.

As Barbara Severin pointed out, "in 1994, for the first time in nearly two decades, few Russian consumers have expressed worry about food shortages."[12] However, the food crisis had not disappeared but actually gone underground. It was no longer evidenced in empty food stores but rather in consumers' diets, whereas in the recent past it was all but the other way around: while food stores stood empty, people's refrigerators and dinner tables often appeared to be cornucopias.

However, Russian agriculture did not benefit from those exorbitant hikes in food prices because prices on agricultural implements, fertilizers, and maintenance grew even faster. Whereas in 1989 one tractor was the price-equivalent to 18 metric tons of wheat, by 1994 its cost had risen to 40 tons.[13]

Russians who are used to pointing to the US as a model for various undertakings are ill at ease when they learn that their increased price disparity is still far behind the respective American ratios (i.e., wheat/tractor and the like). Likewise Russia lags behind the US in terms of the ratio of retail food costs to farm-gate prices. Also, with all the alleged withdrawal of state support for agriculture, Russia's agrarians are still in a favorable position with respect to American farmers in terms of interest on agricultural loans adjusted to inflation. Specifically, the annual interest on 1995 agricultural loans in Russia was 180%, whereas the all-year food price growth index in 1995 was 230%.[14] Russian agriculture's advantage holds true if one com-pares the ratio of the budget subsidies to the agricultural output of the two countries.[15] Nonetheless, these statistical advantages do not cure Russian ills, as the differences in the quality of farming implements and ma-chinery, and especially in market infrastructure and human capital of agri-culture, are definitely in America's favor.

For Russian agriculture, though, one side-effect of the price reform was particularly essential: a switch to barter in many regions. This is largely due to a cash shortage, to the unreliability of financial clearing operations, and to local attempts at subsidizing food prices. All those factors also motivated many provincial leaders to introduce food rationing and, as an unavoidable result of rationing, to isolate one province from another by a kind of in-ternal custom control. Combined with increased transport fares and the termination of state subsidies for transportation of food, these factors result-ed in Russian provinces' beginning to resemble the Thunian isolated states, if not medieval princedoms.

Some provinces, like Orel (in the Industrial Center), Voronezh (in the Chernozem Center), and especially Ulyanovsk (in the Volga region, the homeland of Lenin) revealed particularly glaring cases of self-isolation. The leaders of these provinces preserve subsidized retail prices and food rationing. The difference between artificially low food prices and the pro-duction costs are covered by a special off-budget fund replenished by requi-sitions from local manufacturing. Some of the manufacturers, including *Aviastar*, a giant of the aircraft industry, receive aid from the federal budget, so this local subsidizing has national implications. In effect, aspects of a cradle-to-grave welfare state are being insinuated into the economy behind the facade of reform. Because such subsidizing serves short-term local inter-ests, it boosted local support of communist elements in provincial adminis-trations. Low food prices entail rigid control over outgoing ship-ments of food. Not only does it violate the existing Russian Federation law that secures economic unity throughout the Federation, it also hurts local farmers wanting to sell their produce in neighboring regions. As a result, food is stockpiled in excess of local demand. In 1994, the Voronezh gover-nor even ruled that local collective farms, joint-stock companies, and independent farmers must not only refrain from sending their produce to other provinces but also from processing milk and sunflower seeds, thus forcing them to sell these products to the state-owned local processing plants at prices set below production costs.

Changes in the subsidies and in land ownership have also shown up. The latter changes are difficult to sort out because land legislation has been in a protracted impasse and remains virtually deadlocked upon this writing. Wegren was exceptionally perceptive in pointing out that Yeltsin's October

27, 1993 decree upholding private ownership of land "was only a general political statement and was of no utility in guiding day-to-day activities."[16]

Before we turn to a more systematic coverage of legislation in the area of land ownership let us review views on Russia's agrarian future dominant in the nation's political circles.

Agriculture and the Political Process

Against the backdrop of crisis and reform, the agrarian planks in different parties' political programs were crystallizing in the early 1990s.

The first, constituent Congress of the USSR's Peasant Union took place in 1990. V. A. Starodubtsev, one of the coupsters of August 1991, became the Chairman of that Union, while most of the delegates to its first congress were provincial Communist Party *nomenclatura,* and chairpersons of collective and state farms. The main declared target of the Union was to preserve collective and state farms, and to protect them from the newly emerging elements of the market. The mainstream ideology of the Union did not go beyond habitual statements about exploitation of the village by the city and about the necessity of retaining all forms of centralized budget support for the agrarian sector of the economy. The Peasant Union officially exists today but it keeps silence as its political stance is taken up by the Communist and Agrarian parties. The independent platform of the so-called Agrarian party made sense only during a brief ban on the Communist party in 1991-93. Today the Agrarians have faded into the background simply because all of their objectives became incorporated in the Agrarian plank of the reinstated Communist Party platform.

West European experience has shown that the issues of survival of traditional forms of farming have been among the favorites of right-wing parties' agenda. Not in Russia, where to be *conservative* means to be on the left side of the political spectrum. This is confusing but is grounded in the fact that conservatives of all nations do share in common the yearning for values of the past, the values from which their nations' strength, pride, and other quintessential features reportedly or actually derived. It is thus a crucial dissimilarity between different nations' recent and distant pasts that brings about the above confusion. Russian conservatives, for example, bemoan what no Western counterpart of theirs would ever miss: the breakdown of *shefstvo* (see Chapter 3), i.e., a system of mandatory recruitment of urbanites to assist collective farmers during harvest and other campaigns. "The village is ultimately brought to its knees, it was left to face harvest alone"-- such was the semi-anecdotal reaction of the main communist newspaper *Pravda* on 12 June 1990 to the collapsing system of *shefstvo.*

A powerful block of Communists and Agrarians, enjoying the enthusiastic support of many provincial administrations, is being opposed by a rather disunited market-and-democratic movement. One of its components is the so-called *AKKOR* or Association of Peasant Farms and Cooperatives of Russia; others are the Cooperative Union, the Union of Land Lessees, and local associations of private farmers. These affiliations do not advocate a total removal of collective farming; they simply want to annul the monopoly held by the government and by socialized farms over peasantry. However, there are more radical market advocates, for example, the small Peasant Party rallied around Yuri Chernichenko, a well-known journalist of peasant background.

Although there is a substantial regional variation in this regard, by and large rural Russia stands out for its support of communists; the strength of the above political camps is thus far from equal. Russian peasantry of the 1990s proves to be even less ready to switch to the rules of the game that have long been cherished in the West (like private ownership and economic self-reliance) than it was during the Stolypin reform. The 1995 elections to the *Duma* proved this. Unfortunately, we have no electorate data on rural population only; however, the fact that regional voting is correlated with the percentage share of rural population has become a commonplace in Russian electorate research. Suffice it to compare Figures 6.1 and 6.2. In order to compile the map shown in Figure 6.1 we grouped 43 political parties and movements in three blocks: *anti-reform* (with communists having the leading role); *pro-reform* (for example, the parties headed by Chernomyrdin, Yavlinsky, Gaidar, etc.); and *patriots* (for example, the Liberal Democrats of Zhirinovsky, etc.). The taxons on the map (Fig. 6.1) reflect the ratio of anti-reform to pro-reform votes. First of all, the so-called red belt of Russia shows up in Figure 6.1; it includes Chernozem provinces and is characterized by the strongest rejection of reforms: the votes for communists and the like exceeds those for pro-reform parties by a factor of 1.3-1.5. The same provinces stand out for their heightened shares of rural population (Fig. 6.2). Note that these provinces' countrysides fared well under communists: once having been poverty-stricken and over-populated, they became relatively prosperous. There is, of course, no rigid correlation between the two maps (Fig. 6.1 and 6.2). Thus, the Northern Caucasus is a bit less pro-communist than the Central Chernozem region, and in the non-Chernozem countryside, whose agriculture was almost completely ruined under the communists, the electoral behavior is more reform-oriented. Inside provinces the elections' outcome is a direct match of districts' well-being, which itself is frequently influenced by location: exurban districts are typically more pro-reform, whereas outlying are more pro-communist.

Although upon this writing we do not have a full picture of the 1996 Presidential elections, it looks like the same features of electoral geography that were revealed in December 1995 have been retained. The first round

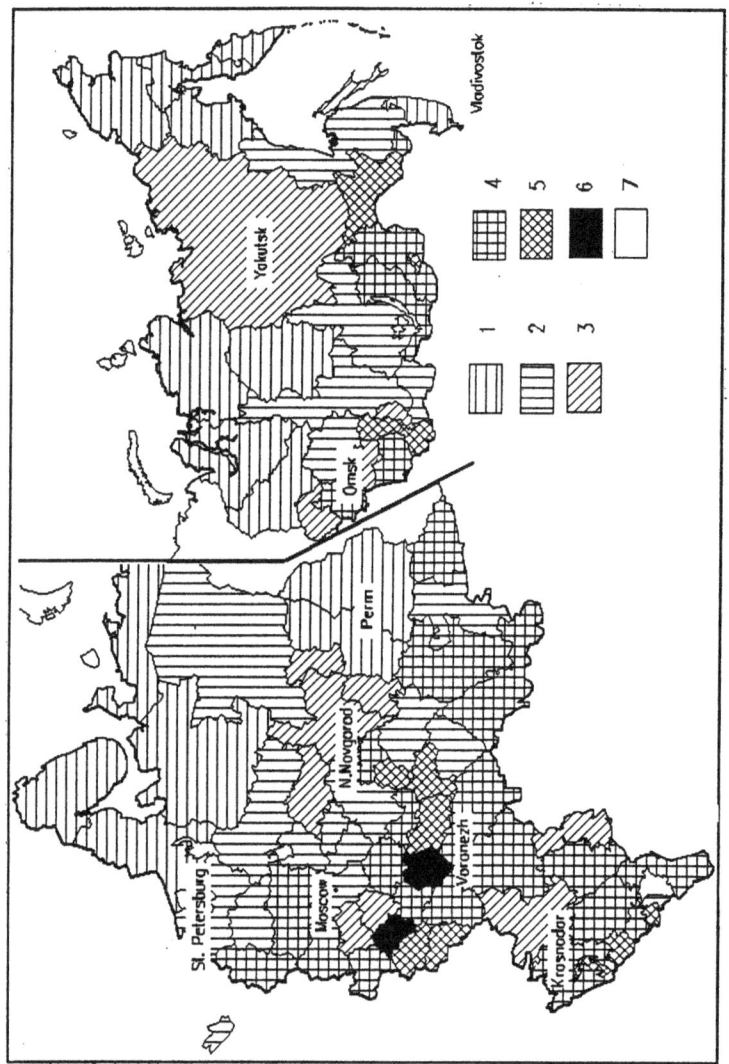

FIGURE 6.1 Ratio of the Number of Seats Gained by Communists to That by Reformers in December 1995 Elections to Duma. 1: 0.1-0.5 (most reform-minded); 2: 0.5-0.75; 3: 0.75-1.0; 4: 1.0-1.5; 5: 1.5-2.0; 6: 2.0-3.2 (most communist-minded).

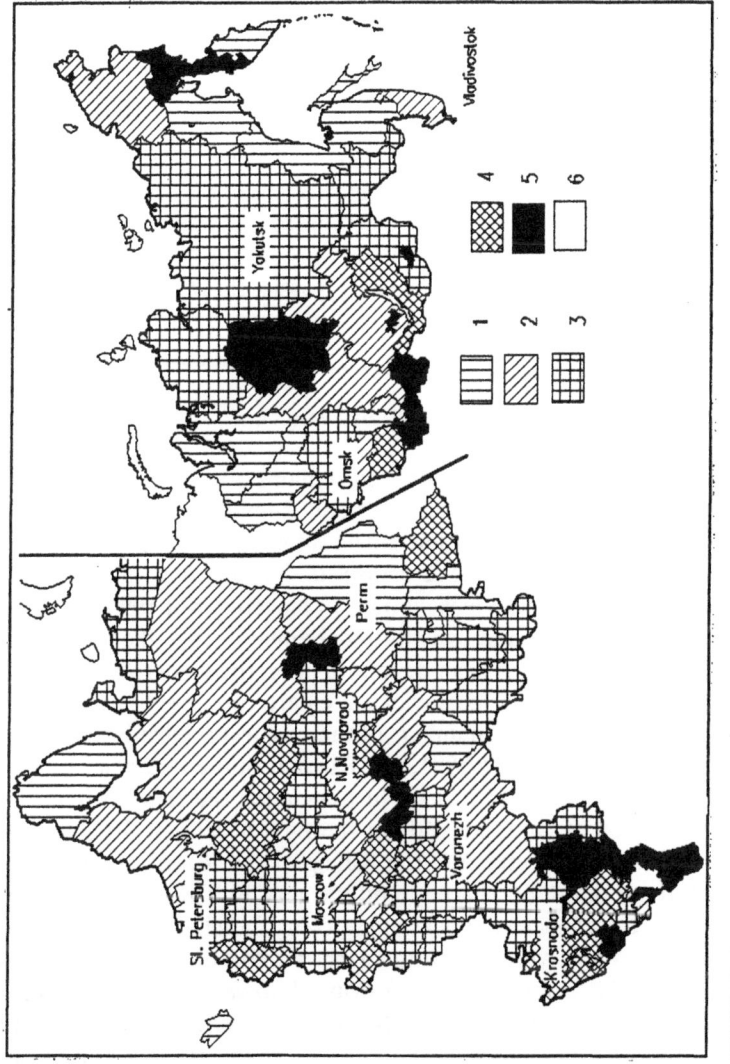

FIGURE 6.2 Rural Population as a Percentage of the Total Population (1995). 1: 8-20%; 2: 21-30%; 3: 31-40%; 4: 41-50%; 5: 51-100%; 6: Data unavailable.

of elections has boosted respect for the Russian domestic opinion polls (because they accurately predicted the number of votes for the two chief rivals), which is why a rural public opinion probe by *VTSIOM,* Russia's leading polling center, two weeks prior to elections may be considered quite revealing. Of 2000 rural respondents from all regions of Russia, 65% believed that the pre-1985 situation in the country ought to be retained. Only 20% support the March 1996 Presidential Decree on private ownership of land (see below). Whereas 39% of urbanites prefer the Soviet political system, in the countryside the respective proportion is 58%. While only 14% said they were intent on voting for Yeltsin, 31% were favoring his communist rival.[17] The preliminary results of the first round of elections (June 16, 1996)[18] largely coincide with the 1995 elections to the *Duma.* Country voters in Central Chernozem and in piedmont provinces of the Northern Caucasus, that is, the premier agricultural regions of Russia, threw their unequivocal support behind Gennady Zyuganov.

Ownership of Land

The issue of private ownership of land remains one of the key issues in the political stand-off between reformers and communists. The major hallmarks of this lingering confrontation can be summarized as follows. In 1989, the Supreme Soviet of the USSR endorsed the "inherited life-time possession of land with a right to lease it out." In November of 1990, the Soviet legislature adopted a law and constitutional amendments allowing private ownership of land. However, just one month later, the Congress of People's Deputies (of which the Supreme Soviet was a part, composed of active legislators on rotation with other members of the Congress) introduced 10-year moratorium on land sales. In 1991, the Russian Supreme Soviet promulgated the Land Use Code of Laws, which included the right to "inherited life-time possession of land" but without the right to lease it. In December 1991, the Presidential Decree and the Ruling of the Russian Government introduced the notion of shared land ownership (*collectivno-dolievaya sobstvennost'*). According to the new ruling, every member of a collective or a state farm could now withdraw from a farm with his own allotment whose size was to be determined at the provincial (*oblast*) level and usually amounted to 5-8 hectares. However, the land sales moratorium was still in place. In 1993, in a new Presidential Decree on the re-organization and privatization of the "Agro-Industrial Complex" and in an associated ruling of the Russian Government, the concept of shared agricultural land ownership emerged once again. Also, urbanites were allowed in effect to privatize land they used for *dacha*s (recreational dwellings), orchards, and vegetable gardens. In the 1993 National Constitution approved by an all-

Russian referendum, the right of private ownership of land was formally endorsed. However, no specific legislative acts following from this endorsement were put into effect by either the Supreme Soviet or, later (after its October 1993 forced dissolution), by the *Duma*. In 1995 a step forward, but one short of passing an operating law, was made when the *Duma* adopted the new Civil Code, which included the right to privately own land and to sell it. This step forward was immediately annulled by the same *Duma's* re-introducing a moratorium on land sales until the adoption of a new Land Use Code to replace that of 1991. In September 1995, this new Code was tentatively endorsed by the *Duma* pending some amendments. The Land Use Code was finally adopted by the *Duma* on May 22, 1996. But on May 24, during a pre-election meeting in Archangelsk, Yeltsin said he would veto the Code.

It needs to be pointed out that the draft of the Code compiled by the Agrarian Committee of the *Duma* and by the pro-communist Agrarian Party virtually emasculates all previous reform-minded efforts of the President and the Government.[19] In particular, the draft does not even mention shared ownership of land. It mentions only two types of ownership, state and private, but the latter is available not to private individuals but only to agricultural production units. Actually, it re-imposes the ownership monopoly of collective farms, whatever their current status. A member of a farm can receive compensation for his land share only if he withdraws from a farm no earlier than ten years following the final endorsement of the Code in question. However, the draft Code gives production units new rights: they may lease out land and may convert it into investment in other production units.

Thus the new Land Use Code is aimed directly at the retention of vast landholdings in the form of former collective and state farms, while granting no real rights to private individuals. To be sure, in due course this draft may be altered. However, its gist is so much against the grain of the policy of the Government that it impelled Yeltsin to issue his March 1996 Decree, "On the Realization of the Constitutional Rights of Citizens in Regard to Land."[20] This Decree has been applauded in the West as a revolutionary breakthrough in the deadlocked land-use legislature.[21] The reaction of the reform-minded Russian press has been much more skeptical.[22] Although the Decree obliges socialized farms to sign agreements with all their members about use of their land shares (upon a member's withdrawal from a farm the member's share can be leased out, presented as a gift, or sold to anybody -- but exclusively for agricultural use and only if other shareholders of the same farm decline buying it), the Decree is in force only until the adoption of the new Land Use Code by the *Duma*. The latter, however, has been quick to denounce Yeltsin's Decree, branding it "unconstitutional."[23] It therefore appears that the main intention of the Decree is once again to indicate the political stance of the President vis-a-vis the pro-communist *Duma*, making it clear that should Yeltsin retain power, the *Duma*'s legislative initiative on land will be vetoed.

Needless to say, the issue of ownership of land is shrouded in uncertainty, which opens the door for unofficial initiatives at the provincial and local levels. Since it is around major urban centers where demand for land is at its highest, a quite real land market has taken shape there, in particular in the provinces of Moscow and Saint Petersburg. But this market serves mostly *dachniks* (actual and prospective owners of recreational dwellings) and has not yet involved farms. Instances of exurban farms selling part of their land do exist (we have come across them oftentimes in our field trips), as a way of helping socialized farms survive, but the land deals involved are being concocted unofficially and are usually not made public.

In regard to agricultural *subsidies* Wegren's insights are again apropos. The "continued egalitarianism undermines the logic of reforms,"[24] he pointed out, and "the poorest farms receive the most help."[25] However, while both statements are correct, the latter does not seem entirely compatible with the former. In reality, the weakest collective farms do receive the most help insofar as the strongest are hit hardest by the deferment of payments for their produce. However, the current egalitarianism in disbursing state subsidies is actually a step forward compared to the earlier distribution in which subsidies were tailored to production costs. Prior to 1991 the principal subsidizing mechanism was the so-called *nadbavki* or farm-gate price mark-ups: the higher the actual production cost of, say, milk, the higher the mark-up set by the state. Needless to say, it effectively penalized economic efficiency. At least now, mark-ups are flat, although regions themselves have a right to modify them when they use funds from their own budgets.

Collective and State Farm Conversions

Even as early as 1990 the state was unable to pay for its own purchase orders for farming produce. The lingering collapse of the government-run procurement and purchasing system pushed farm production units to turn a blind eye to state orders and to pursue new market connections. So covert "privatization" of the collective-and-state-farm monolith had started even before Gaidar took over as acting Prime Minister.

The initial ruling of the government issued December 29, 1991, envisaged transformation of all collective and state farms into joint-stock companies (JSC) and partnerships with limited responsibility (PLR) or their breakup into a set of family-owned farms. Subsequently the option to retain collective and state farms was also added.

The implementation of this ruling resembled a haphazard political campaign and was even likened by some prominent analysts to the collectivization of 1929-35.[26] The objectives of the 1991 ruling were not explained to rank-and-file members. They became formal shareholders, and the signboard "*kolkhoz*" (collective farm) gave way to "JSC" or "PLR." But the actual meaning of the new signs never materialized: profit allocation in these production units continues to be based upon labor input, not on accumulated shares, and their production activity and accounting practices abide by the old rules.

The administration of such new old enterprises often has vested interest in de facto preservation of the collective farm system, so local bosses incite rank-and-file employees against real market changes. Administrators have found that buying up shares while reform has not yet gotten serious is very attractive, all the more so because a collective farm's property assessment is in many cases based on outdated prices, and the real value of property is unknown to current shareholders. Subsequently, according the 1991 ruling (later cancelled), the number of property-related and land-related stocks one receives depends upon one's length of tenure with a farm. In most cases that makes elderly workers larger shareholders than the administrators, especially in backwater, weak production units where leaders used to change frequently. Anyway, in many former collective farms it is the chairman, the chief bookkeeper, and the agronomist who are effectively accumulating shares. This management triumvirate is going to be the owner. In other words, "new" property relationships so far have only been conferring economic legitimacy to power structures that had been previously set up under the Soviets.

Since the activity of large agricultural enterprises and allocation of their profits have undergone little change, the decision to remove or retain the title "collective (state) farm" reveals people's attitude toward the overall economic reforms going on in the nation as a whole rather than toward specific farm reform. For example, the former autonomous republics of the Russian Federation and the ethnic Russian provinces of the Volga region as well as the majority of the Chernozem provinces have preserved the status of about 50% of their collective and state farms. These areas are much less in favor of the general economic reforms than areas like the Moscow and Leningrad provinces where the renaming of collective and state farms has been overwhelming. Interestingly enough, it is only in Chechnya that *all* collective and state farms have been annulled and all the agricultural output is being produced by about 1000 private farms.

In most places, however, the only tangible reform-triggered results achieved at the level of production units in agriculture by 1996 have been the collapse of the state-run procurement and produce distribution system and the removal of the government price control. These changes may have created a new economic environment, but they did not change the internal management practices inherent in collective and state farms; changes to their formal status have had but little real effect.

The Experiment in Nizhni Novgorod

In 1992-94 there were a lot of media reports on the idea of the provincial Administration in Nizhni Novgorod to transform huge and unwieldy collective and state farms into production units able to successfully function in a market economy.[27] For quite some time, however, the stir in the Russian media dwarfed the actual scale of the phenomenon: by the beginning of 1995 the experiment embraced only seven of the seven hundred socialized farms of the province. But subsequently the pace picked up appreciably and by April 1996 about 10% of the existing socialized farms had been affected.[28]

The idea for the experiment belongs to Boris Nemtsov, a young, ambitious, and reform-minded governor of the province, a confidante of a leading Russian economist (and a 1996 contender for the President's post), Grigory Yavlinsky. From the very beginning US experts supervised the implementation of the experiment.

The gist of the idea is as follows. At the first stage of the proposed transformation, a so-called conditional (without cash transactions) auction takes place on a farm. Its objective is to determine each member's size of land and non-land property share. The size of property share depends upon years spent on the farm and one's level of earnings. At the second stage the members-turned-shareholders choose the way to manage the property: they form close-knit collectives in which they accumulate their shares. At the third stage the farm is actually divided among these smaller collectives, openly with careful consideration of the cumulative shares and conflicting interests. The first auction took place in November 1993 at a large (over 3000 hectares) exurban farm. Soon it was divided into seven independent units: five farming comradeships, one maintenance, and one supply-and-procurement association. The land share of each member of the former farm was 6 ha; non-land property shares ranged from 5,000 to 700,000 roubles. The total farm land was divided without argument. One member of a former state farm who managed to successfully sell potatoes produced at his subsidiary plot and buy shares from 60 retirees was able to convert them into 350 ha of land and five tractors; thus an independent private farm was established.

The on-going experiment has already revealed many pluses and some minuses as well. Among the biggest pluses are openness of the privatization process and the lack of any administrative coercion; the smaller collectives in most cases show higher productivity, they use machines more efficiently and have cut back on hard drinking; reorganization has given a chance to energetic personalities to express themselves; eventually, for many bankrupt collective farms, a division of the kind described is all but the only way to achieve an economic cure. Actual and potential minuses include the following: maintenance and supply-procurement associations become local monopolists, inflate prices on their service and, then find that demand for their work plummets; the controlling interests of the farming comradeship are being concentrated in the hands of a few, which in the Russian countryside intimidates people; the tax burden has increased since each transaction between former parts of a socialized farm is subject to taxation, so in many cases such transactions evade documentation; and finally, the fate of the "social sphere" (housing stock, day care, local roads, etc.), formerly taken care of by the collective farm, is unclear.

Private Farming: Does It Have Future in Russia?

In early 1991 there were slightly over 4,000 private farms in Russia, in early 1993 their number had already risen to 183,000, and in early 1994 to 270,000. But by the end of 1994, the dramatic rise had stopped. During 1994, 36,000 units were newly created, but 29,000 then went out of business. Only 279,000 family farms were recorded in early 1995 and 280,000 in early 1996. Despite these imposing looking numbers, in 1994-1995, private farms occupied a relatively small chunk of Russia's agricultural land, 12 million hectares or 5%. But their share in the total agricultural output is about one half that in land.[29]

There are several principal reasons for the initial success of private farmers.

- In 1991-92 private farmers were assigned large landholdings, plus they could take advantage of an 8% bank credit at a time when national inflation figures were at 93% (1991) and 2600% (1992).
- In 1991 and in early 1992 the overwhelming majority of would-be farmers could and did purchase used and even new agricultural machinery at relatively low costs.
- Private farms' output was freed from taxation for two years after registration.
- They had ample reason to expect assistance in selling their produce.
- As a result of official encouragement, highly energetic people opted to become private farmers. These were largely rural intelligentsia (agronomists, livestock specialists, engineers, and machine operators), as well as some urbanites yearning for rural work who were enthusiastic and dedicated to reform.

Not all conditions, however, were favorable for private farmers. Their main problems were rooted in their relationship with collective and state farms. Rarely did this relationship include mutual assistance, let alone cooperation. Hostility to new farmers reigned supreme. In all fairness, however, not all those calling themselves private farmers have been the genuine article. Field observations made in mid-1993, that is, during the heyday of the private farming campaign, revealed that those registered as private farmers break down into three groups.[30]

- The *first* group is farmer-enthusiasts. They did not solicit huge bank loans but worked nearly round the clock. Their livestock is limited: five to ten cows per farm plus a few calves and piglets fattening up on a subcontract with collective farms. They have at least 30 hectares of arable land and have a tractor and some other implements available. A widespread initial arrangement is that several families share a farm. Most multi-family farms subsequently split into single-family units. Private farmers in this group hire temporary help. Usually members of a nearby collective or state farm are willing to hire on despite their generally hostile attitude towards rural "capitalists." The reason: pay is higher and more reliable. Permanent full-time help is not hired because the new farmers do not expect outsiders to share their own zealous attitude to work. Private farmers in this category usually come from the rural elite, but some are former urbanites, and some are resettlers from non-Russian ex-Soviet republics.

  The results of this group's work cannot be matched by collective farms. Their cattle are groomed: milk yields are 3 to 4 times as high as in publicly-run enterprises. Such private farmers pay off their loans by selling meat and milk at the same prices as collective farms. But many individual farmers have a hard time selling their produce because officials in charge of state procurement prefer to deal with large socialized production units. So private farmers in many cases produce for their own consumption. And only a high labor input allows them to have some modest profit. One way to increase it would be to set up some processing operations. Sour cream, cottage cheese, and sausage are much more transportable than meat or, particularly, milk, and easier to sell in a town. However, small-scale processing lines are mostly imported and usually unaffordable. Also, whereas farming produce per se was freed from taxation, food processing is subject to a stifling 30% tax. As a result, this group of farmers is not much different from the semi-subsistence, personal auxiliary farmers (who always existed under Soviets), though their landholding is more spacious and their cattle are a bit more numerous than those of farmers in PAF.

- The *second* group of registered private farmers is either owners of second dwellings in the countryside (so-called "dachniks") or full-time rural residents conducting auxiliary farming that has traditionally co-existed with socialist-style commercial agriculture. This group consists of urbanites and those engaged in service jobs in the countryside. They work the land only during the summer season to support their own families with vegetables. They do not solicit bank loans. For them land is a form of investment that can be leased out, preserved till retirement when one might choose to spend more time in the country, or sold if need be.
- The *third* group is "fixers" (system manipulators), who particularly discredit private farming in the eyes of fellow country residents. Some estimates place the proportion of this group at the level of about one-fifth of all registered farmers. These are people who are personally remote from agriculture and include, for example, urban relatives of some locals. They have moderate-to-small landholdings, but are intent on getting hold of bank loans as large as possible because early on, loanholders were able to profit from inflation and preferential treatment of farmers by banks. Money and the legal status attached to a farm production unit are what these people are actually out for.

However, after the initial rapid spread of private farming, conditions for it worsened. First, interest on bank loans was made equal for everybody, at about 200%. Few are interested in such loans. Secondly, machinery became more expensive. With their favorable treatment terminated, private farmers lost whatever edge they might have had. In 1995 another major privilege was lost: private farmers are now required to pay value-added tax; it is levied not only on their profit from food processing but on their profit from the primary produce itself. It consumes 10-20%; when combined with other tax exemptions, 60% of the profit. Moreover, following the change in status of collective and state farms and the formal introduction of shareholding, new private farmers are no longer assigned land in excess of the per-family allotment set in the area. So if the allotment is, for example, six ha per person in publicly-run farms, as is the case in the Novgorod province, a family of two would be assigned twelve ha whether on a collective farm or a private farm. However, in the non-Chernozen regions of Russia one needs no less than three ha to feed one cow, so a landholding for a decent private farm has to be at least 20 ha. According to other estimates, the profitability threshold for a private farm is 40-50 ha in Russia's North and 10-15 ha in the South.[31] The March 1996 Presidential Decree opens some new possibilities to expand a private farmer's allotment: renting land and buying land shares of former collective farmers (in the case where current members of a collective farm do not want to obtain these shares).

The pioneers' enthusiasm has long been evaporating. Cases of inequitable assignment of land were revealed, and some private farms were legally terminated because they were not engaging in agricultural activity, which favorable loans were supposed to promote. So the wave of private farms setups began to subside. "The primary hindrance to the privatization of land in the Nizhni Novgorod province," said Governor Nemtsov in *Argumenty i Facty* in July 1996, "is the lack of people who want to become owners."

There is a considerable regional disparity in the spread of private farming (Fig. 6.3). Lowland provinces of the Northern Caucasus economic region (Stavropol and Krasnodar/Kuban) and the southern part of the Volga region lead both in terms of private farmers' numbers and their share of arable land. On January 1, 1995, the Stavropol and Krasnodar provinces were the absolute leaders in private farms: about 20,000 units in each. Dagestan, Rostov, Volgograd, and Saratov provinces followed with over 10,000 in each. In the non-Chernozem part of European Russia, the Moscow and Leningrad provinces led with 6.3 and 5.5 thousand private farms respectively. However, in terms of the ratio of private farmers to all agricultural workers, the leading position of the European south is shared by the northwest and by the south of Siberia (Fig. 6.4). Inside the provinces, the exurban areas and areas with intermediate accessibility to provincial centers (i.e., semi-peripheral areas) are the most attractive to private farmers.

The average landholding per private farm varies also depending on the economic profile of an area and on the degree of land abundance. Whereas in the Moscow and Leningrad provinces the average size is 10-11 hectares, in other provinces of the Industrial Center it is 17-51 hectares, in the Chernozem Center 24-57 hectares, and in Siberia a whole lot more (for example, 136 hectares in the Siberian province of Chita and 120 hectares in Altay). The smallest landholdings are in the densely populated republics of the Northern Caucasus.

The Crises: Systemic, Cyclic, and Regional

Crises of different natures are not mutually exclusive. A *systemic* crisis accompanying the transition from central planning to a market economy has afflicted all of Russia's economy. Every sphere faces a shortage of market infrastructure. The very underpinnings of the new system, in which consumer demand and sales drive producers' decisions, are not easily constructed.

Cyclic crises are characteristic of a market economy. However, in Russia the current dramatic downturn in the economy was also preceded by sagging economic growth in the late 1980s. In agriculture, stagnation and a widening gap with the West and the non-agricultural sectors of the domestic USSR economy were quite apparent at that time.

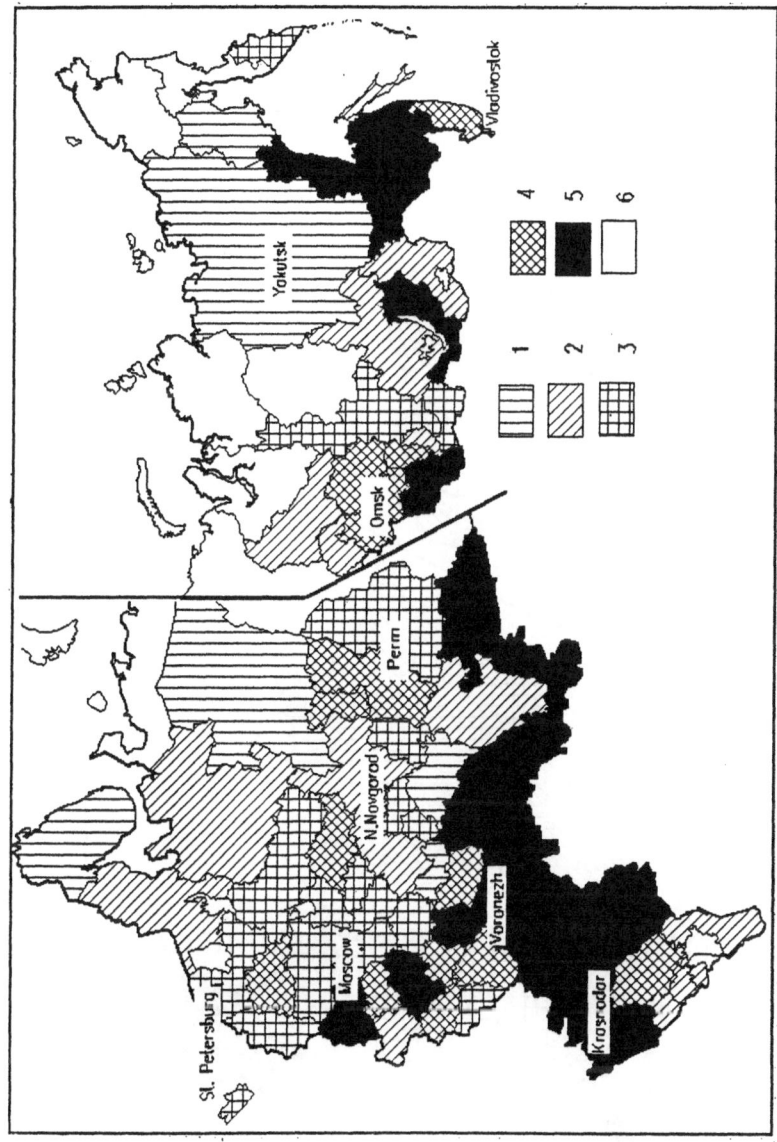

FIGURE 6.3 Private Farmers' Landholding in Arable Land as a Percentage of the Total Arable Land (1994). 1: Below 1%; 2: 1-2%; 3: 2-3%; 4: 3-4%; 5: 4-11%; 6: Data unavailable.

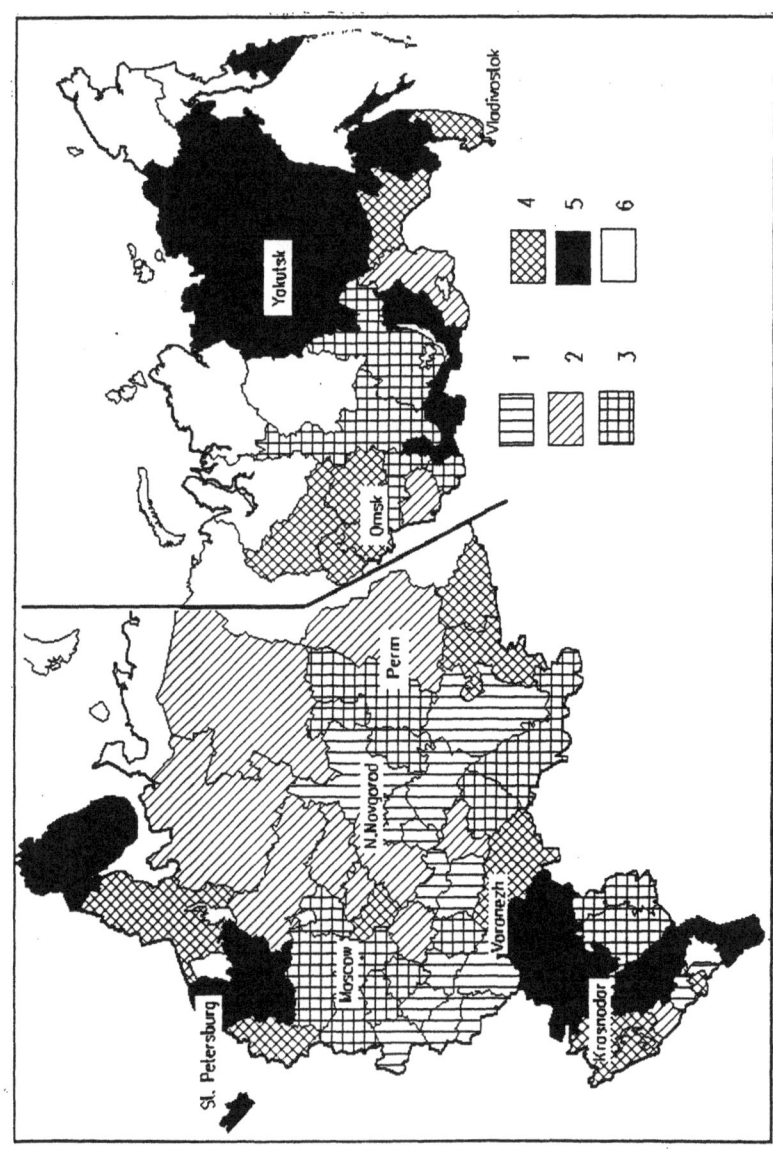

FIGURE 6.4 Number of Private Farms per 1000 Persons Engaged in Agriculture (1995). 1: 3-20; 2: 21-30; 3: 31-40; 4: 41-50; 5: 51-85; 6: Data unavailable.

In the 1990s, the output distribution crisis or, more precisely, the crisis of the old state-run system of food appropriation, became superimposed upon a chronic shortage of farm-related production inputs (resources) fairly typical for a centrally-planned economy. Agricultural investments downsized from 26% of the overall investment in 1991 to 16% in 1992, and to 9% in 1994[32] despite the growing proportion of agriculture in budget spending.

Subsequently, the government failed to pay for the produce it had ordered. After bestowing free gifts in the form of subsidies, discounted loans, and writing off debts, the state then dragged its feet on payments. Payments deferred till winter or even till spring were devalued by inflation two- or fourfold. The strongest production units, that is, those which produced a lot and actually delivered to the state, were among the hardest hit. Abiding by the state mandates, therefore, was no longer in the interest of producers.

Consequently, output began to shrink. However, it is obvious from the above discussion that this shrinkage had hardly anything to do with reforms in agriculture but rather with the overall economic situation in the country.

In Russia as a whole the 1994 output of agriculture was 25% below the level of 1990.[33] The reduction, thus, was not as dramatic as what occurred in industry, where output was halved during the same period. Agriculture's higher inertia and invariably successful fight for subsidies have been among the reasons for the gentler downward slide. But an even more important moderator was the growth in personal auxiliary farming (PAF).[34] It increased its output by 20%, whereas the public sector reduced its produce by 40%.[35]

During five years between 1990-94, the overall agricultural output has dropped most substantially in the Far East, in the republics of Siberia and of the Northern Caucasus, and in some provinces of the Volga region. The Northern Caucasus as a whole and the Chernozem region are being more adversely impacted by the overall crisis than the traditionally ailing non-Chernozem regions. The latter did not sag as badly as the former (Figure 6.5 and Table 6.3). By affecting the major producers the most, the crisis initially led to some levelling of regional disparity in output.

Another trend has been the growing spatial concentration of livestock. Livestock numbers declined everywhere (mostly due to cutbacks on animal feed imports) except in the most economically viable southern and exurban production units. This trend was not lessened even by PAF's growing share of milk output. The opposite, however, occurred in potato farming: with PAF eventually providing two thirds of the output, spatial *de*concentration of potato produce ensued.

Overall the pattern of inter-regional disparity of the mid-1980s described in Chapter 4, has largely survived, although the leaders have lost some of their edge. But many gradients detected in the 1980s are virtually intact. For

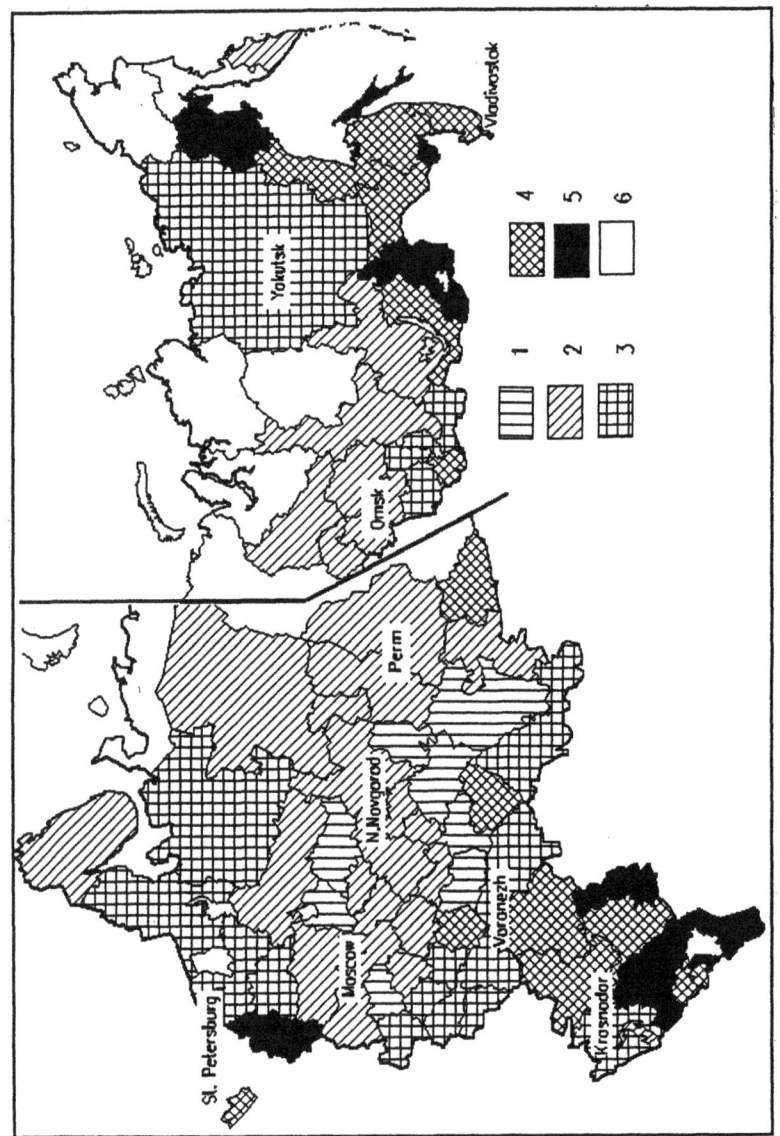

FIGURE 6.5 Output of Socialized Farm Units in 1994 as a Percentage of 1990. 1: 91-95%; 2: 81-90%; 3: 71-80%; 4: 61-70%; 5: 15-61%; 6: Data unavailable.

TABLE 6.3 Russia's Agricultural Output in the 1990s (All Kinds of Farms).

	1990	1991	1992	1993	1994
Area Under Crops, million ha					
Total Cropped Area	117.7	116.5	114.6	111.8	105.2
Grain	63.0	61.8	61.9	60.9	56.2
Potatoes and Vegetables	4.0	4.0	4.3	4.4	4.1
Output, million tons					
Grain	104.3	89.1	106.9	99.1	81.3
Sugar Beets	33.2	24.3	25.5	25.5	13.9
Potato	35.9	34.3	38.3	37.7	33.9
Vegetable	11.2	10.4	10.0	9.8	9.6
Meat	14.8	14.5	12.9	11.9	10.8
Milk	54.2	51.9	47.2	46.5	42.3
Yields					
Grain, c/ha	15.3	14.4	17.2	16.3	14.5
Milk per cow, c	229	256	233	232	219
Number of Livestock, millions					
Cattle	57.0	54.7	52.2	48.9	43.8
Pigs	38.3	35.4	31.5	28.6	25.0
Cattle Plague, % of Total Number	3.0	3.5	4.2	4.9	6.0
Pigs' Plague, % of Total Number	6.9	8.3	10.7	11.9	15.1
Output of Mineral Fertilizers, million tons	16.0	15.0	12.3	9.9	7.5
Application of Mineral Fertilizers, kg per 1 ha of Arable Land	83	79	44	32	12
Production of Tractors, thousands	214	178	137	89	29

Source: Statisticheski Bulletin 1 (APK). Osnovnye Pokazateli Funktsionirovaniya APK v 1994 godu. Moscow, Goskomstat 1995 (computer printout).

example, in 1993-94, at the peak of the crisis, grain yield in the Moscow province was 19-23 centners per hectare, whereas in the neighboring Kaluga province, which is warmer and has a more fertile soil, the output was only 14 centners per hectare. As another example of the persistence of pre-crisis patterns, in 1993-94 the most urbanized provinces and the lowland provinces of the Northern Caucasus (the Russian corn belt) continued to lead in terms of milk yields per cow.

A few large specialized livestock farms using mostly imported technology look especially productive compared with the lamentable situation of the majority of collective and state farms. Such a contrast led in some provinces to a substantial concentration of herds. For example, 40% of all the beef in the province of Nizhni Novgorod in 1994 was being produced by one farm in Tolmachevo (Kstov district) close to the provincial center. More than half of all the pork of the same province is produced by the two gigantic farms in Ilyinogorskoye (Vyksa district). Such farms were doing well up till 1995 when many of them came to a halt because of steep increase in grain prices.

However, the overwhelming majority of collective and state farms were put on the verge of bankruptcy even earlier by the changing economic situation. So far, however, there has been no defined procedure for bankruptcy in Russia. No single instance of bank-imposed confiscation of property has ever occurred in the socialized sector, in part because the failing property legally belongs to the so-called Agro-Industrial Complex, a state-run composite of the Soviet-style agriculture and food-processing industry. But in 1994 even the usual debt write off, this time of one half of the collective and state farms' debts, did not alleviate their financial situation. The Chernozem Center has fared worst in terms of outstanding debts. Conversely, in the Moscow and Leningrad provinces, and in the lowland provinces of the Northern Caucasus, over one half of all the enterprises have no outstanding debts whatsoever. In our view, this statistic accurately reflects the economic situation in agriculture on the eve of 1995 as does the distribution of unprofitable farms. Figure 6.6 reveals the cost of artificially sustaining the well-being of the "islands of socialism" in the Ulyanovsk and Orel provinces, and in the Tatar and Bashkir republics.

Regional Patterns of Crisis and Reform

Farming regions in Russia can be usefully classified according to two characteristics: the depth of their agricultural crisis and the degree of reform advancement in 1990-94.

The *depth of crisis* is reflected by the following indicators: output decline in the public sector, assessed in constant 1990 prices (Fig. 6.5); decline in the number of livestock; 1994 milk yields per cow; milk yield changes in 1990-94; and the proportion of enterprises with outstanding debt in December 1994.

The above indicators[36] do not vary considerably across Russia's provinces. The fifth indicator, a proxy for the potential number of bankruptcies, varies the most with the ratio of standard deviation to the average at the level of 0.19; the milk yield changes vary the least (0.11).

The statistics shown in Figure 6.7 were calculated in the following way. In an 89-by-5 matrix, where rows represent the subjects of the Russian Federation (its provinces and republics) and columns represent the above indicators, each entry was standardized by subtracting the column's average and

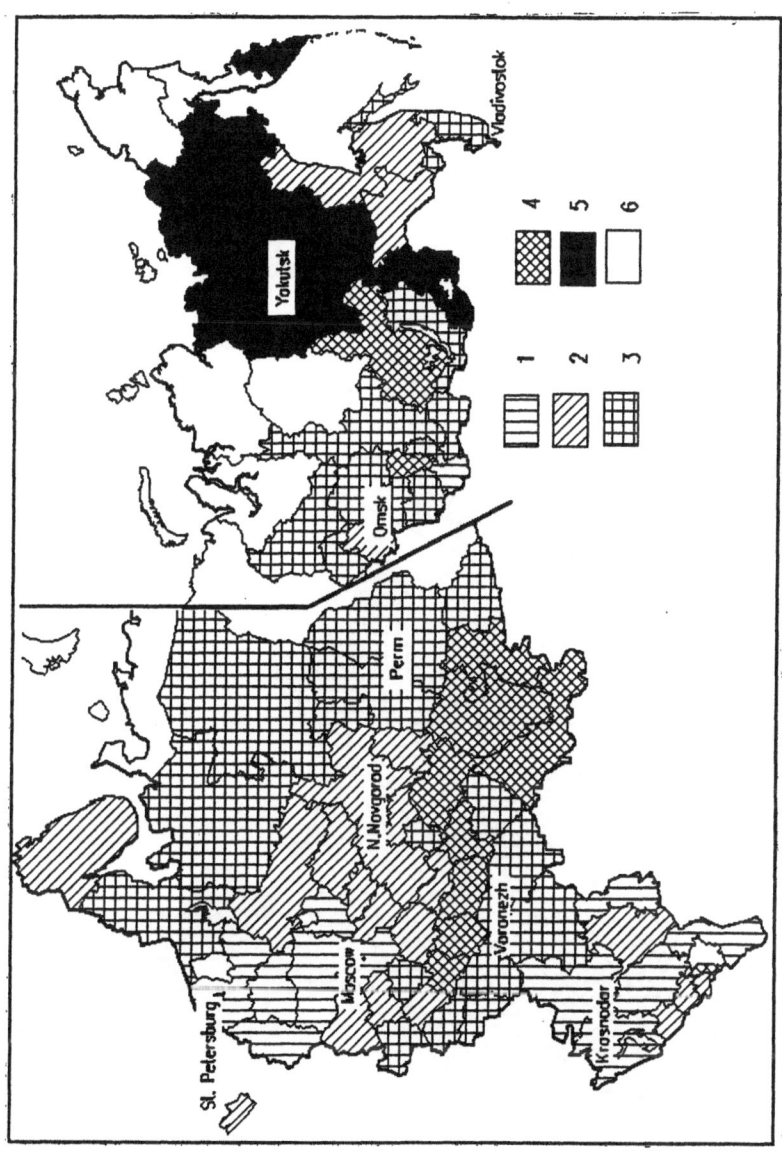

FIGURE 6.6 Outstanding Debt of Socialized Farm Units in Thousands of Roubles per One Employee (1994). 1: 23-70; 2: 70-140; 3: 140-210; 4: 210-280; 5: 280-575; 6: Data unavailable.

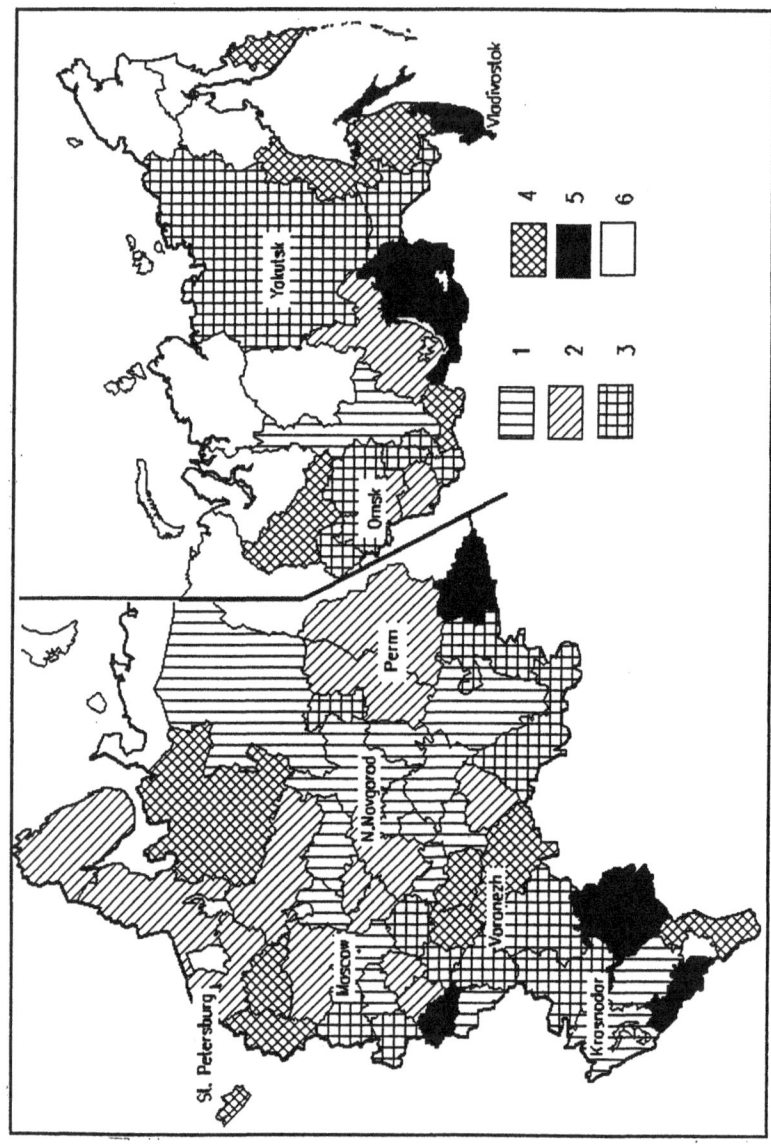

FIGURE 6.7 Crisis in Agriculture (1994). 1-5: Ascending order of crisis depth (see text for explanation).

dividing the resulting difference by the standard deviation in the column. Each row's sum was then interpreted as a composite estimate of the depth of crisis. It is the variation of these estimates that appears in Figure 6.7. The figure shows that the most acute crises are centered in the mountainous republics of the Northern Caucasus, the Kalmyk republic and Astrakhan province, the Novgorod, Pskov, and Archangelsk provinces, and some Chernozem provinces from Kursk to Saratov.

In contrast, several groups of regions stand out as the least crisis-ridden. These are:

- Most of the *non-Chernozem* provinces. Their productivity is falling but not nearly as dramatically as in the area immediately to the south of them. Two factors may account for this. First, these provinces had been experiencing such a profound and protracted regress for at least 30 years preceding current reforms that there was hardly anything left to lose; the downward trend had bottomed up. Secondly, in the non-Chernozem provinces more diversified production, aimed at local self-sufficiency, is more self-contained and, therefore, more viable than many forms of specialized commercial agriculture.
- Provinces with large *urban agglomerations,* also mostly non-Chernozem. These especially stand out because they have retained their 1980s level of productivity; the exceptional position of these provinces, particularly the Moscow one, is, however, nothing new.
- Lowland provinces of the *Northern Caucasus* region: Stravropol and Krasnodar/Kuban'. Not unlike the most urbanized regions, they thrive on long-accumulated fixed assets and more than sufficient labor, but nature itself is of help as well: these are the most fertile areas in all of Russia.
- Agricultural strongholds of *Siberia:* in Altay, Krasnoyarsk, and Irkytsk provinces.
- "Islands of the socialist heaven," that is, regions which have *shut themselves off from reform measures* so that nothing actually has changed there. These regions form a band stretching from the Komi republic to the most outspoken separatists --Tatarstan and the Ulyanovsk province.

It seems clear from the above that essentially *two types of territorial units fare better than the rest: the most open to reforms and the most resistant to them.*

Regarding the regional pattern of the degree of agricultural reform advancement, two preliminary remarks are in order. *First,* no matter how reasonable and all-encompassing the set of relevant indicators might be, the insights that can be drawn from them are ultimately limited by nature of the subject studied. If the subject itself is obscure, vague, and elusive, if it has not yet crystallized, research conclusions are bound to be highly tentative, regardless of the indicators used. In this regard, *crisis* is currently a much more tangible and crisp phenomenon than *reform* is, and so we believe the above regional classification by depth of crisis (Fig. 6.7) is more meaningful and reliable than the following analysis (Fig. 6.8). However -- and this is our *second* remark -- some clear assessment of the still unfolding reform is, of course, desirable.

An informal opinion making the rounds in reform-minded sections of the Moscow public goes as follows: "Our agriculture was slain long ago; to reanimate what had been kept alive only through artificial injections is all but impossible. At this point it is even unclear what the driving force of resurrection might be. In the future, perhaps it will be private farmers. Not now, though. Maybe modified collective farms are what we should wager on, but even their type and size are unclear. *Life itself will show the way.*" While the last sentence is a famous cliché (*Zhizn' pokazhet*) that Russians invariably resort to when they face major uncertainty in their lives, it does not befit researchers to be so non-committal. Especially when various Western commentators' preoccupation with private farming in Russia implies that they already know what Russians themselves have not yet discovered.[37] Moreover, after the research proposal this book is based on had been funded we received a letter from the funding agency alerting us to the necessity of selecting the proper criteria for assessing reform. "How will you evaluate the success of reforms?" was the query. (How indeed. And would it be politically correct to say that at this point we honestly do not know?)

Because of the lack of market infrastructure, because of atrocious roads, the fiscal suffocation of farmers' food processing operations, and the heterogeneity of the emerging private farmers, an indicator like *the number of private farmers per 1000 engaged in agriculture,* or any other similar indicator, cannot serve as a useful measure for reform's progress. Very few private farmers are in the vanguard of reform, most barely feed themselves, and many are farmers in name only. Nor would the *proportion of collective and state farms that have changed their status to JSC and PLR* work well because it would reveal only the attitude toward reform-related code words at best. The *proportion of large enterprises* (both collective and state farms *and* JSC and PLR) *without outstanding debt* would indicate only the degree of adjustment to the new extrinsic economic conditions, not the degree of reform of agricultural production units per se. Yet the *combination of the above indicators* may, in our view, be considered a reasonable approximation of the progress of reform. Like the estimate of crisis (Fig. 6.8), this combination is calculated as the sum of standardized partial indicators. As Figure

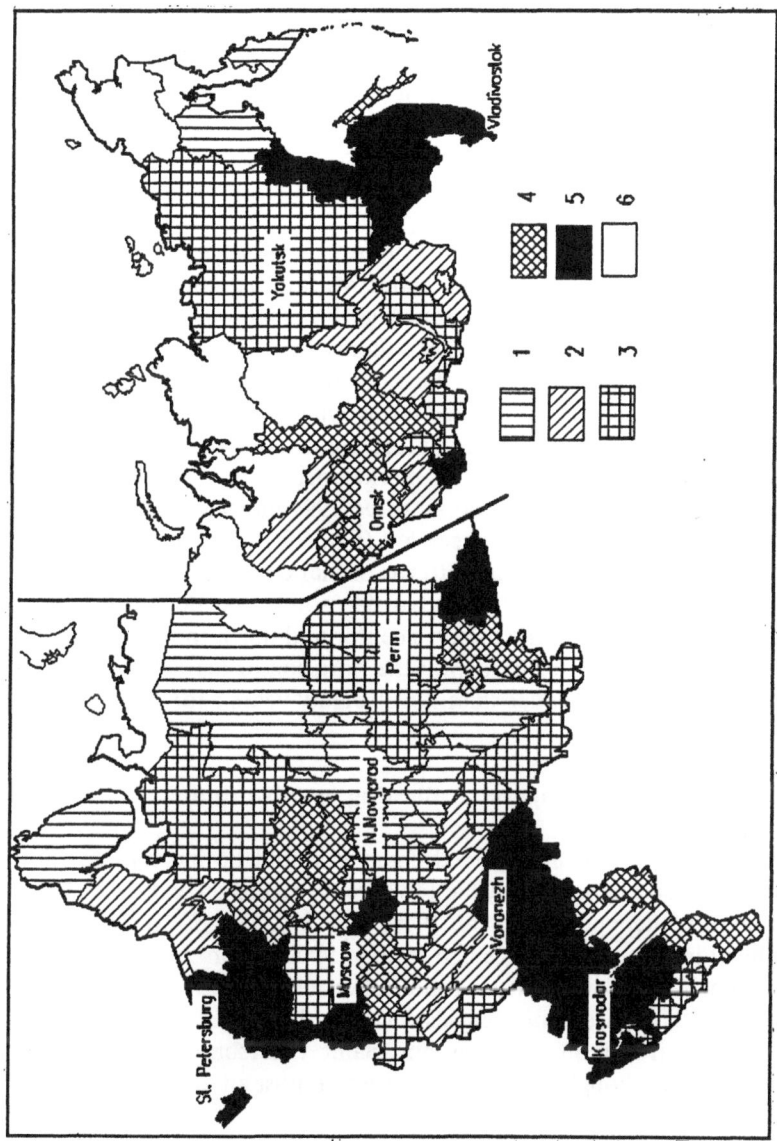

FIGURE 6.8 Reform in Agriculture. 1-5: Ascending order of reform advancement (see text for explanation).

6.8 shows, the band stretching from Komi to Ulyanovsk (including almost the entire Chernozem region along with some northern provinces with tiny agrarian sectors) appear to be the least affected by reform. In contrast, the piedmont regions of the Northern Caucasus and most of the non-Chernozem provinces are in the vanguard of reform.

We also attempted to compare geographically the degree of market conversion of Russian farming back in 1913 with that in 1994, following over seven decades of communism (Fig. 6.9 and 6.10). For 1913 two indicators were singled out: the percentage share of peasants who withdrew from their communes and the level of farming profits per one laborer. It is difficult to select commensurate indicators for 1994, especially since private farmers are not numerous and the erosion of collective farming unfolds along several lines (strengthening subsistence farming, the emergence of private farms, and the conversion of socialized farms themselves). So we limited our choice to two restructuring indicators (the ratio of PAF to independent private farmers and the extent of socialized farms' change of status) and to the indicator of labor productivity in agriculture. Since none of the above statistics is self-sufficient we assumed that market orientation of farming is pronounced if certain thresholds on all three 1994 indicators (Fig. 6.9-A) and on both 1913 indicators were exceeded (Fig. 6.9-B) by a region. It appears that the 1913 and 1994 spatial patterns of market reform are not far apart (Fig. 6.10): in both cases it is the Russian south that demonstrates more readiness for market; in the nation's center and north the market enclaves are mostly confined to exurbia. Still the diffusion of market innovations in 1994 appears to be broader than it was eighty years ago.

Food Imports: A Blessing or Misfortune?

An analysis of reform-triggered changes would be incomplete without focusing on food imports. In Chapter 3 we discussed the causes of Russia's food imports' dependency in the 1970s and '80s. The rise of inflation and crumbling inter-regional exchange in the 1990s underscored the drive for regional self-sufficiency. When food deliveries to centralized federal and regional food procurement agencies declined, the largest urban centers were among the hardest hit. Moscow and Saint Petersburg were affected in particular, even though they used to absorb almost all of Russia's dairy and meat imports.

As to the unprecedented expansion of imported food in the 1990s, there are two viewpoints circulating in Russia. According to one, imports aim at stifling Russia's farming; according to another, domestic farming is simply unable to meet large cities' demands. The real situation may not fit either of these opinions. For example, according to Yevgenia Serova, a chair of the

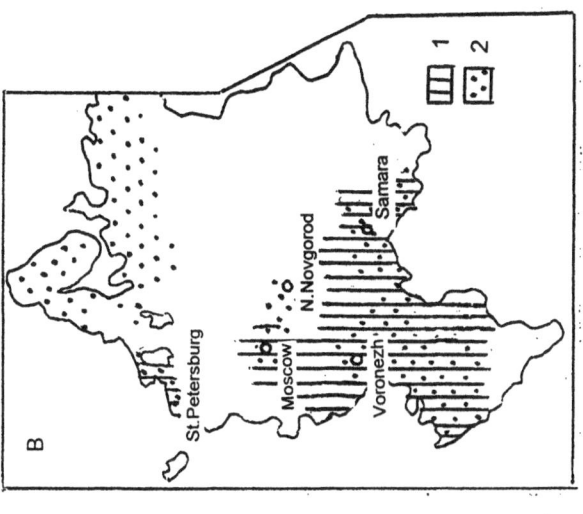

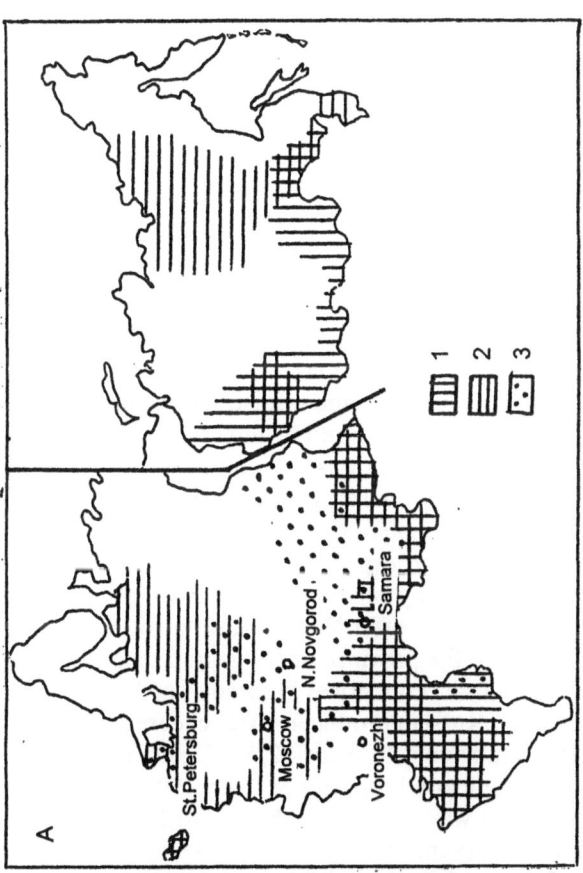

FIGURE 6.9 The Second Coming of Capitalism in Russia: Selected Indicators. A: 1994; 1. Acreage of private farms exceeds that of subsidiary plots; 2. Over 70% of socialized farm units changed their status; 3. Labor productivity in agriculture is above average. B: 1913; 1. Over 20% of peasants households took out title to their land; 2. Profitability of peasant farmers and earnings of hired labor were above average.

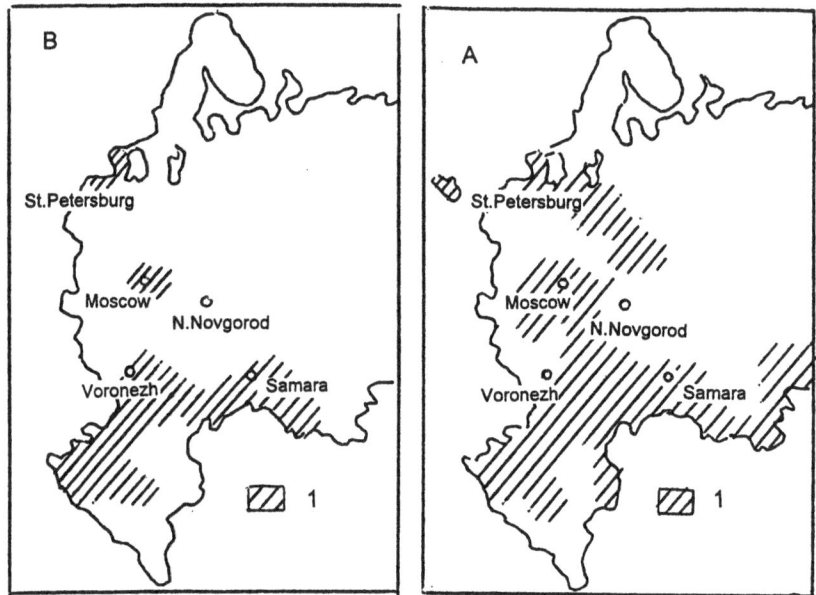

FIGURE 6.10 The Second Coming of Capitalism in Russia. A - 1994; B - 1913;
1 = Areas in which at least two 1994 and two 1913 indicators used in Figure 6.9 are
present.

agrarian lab in Gaidar's Economy-in-Transition Research Institute, in 1994
and 1995 there was an over-production of dairy and meat produce in Russia.
However, the domestic producers could not make it to Moscow and other
large cities because of a lack of information and infrastructure, and because
of racketeers on Russia's roads and thugs hired by importers. At the same
time, the Western food invasion of Russia was and continues to be is promoted by dumping made possible by special export subsidies practiced in the
exporting countries.[38] The Western combination of technological advances,
high labor productivity, low costs, and export subsidies actually beats the
Russian combination of low productivity, high production costs, and the lack
of selling skills. As a result, importers offer Western butter for $2 per kg,
about 9000 roubles in 1995, while the province of Volgograd sells domestic
butter for 20,000 roubles, given that retail prices in summer of 1995 did not
exceed 18,000 per kg. Similarly, Russian producers offer cheese for 18,000-
20,000 roubles per kg, and Lithuania sells it for 12,000-13,000 roubles.[39]

At a pompous government-sponsored conference (May 1995) devoted to
the food demands of Moscow, it was announced that, contrary to Serova's
claim about overproduction, the domestic producers of meat could deliver
only 200,000 tons of 650,000 tons of meat needed by the city of Moscow and

only 9,000 of the 35,000 tons of cheese needed. After all, even during the last decade of the Soviet Union, Russia could meet only one-third of the food demand of its capital city; another one-third was supplied by other Union republics, and the rest came from imports. Today imported food meets two-thirds of Moscow's demand.

However, Russia's overall dependency on food imports may be overstated if one examines to sales or monetary statistics only. Indeed in 1995, imports accounted for as much as 54% of all food sales. But if one switches from dollars or roubles to physical weight of produce consumed, it would appear that only 20% of the meat, 10% of the milk, and 9% of the vegetables consumed in Russia in 1995 were imported.[40] In 1990-1994 there had been a substantial decline in the volume of imported food: from $16,600,00 to $10,400,00.[41]

The structure of food imports changed in the 1990s (Table 6.4) as did their destinations. For example, imports of grain substantially decreased, while meat increased. When livestock weight gains per unit weight of fodder are very low, such a change is a benefit. But it has been decried by domestic meat producers: deprived of imported grain fodder they are obliged to cut back on cattle numbers. In 1995, which is beyond the time frame of Table 6.4, grain imports began to rise again, this time through the efforts of regional importers who set their own contacts with the Ukraine, Hungary, Kazakhstan, the Czech Republic, and Slovakia.[42]

In 1995 a peculiar turning point was reached in food imports: the mass and unprecedented invasion of Russian provinces. Whereas in the summer

TABLE 6.4 Food Imports to the USSR and Russia in the 1990s.

	1990 (USSR)	1992 (Russia)	1994 (Russia)
Imports Per Capita, in kg:			
Grain	111	194	24.7
Meat	3.9	2.2	7.6
Dry milk	0.0	0.3	0.3
Raw Sugar	13.5	14.4	14.7
Butter	1.0	0.2	1.0
Cooking Oil	2.0	2.7	0.8
% Share of Imports in Annual Consumption			
Meat	5.2	3.7	12.9
Butter	16.4	3.4	No Data
Cooking Oil	19.6	40.3	13.8

Source: Adapted from: Serova, Y. "Predposylki i Sushchnost Sovremennoi Agrarnoi Reformy." *Voprosy Ekonomiki,* 1995,1:31-46.

of 1994 imported foods did not compete with domestic produce except in the largest cities, after the "butter crisis," when butter was unavailable in stores for several weeks, food stores in small and medium-sized cities became full of imported dairy produce. It appears, however, that on average food imports in Russia do not supplant domestic produce, rather they make up for deficiencies. It is because of this that regional contrasts in food imports' share of consumption are high and related to degrees of regional self-sufficiency. Thus imported food abound in the north and all across the Non-Chernozem Zone, whereas in Chernozem provinces we rarely came across imports in the summer of 1995 among the most basic and frequently consumed foods (*Snickers* and *Bounty* do no count -- they have taken all of Russia captive). In part this was because of restrictions on imports imposed by provincial administrations.

Russia is not the first nation in Europe that cannot feed itself. Hyper-industrialized England shared the same lot for 200 years, as did Germany between the two world wars and as does Japan of today. However, Russia's distinction in this regard is that it buys food not in exchange for manufactured products but rather in exchange for petrodollars. It is unlikely that these dollars suffice for 150,000,000 people in the long run. But in the meantime, the fate of Russia's manufacturing will have to be cleared up. Until it is, the type of food dependency that exists today serves as an incubator for virulent nationalism and sabre-rattling.

Finally, as in England of the past, in today's Russia the confrontation between protectionists and freetraders runs high, especially in regard to food. This confrontation has a clear spatial dimension: while the premier agrarian regions are bastions of protectionism, the principal manufacturing regions are on the free-traders' side.

Notes

1. Van Atta, Don (editor and contributor). *The 'Farmer Threat.' The Political Economy of Agrarian Reform in Post-Soviet Russia.* Boulder: Westview Press 1993.

2. Wegren, Stephen. "Political Institutions and Agrarian Reform in Russia." In *'The Farmer Threat.' The Political Economy of Agrarian Reform in Post-Soviet Russia.* Boulder: Westview Press 1993, pp. 121-148.

3. See, for example, Bogert, Carrol. "Why Russia Cannot Feed Itself." *Newsweek* 15 October 1990, p. 44.

4. Wegren, Stephen. "Rural Reform and Political Culture in Russia." *Europe-Asia Studies* 1994, Vol. 46, p. 234.

5. Wegren, Stephen. "New Perspectives on Spatial Patterns of Agrarian Reform: A Comparison of Two Russian Oblasts." *Post-Soviet Geography* 1994, Vol. 35, No 8, pp. 455-481; "The Development of Market Relations in Agricultural Land: the Case of Kostroma Oblast." *Post-Soviet Geography* 1995, Vol. 36, No 8, pp. 496-512.

6. Wegren, Stephen. "Rural Reform and Political Culture in Russia." *Europe-Asia Studies* 1994, Vol. 46, p. 219.

7. Tucker, R.C. "Sovietology and Russian History." *Post-Soviet Affairs* 1992, No. 8, pp. 175, 193.

8. Wegren, Stephen. "Rural Migration and Agrarian Reform in Russia: A Research Note." *Europe-Asia Studies* 1995, Vol. 47, No 5, pp. 877-888.

9. Serova, Y. "Predposylki i Sushchnost Sovremennoi Agrarnoi Reformy v Rossii." *Voprosy Ekonomiki* 1995, Vol. 1, pp. 31-46.

10. Shimov, Y. "Tseny na Maslo, Sakhar i Benzin u Nas Vyshe Mirovykh." *Izvestia* 9 August 1995.

11. *Argumenty i Facty* 1995, No 6.

12. Severin, Barbara "Observations on Regional Aspects of Food Availability in Russia." *Post-Soviet Geography*, 1995 Vol. 36, No 1, pp. 41-57.

13. Ovchinnikov, Oleg. "Krizis Selskogo Khoziaistva." *Finansovye Izvestia* 21 December 1995.

14. "Rost tsen v 1995 godu." *Izvestia-Expertiza* 17 January 1996.

15. Ovchinnikov, Oleg. "Krizis."

16. Wegren, Stephen. "New Perspectives." p. 478.

17. Golov, Alexander. Za Kogo Golosuyet Derevnia. *Izvestia* 6 June 1996.

18. We have been finishing our work on this book in the last days of June 1996.

19. Morochenko, D. "Protivniki Chastnoi Sobstvennosti na Zemliu". *Izvestia* 22 December 1995; Virkunen, V. "Sadovodov Zagoniat v Kolkhozy?" *Argumenty i Facty* 1996, No 1.

20. *Rossiyskaya Gazeta* 12 March 1996.

21. Michael Specter's commentary on the Decree in *The New York Times* was especially indicative of such a reaction (see: Michael Specter. "Russian Farmers Get Land. The Sunday News Observer." Raleigh, N.C. March 17 1996:1A,14A; reprinted from *The New York Times*).

22. See, for example, articles by Yelena Yakovleva in *Izvestia* 12 March 1996 and by Yelena Kriviakina in *Finansovye Izvestia* 19 March 1996.

23. *Selskaya Zhizn* 16 March 1996.

24. Wegren, Stephen. "Rural Reform," p. 234.

25. Wegren, Stephen. "New Perspectives," p. 465.

26. Danilov, V.P. "Agrarnaia Reforma v Post-Sovetskoi Rosii." In *Kuda Idyot Rossia.* Moscow: Interpax 1994, pp. 125-35.

27. Among the most meaningful reports are those by Yakovleva, Y. and Yershov, A. in *Izvestia* 13 October 1993 and 25 October 1994.

28. Nizhegorodskiye Kolkhozy Uspeshno Idut s Molotka. *Izvestia* 13 April 1996.

29. *Sravnitelnye Pokazateli Ekonomicheskoi Situatsii v Regionakh Rossiiskoi Federatsii* 1994. Moscow: Goskomstat 1995, pp. 329-334; *Statisticheski Bulletyn 1 (APK). Osnovnye Pokazateli Funktsionirovania APK v 1994 godu.* Moscow: Goskomstat 1995 (computer printout): 36; Kriviakina, Y. "Fermery Prizyvayut Moskvu." *Finansovye Izvestia* 29 February 1996.

30. Nefedova, T. and Pavlova, I. "Fermery i Fermery." *Vash Vybor* 1993, No 3, pp. 35-40.

31. *Food and Agricultural Policy Reforms in the Former USSR.* Washington, D.C.: The World Bank 1992, p. 75.

32. *Statisticheski Bulletyn,* p. 53; *Sotsialno-Ekonomicheskoye Polozheniye Regionov Rossiiskoi Federatsii.* Moscow: Goskomstst 1995, p. 47.

33. *Statisticheski Bulletyn,* p. 56.

34. We resort to *this* literal translation of the Russian "*lichnoye podsobnoye kho-zyaistvo,*" instead of, say, "private subsidiary plots," "smallholders," or "private plot operators," in order to set this entity apart from *private* farming. The latter is an independent business, at least by formal definition; it was not allowed before 1991; whereas the former has always been an inalienable part of the USSR's agricultural production. Even though it transcended its "auxiliary" nature by composing about one quarter of the entire agricultural output, it has been auxiliary in the sense that it was conducted by families whose able-bodied members were full-time employees in the public sector.

35. *Statisticheski Bulletyn,* p. 56.

36. Clearly this set of indicators is not exhaustive and we welcome discussion of other possible criteria. However, one potential criticism, why only the public sector is taken into account, can be addressed now. The point is that the other two components of output do not seem worth including. Private farming's proportion in the output has so far been tiny (2-3% in 1994) and its "progressiveness" has not yet been demonstrated in the Russian context. PAF, on the other hand, looks like an anomaly in an economically advanced nation, which Russia is not, as some may suggest. But even so, PAF's technological level, which was referred to in one recent Russian publication (by Vasily Uzun in *Finansovye Izvestia* 22 December 1995) as reminiscent of ancient Egypt, is far behind that of any other economic activity of today's Russia. PAF's growth, therefore, is a clear *reaction* to crisis. There is no certainty that as such it correlates with the depth of crisis itself, and, therefore, does not fit the purpose of our assessment (the regional variation in the depth of the crisis). That does not mean that we favor collective farming as the optimal form for Russian agriculture. It is simply that no other economically attractive form has yet taken root.

37. This attitude is revealed in the prescriptive nature of researchers' statements such as: "Reforming the existing agrarian order...requires...two...sets of...changes. First, production, land-holding and property relations...*must* be altered... Second, the economic and political environment in which all Soviet farms operate *must* be recreated through the development of new supply, marketing and service arrangements (*The 'Farmer Threat,'* pp. 9-10; emphasis added to the original).

38. Serova, Y. and Yakovleva, Y. "Politicheskiy Privkus Zamorskikh Produktov." *Izvestia* 13 June 1995.

39. Berger, M. "Rossiiskiye Yedoki Otviechayut za Provaly v Agrarnom Sektorie." *Izvestia* 24 February 1995.

40. Yakovleva, Y. "Pravitelstvo Vvodit Kvoty." *Izvestia* 23 April 1996.

41. Latsis, O. "Bolshe Ministerstv v Zakroma Rodiny." *Izvestia* 24 February 1995.

42. Sizov, A. "Rossiiskiye Regiony Narashchivayut Import Zerna." *Izvestia* 1 March 1996.

7

Crisis and Reform in the 1990s: Social Implications

While the post-Soviet states are making their painful transitions to market conditions, they must eat.

--Don Van Atta[1]

How Do People Survive in the Countryside?

In the midst of the total disarray of Russian farming's public sector, galloping inflation, and the scarcity of food items in country stores, people survive as they have always done -- through personal auxiliary farming. All the more so since the slip-knot long strangling peasant subsidiary economy (its access to land, and erstwhile restrictions on the number of cattle and on PAF-related profit) has been loosened. It is now possible to expand personal plots for free (in some provinces by 1 to 2 hectares) using a Special Land Re-Distribution Fund.[2] According to Yeltsin's March 1996 Decree, it is now possible for members of socialized farms to claim up to five hectares of land for PAF from a personal land share. In the 1990s, urbanites in dire straits began to look to the countryside to escape high consumer prices. Everybody in possession of at least a tiny piece of land outside the corporate limits of a town took to growing potatoes and other produce. In a peculiar Russian agro-recreational entity called a *dacha* the agrarian component is clearly on the rise.

As a result, PAF's share of the overall agricultural output has risen from 24% in 1990 to 38% in 1994, a change of truly enormous proportion. Table 7.1 shows the 1990-94 changes in the structure of agricultural output by sector. In European Russia it is in the most depopulated provinces encircling that of Moscow that PAF has increased its output the most. While land used for PAF officially grew from 2 to 5 million hectares, in reality there are now no less than 18 million hectares that PAF uses.[3] This is be-

TABLE 7.1 Percentage Share of Agricultural Output of Socialized Farms, Personal Auxiliary Farms, and Private Farms.

	1990	1991	1992	1993	1994
Total Output					
Socialized Farms	76	72	66	63	60
PAF	24	28	33	35	38
Private Farms	0.0	0.2	1.1	1.9	2.0
Potato Output					
Socialized Farms	40	28	21	16	11
PAF	60	72	76	83	88
Private Farms	0.0	0.2	0.8	1	0.9
Meat Output					
Socialized Farms	76	70	65	60	56
PAF	24	30	34	39	42
Private Farms	0.0	0.1	1	1	2
Milk Output					
Socialized Farms	77	74	68	64	60
PAF	23	26	31	35	39
Private Farms	0.0	0.1	0.5	1	1

Source: Statisticheski Bulletyn 1 (APK). Goskomstat 1995.

cause the pastures, meadows, and sometimes the arable land transferred to rural settlements from official collective and state farms are actually at PAF's disposal (as, of course, is the part of agricultural land still nominally under collective farming). And although it contradicts official statistics claiming that in early 1995 PAF commanded only 4% of Russia's agricultural land,[4] it is safe to say that almost one-third of all of Russia's agricultural land is now being used in PAF one way or another, and in some regions this proportion is still higher. In 1989 in the Russian Federation the average collective farming family earned 58.6% of its income from the collective farm, while only 21.5% from PAF. By 1993 the ratio had become 32.2% to 40.3%[5]: PAF-related income had outstripped that from public sector farming.

Russian peasants, therefore, have de facto opted for PAF even though the level of technology available for PAF is in most cases antediluvian.

In 1994, PAF produced 88% of Russia's potato output, 67% of its vegetables, 42% of its meat, 39% of its milk. PAF's proportion of cattle and pigs grew from one-fifth in 1991 to one-third in 1995; about half of the total number of sheep and goats are also in personal ownership.[6]

Though official statistics underrate the share of agricultural land actually used in PAF, we still use them to distinguish regions that are more PAF-

TABLE 7.2 Percentage Shares of Private Farms and Personal Auxiliary Farming in 1994 in Their Respective Regions.

	PF *Land*	*PAF* *Land*	*PF* *Output*	*PAF* *Output*	*PF* *Cattle*	*PAF* *Cattle*
North	2.1	10.5	1.2	35.0	1.3	16.4
Northwest	3.9	13.0	1.4	43.0	1.3	18.3
Center	3.5	6.5	1.1	39.0	0.8	14.9
Volgo-Viatka	1.9	5.3	0.7	42.0	0.4	21.9
Central-Chernozem	3.9	6.8	1.4	36.0	0.3	19.3
Volga	7.3	1.3	2.3	35.0	1.6	26.6
N.Caucasus	4.8	3.0	2.8	35.0	2.2	32.0
Ural	4.5	3.2	1.8	39.0	1.2	32.0
W.Siberia	5.0	1.8	2.1	39.0	1.4	28.3
E.Siberia	3.8	3.3	1.5	46.0	2.4	40.4
Far East	7.4	7.2	4.9	41.0	5.2	35.1
Russia	4.6	3.6	1.9	38.0	1.4	26.7

Sources: Economicheskoye Polozheniye Regionov Rossiyskoi Federatsii. Moscow, Goskomstat 1995; *Statisticheskii Bulletyn 1 (APK)* Moscow, Goskomstat 1995.

prone from regions that lean more toward independent private farming.[7] Table 7.2 shows that in Russia the (European) North and the Northwest top other regions in terms of relative significance of PAF, whereas in the Northern Caucasus and in Siberia the position of private farmers is relatively stronger. At first sight this contradicts Figure 6.4, according to which both the Northwest and the southern provinces of European Russia stand out for the number of private farmers per 1000 agricultural population. The latter, however, is linked to a very considerable reduction in the Northwest's agricultural population. It thus appears that the Northwest stands out in terms of the heightened role of both PAF and private farming in the area, but much more so for the heightened role of PAF. As for the comparison of the shares of *output* produced by PAF and by private farmers, the result is so much in favor of PAF that a systematic flaw in statistical records of *land* used in PAF suggests itself immediately. Yet even these inherently flawed comparisons (underestimating the role of PAF) corroborate that at least in the first half of the 1990s, Russia's agriculture has been evolving not along capitalist lines but rather by emphasizing semi-subsistence, archaic family farming based upon manual labor.

Scores of middlemen have emerged, effectively linking PAF with retailers, public catering, and food processing. These intermediaries are still branded "speculators" by the man-in-the-street, but they appear to close a lot of loopholes. Aside from the higher quality of produce that they peddle, they do not require advance payment in wholesale transactions; however, when buying produce they themselves pay at the time of the transaction in cash. Also their mistakes in strategy, such as assisting the growth of uncompetitive small processing plants, are not as costly as those made by central planning. Yelena Yakovleva, whose essays on Russian agriculture appear in *Izvestia* and are particularly worth reading, asserts that whereas "from the perspective of ideal economic calculations a large meat processing plant is more productive and competitive than small sausage shops and a couple of bumpkins peddling piglets, in actual economic situations the plant languishes because of capricious deliveries while bumpkins with piglets fare OK."[8]

Gennady Lisichkin, an unchallenged pioneer in Russian pro-market discourse[9] and also one of the very first open preachers of independent commercial farming in Russia, now eulogizes regional authorities who have consciously supported PAF and its symbiotic relationship with what is left of collective farms. He does so more assertively than Yakovleva but for essentially the same reason, acknowledging that under current conditions independent farming would require expensive state-of-the-art agricultural machinery which is beyond the reach of the average farmer. Having farmers buy such equipment, says Lisichkin, would be like forcing a concert grand piano on the owner of a one-room apartment. He believes collective farms will be related to peasant PAF in the way that the so-called Machine-Tractor Stations[10] were once related to collective farms before the latter were ordered to buy out the former in the late 1950s. In other words, he expects that existing collective farms will provide the technology that PAF will use.[11]

The role of PAF in today's Russia shows up, indirectly, when one compares the earnings of urbanites and country folk and the patterns of their spending. By 1990 average rural earnings had grown to 80% of urban earnings (the gap between the two was thus closing which was considered a great achievement of the Soviets); however, by 1994 rural earnings had slumped to 54% of urban earnings. Yet food accounted for 41% of urbanites' spending in 1994 (25% in 1990), but only for 33% of spending in the countryside (24% in 1990).

The Presidential Decree of March 1996 discussed in Chapter 6 in conjunction with land ownership also expands opportunities for subsidiary farming. Specificaly, limits to land allotted for a rural household have been extended to five hectares. This means that to establish a farm that feeds the owner and produces surplus that can be sold one does not necessarily need to register as a farm owner; one need only apply for extra land (in many cases one is already uses it) within the confines of one's land share. Not very many are likely to choose this option since working the expanded parcel without machines is impractical, and the illegal services of tractor operators from nearby socialized farms may be very costly. On the other hand, many will be willing to get the extra land to use it as a hayfield or a pasture. One point seems important to underscore in this regard: one way or another the ways of PAF and independent private farming in Russia appear to be converging.

This convergence in part explains a fairly new phenomenon, rural unemployment. That the countryside suffers from a shortage of labor was a well-established truism used to justify dispatching urbanites to assist villagers for free during harvests. In the 1990s this practice has continued but on a much smaller scale than before and not free of charge. Simultaneously the first rural unemployed were recorded and began to receive state benefits. According to some unpublished data of Goskomstat, by the summer of 1995, officially recorded rural unemployment (2.8% of the economically active population) had exceeded its urban counterpart (2.7%). The highest percentage of rural unemployed is in the Ingush republic (20%); in the provinces of Vladimir and Ivanovo it is over 9%, in the provinces of Pskov and Yaroslavl, over 7%. At the same time, in the province of Smolensk official rural unemployment did not exceed 0.7% and in Tula and Riazan, 1%. We will focus on this phenomenon once again in Chapters 10 and 11 devoted to concrete regions, but some remarks are in order at this point. First, such a range of variation in unemployment statistics within a fairly homogenous group of central provinces makes us suspect that the very definition of unemployment is subject to local interpretation. Secondly, our field observations have shown that the rural unemployed consist of two categories. One is derelicts, especially alcoholics, who were discharged by farm chairmen began in the 1990s (prior to the 1990s it was next to impossible to fire a worker for any reason). The second, larger category is those who voluntarily resigned from socialized farms. Such people live off PAF, occasionally selling surplus food, and are not averse of taking advantage of unemployment benefits however modest, even though they actually work their subsidiary plots; PAF and benefits combined exceed their previous earnings. Local administrations' attitudes toward such people vary from place to place, explaining the differentials in unemployment statistics, which thus do not reflect the real picture.

However, despite all the new opportunities opened for PAF, it seems to have almost exhausted its growth potential by 1994, not unlike private farming, and in 1994-95 the output of PAF in most provinces did not increase for the first time in the 1990s, while in quite a few provinces it declined. The trend suggests research attention may have to shift once more to collective farming, likely to remain the mainstay of Russian agriculture.

Rural Population Growth:
The Reversal of a Long-Term Trend

Russia's rural population began declining in the late 1920s to early 1930s. Change in the rural population of European Russia was for a long time an integral part of two processes: urbanization and colonization. Rural migrants flocked to growing towns and, up until the Second World War, scores of villagers migrated to the east. After the war villagers played a decreasing role in the colonization of Siberia and in the settlement of non-Russian republics.

East-bound migration waves had subsided by the end of the 1960s. In the 1980s in all of Russia few regions were attractive to migrants. In fact, Moscow and Leningrad and their provinces and the West-Siberian province of Tiumen with its gigantic oil-fields absorbed about two-thirds of all incoming migrants, urban and rural. Aside from these, positive net rural migration in the 1980s occurred in Murmansk, north of Krasnoyarsk, in Kamchatka, and Primorye. That is, almost all of European Russia and the area south of Siberia suffered losses in rural population.

The 1990s have brought about a sweeping change: all of European Russia, with the exception of its north, began to experience a growth in rural population, while the north, both European and Asian, and the Far East began to lose their rural populations sharply. The heaviest exodus has been from the province of Magadan: according to the data generously shared with us by Zhanna Zayonchkovskaya, the most reliable Russian authority on migration, every sixth rural resident has abandoned this province in 1994 alone. This landslide out-migration resulted from the combination of circumstances related to crisis and reform: the removal of budget supports and of the mark-ups of the "northern" salaries, vanishing personal savings due to inflation, etc. One way or another, the introduction of market elements into the economy has revealed that the erstwhile labor-intensive colonization strategy for the areas with harsh natural conditions can no longer be sustained. Most migrants from the north and Far East are headed to European Russia, as are re-settlers and refugees from ex-Soviet republics. The latter flow is nothing new, though; it has been under way since the mid-1970s and its insignificant numerical increase in the 1990s has been far short of the splash in publicity it has occasioned. What *has* been new, that is, originating in the 1990s, is the drastic reduction in migration *from* Russia to other ex-Soviet republics.[12] In other words, the colonization process has come to a halt.

Figure 7.1 shows that in many European provinces the *rate* of migratory inflow to the countryside exceeds that to urban areas. At the same time actual migrants heading to the countryside outnumber those heading to towns only in a few provinces: in 1989-94 these provinces were Orel, Kras-

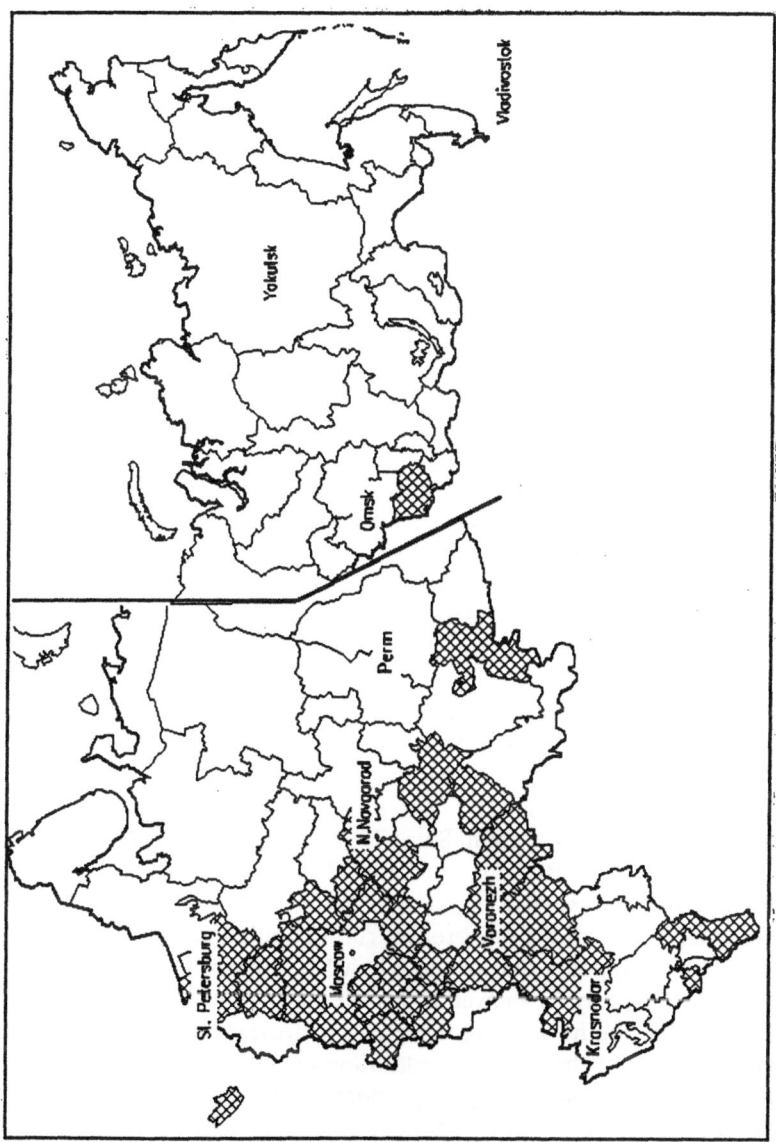

FIGURE 7.1 Provinces in Which 1994 Net Migratory Inflow in the Countryside Exceeds That to Urban Areas.

TABLE 7.3 Population Changes in the Russian Federation, in Thousands.

'59-69	'70-78	'79-88	'89-94	'89	'90	'91	'92	'93	'94
Rural									
-6825	-6921	-3202	937	-173	-58	288	721	151	65
Urban									
19370	14393	13052	43	813	560	-127	-752	-458	-125

Source: The Demographic Yearbook of the Russian Federation. Moscow, Goskomstat 1994.

nodar, Stavropol, Chelyabinsk, Altay, Kemerovo, and Dagestan. According to Zayonchkovskaya, in 1989-94 the distribution of rural migration gains within European Russia was as follows. The biggest gainers were the piedmont provinces of the Northern Caucasus (net migration amounted to 642,000 persons). The second biggest gainer was the Industrial Center: 470,000 people. In 1979-1989 net migration here was only 98,000 people with virtually all of them settling in urban areas. Now the flow is more rural-oriented and far less bound for the province of Moscow, whose niche for new settlers is not big. The Central Chernozem region received 324,000 rural net migrants, whereas in 1979-1989 it lost 14,000.

Seventy-seven percent of migrants from non-Russian ex-Soviet republics are ethnic Russians. In 1990-1993 alone the Russian countryside received about one million people of working age, about 5% of its entire 1989 working age population.[13]

In 1991 the rural population of Russia showed an increase (see Table 7.3) for the first time since the 1930s. Population growth in the republics of the Northern Caucasus outweighed the continuing decline in other regions. And since 1991 the rural population has been growing in most of the peripheral regions of the Russian Federation. Since 1992-93, the rural population has also been growing in the Russian heartland (Table 7.4).

Available sources on Russian population[14] provide the basis for our analysis of the components of the most recent rural population change (Fig. 7.2 and Table 7.5). It appears that of the 89 civil subdivisions of the Russian Federation, in 1993 only 23 showed a positive natural increase, i.e., an excess of births over deaths. In 1994 these 23 territories contained 17.5% of the rural population of the Russian Federation in 1994. Only three of them (Astrakhan, Orenburg, and Omsk) are predominantly ethnic Russian oblasts with agriculture being a sizable component in employment composition; 13 are thinly populated civil subdivisions of the Russian North and Eastern Siberia, while the remaining six (accounting for 70% of the excess births over deaths) are the republics of the Northern Caucasus. Note that all the above entities are located outside Russia's heartland.

TABLE 7.4 Rural Populations, in Thousands

	1961	1971	1981	1986	1989	1990	1991	1992	1993	1994
North	2006	1799	1484	1407	1435	1424	1415	1423	1486	1465
Northwest	1663	1371	1153	1115	1107	1100	1097	1088	1086	1080
Central	9746	7742	6025	5372	5307	5263	5229	5158	5189	5185
Volgo-Viatka	4812	3858	3013	2665	2627	2595	2578	2559	2564	2550
Central Chernozem	5536	4708	3550	3150	3076	3043	3014	2993	3024	3031
Volga	6463	5838	4830	4457	4389	4373	4361	4404	4505	4546
N. Caucasus	6914	7149	6959	6989	7154	7188	7207	7346	7564	7688
Ural	7322	6617	5477	5162	5124	5089	5087	5142	5223	5225
W. Siberia	5432	4565	4082	4038	4082	4079	4068	4190	4353	4386
E. Siberia	3039	2809	2517	2486	2572	2564	2591	2646	2657	2651
Far East	1520	1641	1739	1788	1920	1901	1911	1926	1907	1896
Russia	54668	48291	41012	38812	38975	38802	38744	39032	39753	39904

Source: The Demographic Yearbook of the Russian Federation, Moscow, Goskomstat, 1994

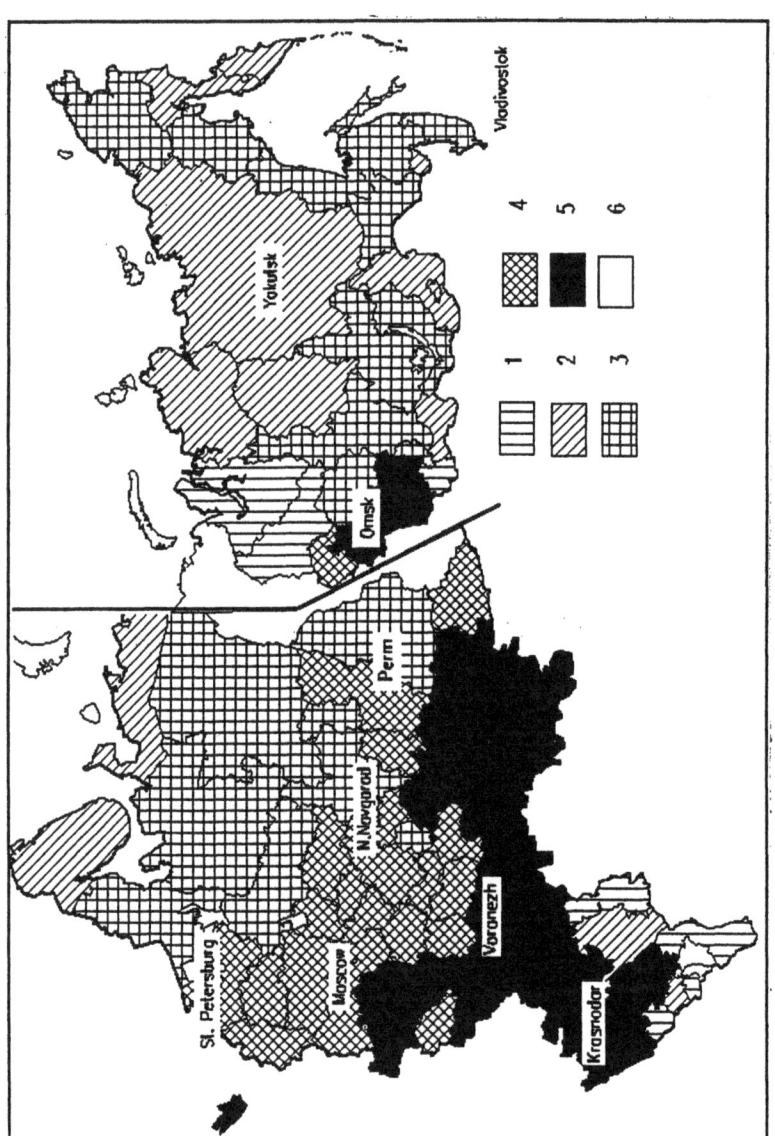

FIGURE 7.2 Components of the 1994-1995 Rural Population Change. 1-5: See Table 7.5; 6: Data unavailable.

TABLE 7.5 Percentage Distribution of Russia's Rural Population Among the Types Featured by Figure 7.2.

Sub-divisions with birth rate in excess of death rate	*(17.5)*
1. those with net migration inflow	8.2
2. those with net migration outflow	9.3
Sub-divisions with death rate in excess of birth rate	*(82.5)*
3. those with net migration outflow	15.9
4. those with net migration inflow short of the population decline due to excess of deaths over births	24.3
5. those with net migration inflow offsetting the population decline due to excess of deaths over births	42.3

Source: Demographic Yearbook of the Russian Federation. Moscow, Goskomstat 1995.

In 1993-94, in-migration more than made up for the overall negative natural increase (i.e., excess of deaths over births) of the rural population. Analysis of the spatial layout of both processes reveals that in-migration overcompensated for the excess of deaths over births only in 16 civil sub-divisions of the oblast level, all of which -- with the single exception of Novosibirsk -- are located west of the Urals. Almost one-third of all migration inflow was absorbed by just three civil subdivisions of the Northern Caucasus: Krasnodar, Stavropol, and Rostov -- long the migration magnets of Russia. While accounting for merely 12% of the rural population of the Russian Federation, these three subdivisions absorbed 35% of the migrants. Together with three other Chernozem provinces -- Belgorod, Voronezh, and Volgograd -- they received 50% of all the migration inflow into the Russian countryside (while containing only 17% of the rural population).

This process has spilled over to the Industrial Center, whose four southernmost subdivisions (the Briansk, Kaluga, Orel, and Tula provinces) have also shown a migration inflow in excess of natural losses.

According to Figure 7.2, the interplay of the major components of rural population growth has assumed a distinctive latitudinal pattern in European Russia. The demographic revival of the countryside seems to be northbound.

Stephen Wegren has recently hypothesized that the reversal of migratory trends might in the long run make the demographic situation in the countryside healthier and it also might "liberalize" generally conservative rural attitudes.[15] While that may turn out to be the case, the new situation has to be interpreted with caution. First, the components of the population change -- that is, natural increase and migration -- are rooted in different factors that have different degrees of stability/volatility. If the on-going migration is not an easily controllable factor, birth and death rates -- with the *natural increase* being a function of these -- lend themselves to social control even less. In fact, in large measure the natural increase reflects the age composition of the population, which, in turn, bears the indelible imprint of a *long-lasting out-migration* of predominantly younger people, which predated the current situation.[16] Age composition has a distinct character in areas subjected to protracted out-migration in contrast to those areas not subjected to its lingering effect. As Table 7.6 shows, until recently the latter areas had more effect on the overall demographic situation in rural Russia, so the overall balance of births and deaths in the countryside was positive.

Ironically, only when the migration flow was favorably reversed did the accumulated effect of previous migration come to exert its full impact, that is, a landslide excess of deaths over births in 1993 (Fig. 7.3). But as a matter of fact, in the Russian Northwest and in the major part of the Industrial Center, an excess of deaths over births in the countryside has been documented annually since as early as 1966-70. *Depopulation,* which seems to be at the heart of the whole issue of rural population change has, therefore, not been reversed. It has in fact gotten even worse.

Secondly, while some urbanites from nearby cities are returning to the countryside, the largest contingent of rural in-migrants has so far been made up of refugees, forced migrants from the non-Russian ex-Soviet republics, and voluntary migrants from the Russian North. In 1993 alone, 493,000 people arrived in Russia from Central Asia and 54,000 from the Baltics. Even though this flow is mostly urban in origin, 53% of the newcomers have been assigned to the countryside (38% in 1989-93).[17]

For example, in 1991-93 about 50,000 forced migrants arrived in the province of Pskov, one of the most depopulated provinces of Russia. About 20,000 of them (equivalent to 7% of the province's entire rural population) were allocated to the countryside. Nearly half of these migrants were from the Russian North, one-quarter were ethnic Russians from the Baltics, and the remaining one-quarter were ethnic Russians from Central Asia and the Transcaucasus.[18] In most of the other parts of the Russian heartland the last group dominates incoming migrants.

Thirdly, one factor contributing to rural population growth has nothing to do with migration: the reclassification of settlements, i.e., a change in settlement status. For years it was quite common for some villages to be given urban status as they became foci of population concentration and developed a non-agricultural employment base. However, when depopulation affected not just rural areas but myriads of small towns as well, many lost their urban status. Table 7.5 shows that in 1991-93 reclassification actually outweighed migration as a cause of rural population growth by a factor of 1.9.

The repopulation of the Russian countryside is not always what it seems to be at first glance.

TABLE 7.6 Components of Rural Population Change in Russia, 1961-1993, in Thousands.

	Total Change	Natural Increase	Net Migration	Reclassi- fication
1961-65	-2176	3483	-4202	-1457
1966-70	-4200	1555	-5171	-584
1971-75	-4668	998	-5073	-594
1976-80	-2343	719	-2591	-471
1981-85	-1591	788	-1836	-542
1986	-272	243	-392	-123
1987	-199	233	-391	-41
1988	-208	180	-331	-57
1989	-208	145	-275	-78
1990	-58	87	-72	-73
1991	285	44	57	184
1992	721	-33	289	465
1993	151	-184	264	71
1994	65	-227	272	20

Sources: Otsenki Chislennosty i Migratsii Naseleniya v Rossiiskoi Federatsii v 1993 godu. Moscow, Goskomstat 1994; Naseleniye Rosii, Moscow, Eurasia 1995:29.

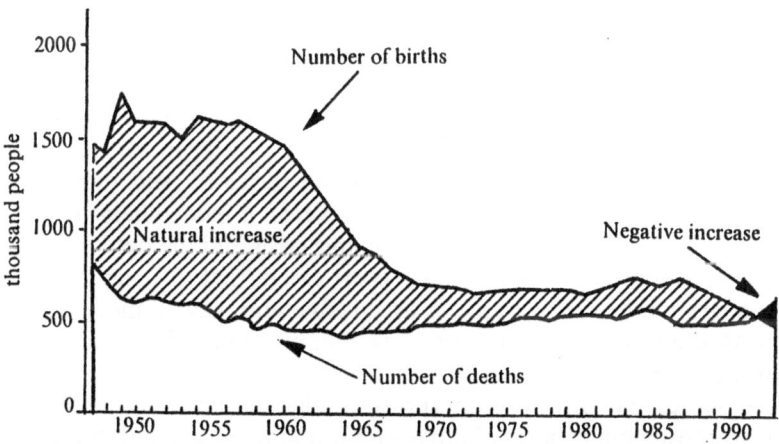

FIGURE 7.3 The Natural Increase of Russia's Rural Population.
Source: Adapted from: *Naseleniye Rossii.* Moscow, Eurazia 1994:31.

Will Recent Migration Trends
Overturn Rural Depopulation?

It stands to reason that only a drastic change in age composition could effectively reverse the countryside's on-going depopulation. But on-going and future in-migration, upon which this change depends, appears not to be conditioned by *pull* factors (i.e., forces coming from within the countryside itself). As one analyst recently claimed, only a crisis in large cities can help revive the Russian countryside.[19] It remains to be seen whether *push* factors (forces pushing people out of urban areas) will be stable over the long run. While net in-flow into the countryside has been on the rise since the early 1990s, it could stop as abruptly as it began, if only because it contradicts global and long-term national trends of population redistribution.

To the extent, however, that migration is motivated by the opportunities created by economic reforms in the countryside itself (i.e., plausible pull factors), it might be more or less long-lasting. Such a trend might suggest that previous, Soviet-engendered-style economic conditions in both urban and rural areas were abnormal, and that a return to "normalcy" will remove the effects of the earlier conditions, excessive urbanization being one of them. In this scenario, global population trends will resume as soon as the new economic order takes root. However, the conclusions about reform in our previous chapter do not support this line of reasoning.

Moreover, it can be argued that rural depopulation notwithstanding, Russia still has relatively more people in agriculture than all economically advanced countries do and that the quality of labor and lack of economic incentives have been more of a hindrance to the performance of Russian agriculture than simply the quantity of laborers has been. If this is true, the ability of even more aggressive and consistent economic reforms to generate substantial and lasting migration into rural areas may be called into question. Note that unlike in the West agriculture remains the most important and in many cases the sole economic activity in the Russian countryside.

It is unclear at this point which trends will develop. And there is a growing evidence that the recent migration reversal is indeed being caused not so much by the reforms as by the economic crisis and the deterioration of the living conditions in *urban areas*. To the extent that this may be true, rural in-migration becomes even less predictable and may be more tightly linked with the situation in the urban-based industry than the situation in the countryside itself.

Indeed, scores of relatively recent rural out-migrants have been employed in the least prestigious spheres of urban employment: on assembly lines and construction sites. It is these people to whom the economic crisis has dealt the severest blow. This is especially true for those who had not managed to obtain urban apartments prior to 1992. Even those who had by that time had their names registered on the waiting lists are now facing years of delay since a growing number of new apartments are being sold for cash, and modest savings have evaporated after the removal of price controls in 1992. That these people, who have not yet lost rural habits and skills -- without which the vicissitudes of rural life in Russia would be insurmountable -- are the most natural candidates for return to the countryside has long been indicated by experts like V. Perevedentsev.[20]

Some of these urban outcasts are now replenishing the rural population pool. They are mostly heading to areas whith fertile soil that they may be able to take advantage of. But nobody knows for sure how many of them are just biding time, waiting to return to the city. It is indeed highly ironic that an outright crisis was required to accomplish what government projects of the improvement of rural life were supposed to deliver. Billions of roubles had been spent in rural areas since 1965 in order to reverse the population outflow. This reversal used to be the number one objective for the Non-Chernozem Zone of Russia, where 98 billion of not-yet-devalued rubles were invested between 1976 and 1987 alone.[21] Needless to say, the Soviets' objective was never realized, and it never would have been, had it not been for the unsolicited "remedy" created by the recent overall economic crisis.

It is unlikely, however, that a substantial number of former urbanites from other ex-Soviet republics will take root in the countryside. They may abandon it as soon as they manage to cast anchor in the nearby cities. Indeed, this process has already begun.

In European Russia substantial portions of the rural population and especially of the land area are located outside a two-hour accessibility range to big cities. The recent spread of private vehicles and road construction and the depopulation of outlying areas have, of course, improved accessibility conditions but not drastically. And it is beyond this accessibility range that most depopulated areas are located. The magnitude of the depopulation, as demonstrated by Figure 7.2, and its powerful impact in terms of abandoned villages and land are gigantic. However lasting the on-going repopulation may be, it would be unrealistic to think it could make up for a sizable share of this impact.

In-migration can hardly change the spatial pattern of the rural population that has resulted from a long-term population redistribution. Current migration in-flow may, in fact, make this pattern even more pronounced.

Summary

Because some major points made in Chapter 6 and 7 are among the most important for this book and because they are inter-related, we will briefly re-emphasize them here.

First of all, collective farming is still a predominant mode of production in Russia's countryside. While collective and state farms' nominal change in status reveals a general attitude towards reforms, the change has been inconsequential in terms of management, the pattern of personal earnings, internal governance, and accounting practices.

The real changes, however, have been the increase in the variety of options in selling produce, and, especially, a stunning rise in personal auxiliary farming's proportion and significance. As far as forms of agriculture are concerned, it is PAF, not Western-style private farming, that has become the prime beneficiary of reforms, even though the rise of PAF is related to these reforms more indirectly than it is to crisis. From a technological standpoint, the increase in PAF is a setback for Russian agriculture, relegating this already backward activity to an even earlier epoch in relation to other economic activities in Russia.

Another real change has been a reversal of rural population dynamics, which, however, is not always as demographically favorable as it seems at first glance. In particular, the rural depopulation in many areas is still underway, barely disguised by migration inflow.

While private farming along Western lines supposedly has emerged in Russia, Western analysts' attention to it clearly outstrips its significance both in terms of its proportion in output and of its potential. Potent obstacles, cultural, economic, and political, stand in the way of private farming in Russia. Private farms are not a uniform category, many of them "exist" only on paper, while many others are not very different from PAF, long a "fringe benefit" for collectivized Soviet peasants.

Agricultural production has been shrinking in Russia in the 1990s, but not as a result of reforms in agriculture but rather as a result of the overall economic situation in the nation. Hardest hit have been the mountainous republics of the Northern Caucasus, the Chernozem belt from Kursk to Saratov, and a few non-Chernozem provinces.

Finally, it seems likely that the changes that befall socialized farm production units will have the greatest impact on the future of Russian agriculture. Despite the dwindling proportion of collective farms in overall output and despite the growing significance of PAF, the latter can exist only in symbiosis with the former. At this point predictions about the changes yet to come are largely speculative. However, our observations point to six *major patterns* that the transformation of various public enterprises seems to follow and, possibly, will continue to follow:

- Some strong production units, fairly stable in the past, will continue to thrive on their accumulated potential and the enthusiasm of their leaders. These units strive to adjust to new economic conditions, pursue new markets, and develop specializations accordingly. Economic survival prompts leaders to fire heavy drinkers and other unreliable workers. Farm units of such a kind exist predominantly in exurban areas; however, in Russia's south they are not confined only to exurbia and are more widespread. Factors that work against these farms today are that fixed assets wear out as does enthusiasm. Also since fired workers do not change their place of residence, their private plots sometimes act as parasites on the collective farms. Overall the production units of this type have a good chance of survival as socialized farms.

- In some not-so-strong production units power will gradually concentrate in the hands of a few farm administrators who become de facto owners. These administrators also fire ineffective workers thus contributing to rural unemployment (especially if a farm is located at a substantial distance from a city). However, because formal ownership rights, as well as property transfer rights, have until recently been obscure, the situation is quite uncertain. Such units abound all around non-Chernozem regions; they also exist in the less dense and less land-use-intensive provinces of the south.

- Some other units will opt for the status quo, pinning their hopes not on the market but on the support of provincial power elites, internal accounting practices, and the overall strategy a la 1980s. Such production units abound in the non-Chernozem and in the Volga regions. This way of survival, however, even more than the preceding, runs the risk of collective farms' being gradually cleaned out by their workers.

- Many farm units will split into separate family and smaller collective units, so called "partnerships based on faith." The most civilized version of this strategy has been offered in the province of Nizhni Novgorod. Very few of these newly emerged entities, however, survive while holding onto commercial agriculture. Those in exurban areas have more chances; those on the periphery switch to subsistence farming.

- In quite a few places there will be a purely nominal retention of collective farms which would act as an umbrella for numerous PAF units. This strategy leads to semi-subsistence farming in which part of the produce, usually milk, is sold through the collective farm, which owns the milk-related machines and implements. Such a situation is particularly widespread in non-Chernozem and Chernozem districts in the hinterlands.

- In some areas outright cessation of agricultural activity will occur; leasing land to *dachniks* and workers seeking non-agricultural employment are becoming increasingly popular in the exurban areas in the North.

Not for the first time, even in this century, Russian agriculture is at a crossroads. It remains to be seen what new forms it will take.

These first seven chapters have addressed the evolution of and reforms in Russian agriculture in general. In the following four chapters, Chapters 8 through 11, we will examine the same issues while focusing on specific regions of European Russia: the Central Chernozem economic region; one of its provinces, Belgorod; the Non-Chernozem Zone; and one of its provinces, Yaroslavl.

Notes

1. Van Atta, Don (editor and contributor). *The 'Farmer Threat': The Political Economy of Agrarian Reform in Post-Soviet Russia.* Boulder: Westview Press 1993, p. 3.

2. See about this fund in: Wegren, Stephen. "Rural Reform and Political Culture in Russia." *Europe-Asia Studies*, 1994 Vol. 46, p. 223.

3. Uzun, Vasily. "Agrarny Sector Ischet Puti Formirovania." *Finansovye Izvestia* 22 December 1995.

4. *Ekonomicheskoye Polozheniye Regionov Rossiiskoi Federatsii.* Moscow: Goskomstat 1995, p. 116.

5. Uzun, Vasily. "Agrarny Sector."

6. Sizov A. "I v Etom Godu Spad v Zhivotnovodstvie." *Finansovye Izvestia* 27 April 1995.

7. First, official statistics are irreplaceable and, despite all their flaws, are still more reliable now than in the past; second, we assume, perhaps arbitrarily, that Russian statistics treat and/or mistreat the phenomena they target in an evenhanded and spatially equitable way.

8. Yakovleva, Yelena. "Ue 'Novykh Spekulantov' Tovar Deshevlie i Kachestvennee." *Izvestia* 14 December 1995.

9. Lisichkin's 1965 book *Plan i Rynok* (Plan and Market) was an eye-opening bestseller describing advantages inherent in a market economy in a way digestible by a relatively broad audience. It was immediately translated into Czech and Hunarian, since in these countries reform thinking was markedly on the rise at that time. However, in the USSR the official reaction was such that Lisichkin was banned from the media up until perestroika and his doctoral dissertation on agricultural eco-omics was denied, on orders from the Central Committee of the Communist Party, a chance to be defended.

10. See about MTS in: Van Atta, Don. *The 'Farmer Threat,'* p. 11.

11. Lisichkin, Gennady. "Zhizn Fiodora Kuzkina, Prodolzhennaya Airatom Gafiatovym." *Izvestia* 30 November 1995.

12. Zayonchkovskaya, Z.A. *Migratsiya Naseleniya i Rynok Truda v Rossii.* Moscow: Institut Narodnokhoziastvennogo Prognozirovaniya 1994, p. 6.

13. Ibid., p. 16.

14. *The Demographic Yearbook of the Russian Federation.* Moscow: Goskomstat, 1994; *Nasselenyye Rossii.* Moscow: Eurasia 1994.

15. Wegren, Stephen K. "Rural Migration and Agrarian Reform in Russia: a Research Note." *Europe-Asia Studies,* 1995 No 5, 1995, pp. 877-888.

16. The lay public is tempted to believe and journalists are prone to propagate the idea that the ongoing demographic winter in Russia results primarily from the deterioration of living conditions. Unfortunately for the conventional wisdom this idea seems rooted in, it cannot be convincingly corroborated even with regard to Russia's total population, 73% of which is urban. In fact, the "baby bust" of the 1960s (when living standards were on the rise) and the reduction in the number of people of childbearing age in the same decade (an echo of the war) account for no less than 90% of the current excess of deaths over births. The above idea is even less applicable to the rural population by itself. First, the total fertility rate, i.e., the average number of children per one woman of childbearing age, has been substantially higher in the countryside of all Russian provinces than in their urban areas (today it is 1.6 times higher on average). The same, naturally, holds true in regard to age-specific fertility rates. Secondly, the exact timing of the rural natural increase reversal in specific provinces is closely correlated with the duration of the rural exodus. Thus in areas with the longest history of rural net out-migration (like Pskov, Novgorod, Tver, Smolensk, and Yaroslavl), the excess of deaths over births has been documented annually since the late 1960s. In contrast, in regions where a vigorous outflow from the countryside commenced later (like Briansk, Riazan, Orel, the provinces of the Chernozem Center, etc.), the switch to a negative rate of natural increase occurred at later dates.

17. *Migratsionnye Protsessy Posle Raspada SSSR.* Moscow: Institut Narodnokhoziaistvennogo Prognozirovaniya 1994, p. 3.

18. Pskov Provincial Migration Service. Unpublished report.

19. Gurkov A. "Rossiya Nie Perestaiot Udivlyat." *Argumenty i Facty* 2 January 1996 (an account of views of Karl Schlegel).

20. Victor Perevedentsev. *Plachu Dolgi, Dayu Vzaymy.* Moscow: Sovietskaya Rossiya 1983, p. 135.

21. Ioffe, G.V. *Selskoye Khozyaystvo Niechernozemya: Territorialnye Problemy.* Moscow: Nauka 1990, p. 6.

8

The Chernozem Countryside

Soils in the Chernozem region are truly black (*Chernozem* means "black soil"), their humus content is 8-12%, and they are over one meter thick. During the Second World War, German occupiers loaded Ukrainian and southern-Russian Chernozem onto cargo platforms and shipped whole trains of it at a time to the *Vaterland*. Foreign visitors are still carried away by these soils: "Here is *soil!*," they used to exclaim, "One doesn't need to fertilize it." We remember how in the late 1970s, American participants of an international symposium visiting the Biospheric station (of the Geography Institute) near Kursk demanded that their bus stop in the midst of a farming area; they stepped onto a field and kneaded fat damp lumps of soil in their hands for a long time. "Grain yields should be high, shouldn't they?" they asked. We were ashamed to answer then that yields did not rise above 15 centners per hectare. Even now they rarely exceed 25 centners.

Officially, five provinces belong to the Central Chernozem economic region, those of Belgorod, Voronezh, Kursk, Lipetsk, and Tambov; together they comprise 1% of Russia's total land area. But the real spread of Chernozem is broader, involving the bulk of the province of Orel (part of the Central economic region) and a few provinces to the east -- down to the Volga River and beyond (Fig. 8.1). Actually, Chernozem is typical for the whole vegetation region of forest-steppe stretching from Kursk and Belgorod to the city of Ufa, the capital of Bashkortostan. So-called fat (*tuchnyi*) or classy Chernozem gradually turns into leached Chernozem south of the forest-steppe and into podzol-tainted varieties in the north. Climatic conditions are also favorable in forest-steppe, especially for wheat, sugar beets, and sunflower. Moisture, however, can be deficient from time to time.

The Central Chernozem region is a typical agrarian belt of European Russia with arable land accounting for 60-70% of the region's total land area and mostly situated on watersheds. Extensive fields intersperse with picturesque green valleys and gullies. Villages are threaded like beads on rivers and creeks, occasionally forming continual settled bands. Forests on

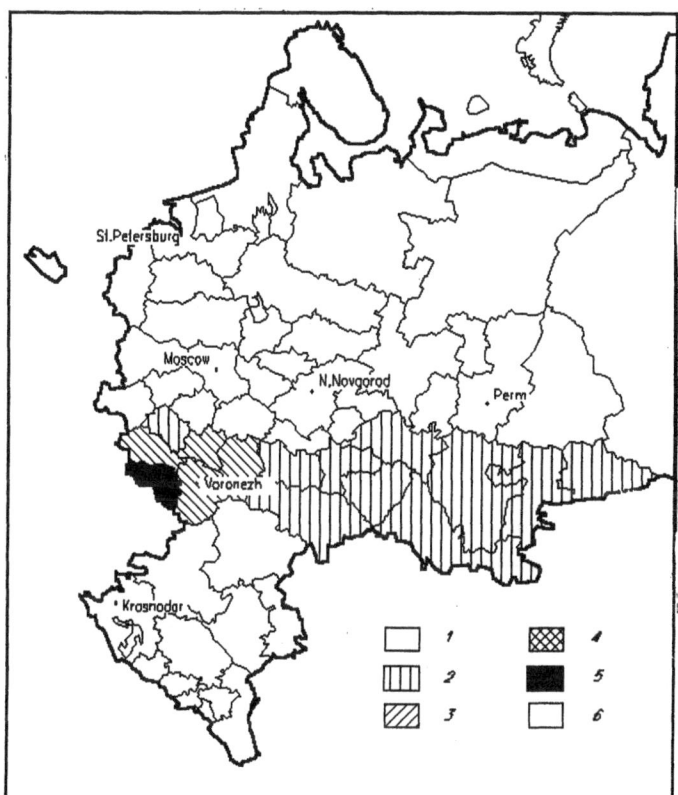

FIGURE 8.1 Chernozem Belt of European Russia. 1 - Approximate spread of classic Chernozem (black soil); 2 - Central Chernozem economic region; 3 - Province of Belgorod (case study province to be discussed in Chapter 9).

average account for 10% of the land area, hayfields and pastures for 12-15%. Since the Chernozem region is transitional from steppe to forest, the provinces differ: the more northern have more forest, the more southern are almost devoid of forest and have the richest soils.

We have mentioned Chernozem provinces on many occasions in the previous chapters because of some specific features of their productivity and of the crisis and reform there. In terms of history this region has been peculiar as well. Bondage was more rigorous and tenacious, and landlords controlled higher portions of the land here than elsewhere. The region has differed from both the more economically advanced Industrial Center and from its southern neighbors not as affected by vestiges of bondage and peasant communes. The Chernozem region had been one of the most ailing regions of Russia, and then became a relatively successful one. How did agriculture evolve on these lands generously endowed by nature but detached from large industrial clusters?

From Kievan Rus to the 20th Century

Colonization of the land traces back to the 10th and 11th centuries, that is, to the Kievan Rus. It is at that time that the oldest towns, like Kursk and Rylsk, were founded. They were then devastated during the Mongolian invasion in the 13th century, while the whole surrounding area became, as it were, a no-man's land, the so-called Wild Field (*Dikoye Polye*) situated between Muscovy to the north, Poland and Lithuania to the west, and the Golden Horde to the southeast. Following 400 years of desolation, a second wave of colonization commenced in the late 15th century when the area joined the Russian state and new towns were created. Yet even then the marginal geographic location of the region inhibited its development: nomadic incursions continued up to the 18th century. In order to thwart them, so-called notch lines (*zasechnye linii*) were created, that is, obstructing bands of lumber, paling, etc., with fortresses as strong points. The government allocated its garrisons, here assigning generous chunks of land to their commanders and mercenaries.

Gradually a fringe of the Russian state was moving southward, to the Don's basin, to the Volga, and to the Dnieper's rapids.[1] However, a Cossack population already lived at the outposts of these areas. Because these Cossacks were free their presence nearby created a kind of a barrier effect causing bondage in adjacent Chernozem lands to be more severe to prevent peasants from fleeing their lords.

During the 17th century an all-Russian market was taking shape, fixing a division of labor between the northern (forest) industrial belt and the agricultural Chernozem south. The latter became Russia's granary. However, output grew mainly through plowing up extra lands, as the returns to labor inputs under bondage were low.

In the pre-railway epoch grain was shipped via rivers and cart-tracks. Most rivers of the region, though, flow south. Only the Oka River offered a convenient way north, thus favoring commercial agriculture in the northeastern corner of the region. The situation changed when the first railway lines came to the region: grain fever rolled over the area in the second half of the 19th century, moving south and south-east where it reached the so-called New Russia along the northern coast of the Black Sea. There true capitalist farming took shape with grains of higher quality and with proximity to sea ports. The grain-oriented agriculture of the Chernozem region did not survive this competition and declined.

The economic development of the Chernozem region was greatly hindered by bondage. Only landlords could respond to the growing demand for grain, which they did by expanding their plow-land. Thus it became inevitable for them to favor corvee (*barshchina*), the most archaic feudal obligation. It required that peasants work more for the lord, unlike in more northerly regions where peasants were mostly liable for quit-rent (*obrok*) and thus more inclined to work on their "own" parcels. With *barshchina* taking up to six days a week, both feeding one's own family and working efficiently remained problematic.

Because of the vestiges of bondage and the competition from the Industrial Center and the *Donbass*, industry and towns of the Chernozem region stagnated. The region was thus not only overwhelmingly agrarian but becoming exceedingly backward as well. Even landlords' estates were primitive economies based on archaic technology. The new crop-rotation systems with mineral fertilizers that appeared in the western and central industrial regions of European Russia did not make their way here.

The region was almost entirely plowed up; by the 1860s arable land acreage was almost the same as it is today. The moderately rugged terrain promoted erosion, which cut back on already low yields. Droughts caused crop failures on many occasions. And fallowing alone was insufficient to retrieve full-scale fertility. Here is how Alfred Hettner described it in a 1905 book: "The predominant economic system has remained the three-field system with unfertilized fallow. The cultivation of fields is mostly surface tilling... The total economy is extremely one-sided and bears the character of plundering exploitation of soil... It has been a terrible mistake of Russian farmers to consider Chernozem to be perpetually fertile... The land which is so suited for agriculture as few other countries of the Earth are and can be compared with the best grain regions of North America has been economically degraded. What is more, it is exporting grain even today but only at the cost of feeding its own population."[2] Backwardness and poverty in the Central Chernozem provinces were also pointed out by Russian researchers of different schools. The provinces between the Chernozem and non-Chernozem belts, that is, those of Orel, Riazan, and Tambov, were among the poorest. Veniamin P. Semionov called them the economically and spiritually ailing part of Russia.[3]

And yet the region was exceedingly over-populated. At the turn of the century, the population density amounted to 51 persons per square kilometer, which, according to the estimates of the time, exceeded the capacity threshold for regions with low land-use intensity. The highest density (62.5 per sq km) was in the province of Kursk, which at that time included the bulk of today's provinces of Kursk and of Belgorod, and the lowest (47 per sq km) was in the province of Penza. Peasants accounted for 94% of the region's population. Peasant holdings were small, 1.5-4.5 hectares per person.[4] All the problems of agrarian, poor, and over-populated European Russia that were discussed in Chapter 2 were entirely typical in the Central Chernozem region.

The industrialization of the 19th century skirted the region. Only a food processing industry developed, sugar refineries in particular. Many of them are still working. For example, nine of the region's thirteen contemporary sugar refineries were built in the 19th century.

Thus the region approached the major cataclysms at the beginning of this century as a backward and ultra-conservative area. Its mainstay, agriculture, was depressed; over 85% of its cropland was under grains and had low and unstable yields. Every tenth resident of the region took part in temporary labor migrations to the north.

The wave of innovations associated with Stolypin reform caused a strong migratory outflow from the region to Siberia and Central Asia. However, the reform failed to destroy the pattern of social relationships in the region, awash as it was with vestiges of bondage. Capitalism made its way to the area extremely slowly. Even the fivefold increase in the region's number of agricultural machines between 1900 and 1913 sidestepped the bulk of peasant households.

Of all regions, the Central Chernozem peasantry seems to have initially benefitted the most from the post-revolutionary re-distribution of land in proportion to the number of mouths in a household. Peasant landholdings became more than one-third larger. However, when the Bolsheviks took power, the so-called food detachments of industrial workers were sent to collect food stored by peasants. Their expropriating zeal peaked in 1918-20 when Bolshevik Russia was separated from the Ukraine and the Volga region, formerly Russia's principal granaries. This situation led to the peasant rebellion in the province of Tambov that was bloodily crushed by the Red Army.

Only in the years of the New Economic Policy *(NEP)* did the situation stabilize. By 1924-25 the agricultural income per one peasant household had risen to the 1913 level, and the region began to recover. However, the most commercialized branches of pre-revolutionary farming (producing sugar beets, hemp, sunflower, and pork) were the slowest to recover. In addition there were fewer opportunities to earn extra money outside farming because of major dislocations in the nation's industry and transportation. And as was pointed out in Chapter 2, the Central Chernozem provinces were the most crisis-ridden areas of Russia in the 1920s with the lowest peasant earnings and high land costs. In the 1920s, a new wave of out-migration affected the region. In 1925-26 alone, 133,000 people abandoned it.[5]

During the first years following the completion of the coercive collectivization campaign, an expansion of the cropland was attempted. The region began to ship more produce to the rest of the country. Regional agriculture diversified; for example, the area under non-food crops increased sixfold between 1930 and 1940. The supply of machines was given a boost. And yet yields remained even lower than before collectivization. Even by 1940, the 1913 level of grain yields had not been reached.

In 1926-39, the population of the Central Chernozem region increased by

211,000, but exclusively because of urban growth; the rural population decreased by 246,000.[6] However, the directions of its migratory flows changed, now moving to cities, not to agricultural lands in the east.

Achievements and Losses

The first wave of industrialization came to the region in the 1930s. The mines and blast furnaces of Lipetsk, the machine-building factories of Voronezh, and the building material factories of both areas were put in operation at that time. In other Chernozem provinces only the centers developed some industry, the remaining territories retaining their purely agricultural character. Another wave of industrialization was associated with the discovery of a rich deposit of iron ore in the 1950s (the so-called *KMA*, or Kursk Magnetic Anomaly), which led to the construction of three large ore-dressing plants: in Gubkin and Stary Oskol in the province of Belgorod and in Zheleznogorsk in the province of Kursk. These developments boosted urbanization. Cities began to absorb more rural migrants. In most provinces industrial growth accelerated, mostly in heavy industry, while traditional food-processing operations stagnated. Whereas in 1940, the region contributed 70% of all sugar produced in Russia, by the end of the 1980s its proportion had dropped to below 60%.

The structure of the labor force in the region testifies to its lagging development compared to the national average, let alone to development in the country's most advanced regions. Thus, by 1990 the primary sector still employed about one-fifth of the labor force even though it had substantially reduced its share in recent decades (Table 8.1). The region is still predominantly agrarian, since its percentage share of agriculture is two times as high as Russia's average. Employment in tertiary and especially quaternary (research and development, management, and finances) sectors

TABLE 8.1 Percentage Shares of Employment Sectors in the Central Chernozem Provinces.

	Primary		Secondary		Tertiary	
Provinces	1959	1990	1959	1990	1959	1990
Belgorod	64	22	16	40	20	38
Voronezh	52	19	23	40	24	41
Kursk	62	23	16	40	21	37
Lipetsk	52	18	23	44	24	38
Tambov	56	22	19	38	25	39

Source: Unpublished data of provincial statistical bureaus.

TABLE 8.2 Productivity of Agriculture in the Chernozem Provinces.

| | Yields in 1986-90 in centners per hectare | | | Milk Yield per Cow in 1990 in kg |
	Grains	Sugar beets	Sunflower	
Russia	15.9	225	12.7	2781
Central Chernozem				
Region	22.4	236	13.3	2767
Belgorod	26.6	273	16.7	3099
Voronezh	22.8	236	13.2	2798
Kursk	23.9	245	10.8	2627
Lipetsk	20.9	221	13.1	2729
Tambov	18.5	188	11.6	2525

Source: Narkhoz RSFSR 1990. Moscow, Goskomstat 1991.

has lagged behind that in many other regions; these sectors' share in employment is one-fourth of that in the Industrial Center.

Though agriculture has expanded during the last 50 years, including a doubling of grain output, its development has had mixed results. In 1986-90, grain yields in all the provinces except Belgorod were 1 to 5 centners per hectare short of the bio-climatic norm (discussed in Chapter 4). Animal husbandry, however, has expanded appreciably and by 1990 was at the average Soviet level in terms of productivity and density of livestock; in this regard it was substantially behind only the Baltics and Ukraine. By 1990, five Chernozem provinces accounted for 5% of the total population and 8% of the rural population of Russia but contributed 12% of its grain, 54% of its sugar beets, 16% of its sunflower seeds, 8% of its potatoes and vegetables, and 9.5% of its meat (Table 8.2). The total output per unit of land is thus well above Russia's average. In large measure this is due to the heightened natural fertility of soils and favorable climatic conditions. Given that the rural population density in the region is also higher than in areas north of the region, agricultural investments have generated relatively high returns. One could argue that under a different system of management the returns would be still higher. And yet it is hard to escape the feeling that socialist-style collective farming has been more accepted here than almost anywhere else in Russia, certainly much more than in the Non-Chernozem Zone. After all, agriculturally speaking, the region has grown from one of the most backward into one of the most successful in Russia.

As Tables 8.1 and 8.2 show, the provinces of the region do not differ much from each other. The provinces of Lipetsk and Voronezh are still the most industrialized, while those of Belgorod and Kursk are the most agrarian. In the latter provinces almost two-thirds of the working rural popula-

TABLE 8.3 Changes in the 20th Century in the Countryside and Agriculture of the Province of Belgorod.

	1885	1915	1940	1954	1970	1980	1990	1994
Rural population density								
per sq km	60	75	50	39	30	22	18	19
Number of cattle								
(heads)	ND[a]	ND	398	451	762	817	937	686
Number of pigs								
(heads)	ND	ND	261	330	788	902	984	664
Number of private cows per 100 rural								
residents	ND	ND	12	15	ND	ND	11	15
Percentage of cows								
in PAF	100	100	83	68	ND	ND	18	26
Cropping area								
in 1000 ha	ND	ND	1397	1428	1698	1712	1586	1483
Grain yield								
in c/ha	11.0	10.6	9.9	8.7	20.4	18.0	26.9	29.0
Sugar beet yield								
in c/ha	120	ND	97	114	155	200	213	134

[a]ND -- no data.

Sources: Narkhoz RSFSR for different years; *Selskoye Khoziaistvo Rossii v XX viekie.* Moscow, Novaya Derevnia 1923.

tion is engaged in agriculture -- the highest share of all Russian provinces. The province of Voronezh appears to be the most representative of the region, the province of Belgorod the most advanced, while the province of Tambov the most backward. Some achievements in agriculture over 100 years are shown in Table 8.3 based on the province of Belgorod.

Rural Population and Settlement Dynamics

The Central Chernozem region was not spared the rural depopulation characteristic of most of Russia. As in the Non-Chernozem Zone, by 1989 the region's rural population was only about one-third of its 1926 level (Table 8.4). However, in distinction from almost all the non-Chernozem regions (Volgo-Viatka is the exception), the net result of the migration of the *total* regional population, that is, urban and rural, has been negative for the region throughout all the post-war period. This is because the region's urban centers were stagnant for most of the period, up until the 1970s, so most rural migrants used to move out of the region. In 1979-1988 the rural

TABLE 8.4 Population Percentage Change, Natural Increase, and Migration per 1000 Rural Population in the Central Chernozem Region.

	Rural Population					Total Population in 1994 as a % of 1926	Natural Increase		Net Migration	
	In 1994 as a % of 1926	In 1939 as a % of 1926	In 1959 as a % of 1939	1989 as a % of 1959	1994 as a % of 1989		1990	1993	1979-88	1989-93
Russia Total	52	94	77	70	102	160	2.2	- 5.1	- 98	14
Central Chernozem										
Region	35	91	72	54	99	82	- 1.8	- 7.4	- 132	49
Belgorod	33	86	73	52	100	86	0.1	- 5.8	- 102	83
Voronezh	43	96	72	62	100	99	- 2.4	- 8.1	- 116	78
Kursk	32	95	74	48	96	73	-2.1	- 7.6	- 182	8
Lipetsk	34	86	68	58	100	84	-0.7	- 6.8	- 108	39
Tambov	31	89	71	50	97	65	-3.2	- 8.6	- 159	17

Source: Demographic Yearbook of the Russian Federation. Moscow, Goskomstst 1994:406-407.

population's out-migration amounted to 1.5% per year, exceeding the average Russian level. The highest out-migration intensity occurred in the provinces of Kursk and Tambov, and the lowest in the province of Belgorod (Table 8.4).

That rural depopulation has, nevertheless, not caused land and settlement abandonment on the scale typical for more northerly provinces is solely due to the region's high rural population density on the eve of industrialization: 50-60 residents per square kilometer versus 20-30 in the non-Chernozem provinces. So even by 1990, the average rural population density in the region was 18 residents per square km, much higher than north of the region.

The rural population decrease was marked by a centripetal gradient, that is, population was declining everywhere but in the vicinity of large urban centers. The pace of the resulting spatial concentration is illustrated by the fact that while in 1959, 21% of the rural population in the province of Kursk lived within the scope of the Kursk urban agglomeration, by 1987 the figure had grown to 41%.[7] Many old towns in the region, however, are so feeble that proximity to them does not affect rural population change. Even relatively large towns are not surrounded by attractive exurban bands unless the town's growth proceeded for a long time. Examples of large but new towns are the centers of metallurgy: Zheleznogorsk in the province of Kursk and Stary Oskol in the province of Belgorod. The case of Zheleznogorsk (along with Naberezhnye Chelny, an automobile-producing center in Tatarstan) was taken up by Zhanna Zayonchkovskaya and is particularly revealing. It appears that the intrusion of a newly-emerging urban center into a long-settled countryside is a painful process. Initially a new town creates a funnel-shaped demographic depression in the surrounding area: that is, close to it the rural population diminishes faster than at a distance. Only after achieving a certain population number (a spatially variable threshold), does the new town become a magnet for the surrounding rural population as well. However, Zheleznogorsk has not achieved that threshold even after reaching a population of 90,000; it still looks like an alien intrusion in the Chernozem countryside.[8] Significant paved roadways do not exert a powerful stabilizing influence on the rural population of the region, either. In part this is linked to the historical gravitation of rural settlements to river valleys and gullies; in contrast major thoroughfares crossing the province run through watersheds.

Thus, spatial gradients in rural population dynamics do exist in the region. Apparently they depend upon two factors. The age and size of a town are just one of them. The type of the surrounding natural vegetation is the second: gradients appear to be steeper in the northern part of the region, in a typical forest-steppe, and very gentle or absent in the south where classic steppe occurs. Given that the spatial polarization of the countryside is most vivid north of the region (see Chapters 1, 4, and 10) and much less apparent south of it, the Central Chernozem region appears to be truly transitional in this regard.

As was pointed out in Chapter 7, in the 1990s the tide of migration was turned, and the countryside began to receive more people through migration than it lost to it. The Central Chernozem region is one of the regional leaders in this regard: whereas in Russia as a whole, net rural migration in the period 1989-1993 amounted to plus 14 per 10,000 rural residents, in the Central Chernozem region it was 49. The most attractive provinces have been that of Belgorod and that of Voronezh: 83 and 78 per 10,000 population respectively (Table 8.4).

New Trends in Farming

Having retained over half of its late-1980s public sector level of agricultural output (Table 8.5), the Chernozem Center has fared average in the crisis of the 1990s. That is, it has done somewhat better than the piedmont provinces of the Northern Caucasus, the premier Russian producers of grain, but worse than the non-Chernozem provinces. The region has retained its leading position in crop harvesting productivity and, along with the Northern Caucasus and the province of Moscow, has the highest grain yields (see Chapter 6).

And yet financially socialized farms are in dire straits, especially in the northern part of the region (see Figures 6.6 and 6.7). After the removal of some of the welfare-style management practices of the 1980s, characteristic distinctions with long historic roots showed up. The northern part of the region (around Tambov and Lipetsk in particular), the most ailing early in this century, once again fares the worst. The southern part, long influenced by the large cities of Voronezh and Kharkov,[9] is in better shape.

The region's total agricultural output was slashed by 28% in 1991-1994, including 14% in 1994 alone. The production cutback in the socialized sector, about 40%, does not vary much across the region. However, the growth of PAF has been highest (42%) in the most crisis-ridden province of Tambov and lowest in the provinces of Belgorod and Kursk (Table 8.5). As everywhere in Russia, the activation of PAF has been the first reaction to crisis. However, by 1994 PAF's growth potential seemed already to have been exhausted.

Nonetheless, even after dropping 40% of their former public sector output (including 16% of its grain, 50% of its sunflower, and 66% of its sugar output), the Chernozem provinces have not suffered from food shortage. They would not have suffered from it in the past either had it not been

TABLE 8.5 Growth Indices of Agricultural Output: 1994 as a Percentage of 1990 and 1994 as a Percentage of 1993.

	All Farm Categories		Socialized Farms		Personal Auxiliary Farming	
	1994/90	*1994/93*	*1994/90*	*1994/93*	*1994/90*	*1994/93*
Russia	75	91	59	86	119	98
Central Chernozem						
Region	72	86	57	83	118	92
Belgorod	69	85	59	83	108	89
Voronezh	69	86	56	82	111	93
Kursk	69	86	56	87	107	85
Lipetsk	74	86	57	82	128	94
Tambov	84	87	60	79	142	98

Source: Statisticheski Bulletyn 1 (APK) 1995.

for the fact that Moscow used to grab off a sizable portion of its produce. Now the region's agriculture is facing a different problem: where to sell its produce. Strangely, communist-minded provincial powers are resorting to protectionist practices, internally thwarting their markets. While it would make some sense to protect these markets from incoming goods (as the impoverished Ukraine is next door), the restrictions imposed on out-going shipments (see Chapter 6) appear bizarre. The province of Voronezh has the toughest of these restrictions; Belgorod followed suit by introducing a special tax for selling meat outside the province. Such measures undermine the economic stability of regional farming.

There are very few independent private farms in the Central Chernozem region. Only the province of Tambov, with 34 private farmers per 1000 persons engaged in agriculture, is on a par with the average Russian level; in other provinces there are only 12-15 private farmers per 1000. Moreover, the Central Chernozem region is the only one in which the number of private farmers substantially declined in 1994 (from 14.3 to 12.9 thousand). The biggest decline (by one-third) was in the province of Belgorod.

In summary, despite some indisputable achievements in agriculture during the Soviet period, or perhaps because of them, the region has preserved the conservatism so vivid in it early in this century when the region was so poor. This conservatism now shows up in the peculiar reactions of local administrators to on-going economic changes and in attempts to challenge them.

Let us take a closer look, in the next chapter, at the processes unfolding in this region using the example of the province of Belgorod.

Notes

1. See a more detailed description by Shaw, D. J. B. "The Frontier Experience in Romanov Russia: The Settlement of the Central Black Earth Region in the Seventeeenth Century" in Pallot, J. and Shaw, D.J.B. *Landscape and Settlement in Romanov Russia 1613-1917.* Oxford: Clarendon Press 1990, pp. 13-32.

2. Hettner, A. *Das Europaische Russland.* Leipzig: B. G. Teubner 1905, pp. 184-186.

3. Semionov-Tianshanski, V.P. "Gorod i Derevnia v Yevropeiskoi Rossii." In *Zapiski Russkogo Imperatorskogo Geograficheskogo Obschestva. Departament Statistiki* 1910, Vol. 10, No 32, p. 36.

4. Semionov, V.P. (ed.) *Rossiia. Polnoye Geograficheskoye Opisaniye Nashego Otiechestva.* Volume 2. *Srednierusskaya Chernozemnaya Oblast.* Saint Petersburg: Devrien 1902, p. 195.

5. Chelintsev, A. "Selskokhoziaistvennye Raiony RSFSR." In *Ekonomicheskaya Geografiya.* Moscow: Sverdlov Communist University 1929, p. 447.

6. *Demographic Yearbook of the Russian Federation.* Moscow: Goskomstat 1994, pp. 17, 20.

7. Zayonchkovskaya, Z.A. *Demograficheskaya Situatsiya i Rasseleniye.* Moscow: Nauka 1991, p. 117.

8. Ibid., p. 117

9. Before custom control was set up at the Russian-Ukrainian border, the province of Belgorod was clearly a part of Kharkov's economic and social backyard. It remains to be seen whether these neighborhood relations will be fully restored or fall into oblivion.

9

The Province of Belgorod

The information for this chapter was collected in June 1995 when co-author Tatyana Nefedova made a series of field trips in the province of Belgorod for the Moscow-based magazine *Vash Vybor*. Such field trips usually make it possible not only to obtain a lot of local statistics unavailable in Moscow but also to interview scores of local officials and rank-and-file people. As early as in the 1970s it had become abundantly clear to us, then graduate students at Moscow State University, that interpreting Russian statistics *without* talks and interviews with *real* people is all but senseless.

The province is one of the most densely settled in the country. Even though it does not have many large cities, its average population density is ranked 16th among the 89 sub-divisions of the Russian Federation; the province is among the top ten sub-divisions in terms of paved roadway per 1000 square kilometers; and it is third in terms of the value of its fixed assets related to farming (per 100 hectares of agricultural land).[1]

One specific feature of the province is that its center does not dominate it. By the end of the 19th century, the city of Belgorod was 2.5 times smaller than Kursk, four times smaller than Voronezh, and 25 times smaller than Kharkov. The latter is next door, on Russia's scale, so for a long time the city of Belgorod developed as a satellite of the Ukrainian city of Kharkov. Even now, with a population of 322,000, Belgorod accounts for only 22% of the province's total population. For comparison: in the neighboring province of Voronezh, the respective share is 35%; in the non-Chernozem Yaroslavl (Chapter 11), 44%; and even in the bi-central province of Pskov (Chapter 14), the share of the major center is slightly higher than in Belgorod.

Having a relatively weak center and habitable space throughout all the 21 districts of the province (Fig. 9.1) results in the area's population density contrasts being subdued. In 1994, the population ranged between 12 persons per sq km in the remote south-east to 20 in the districts located on major thoroughfares. The district of Belgorod is the one exception with 33 rural residents per square kilometer.

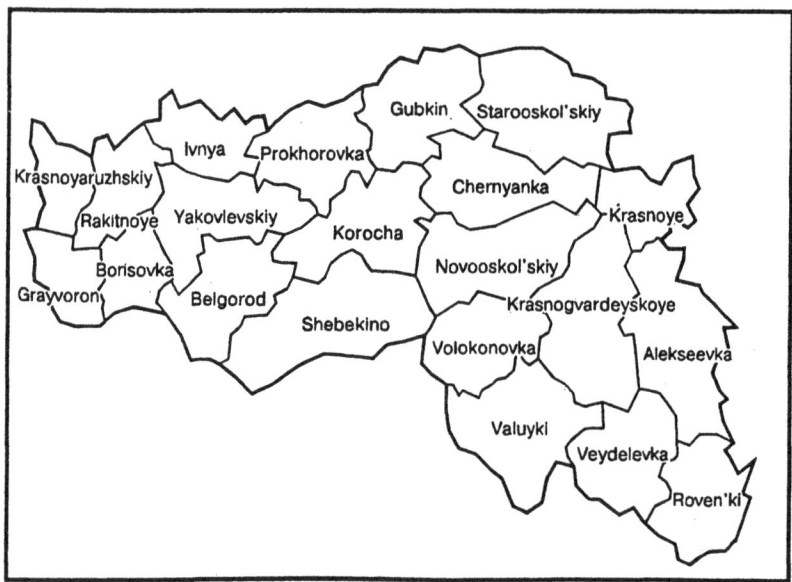

FIGURE 9.1 Districts of the Province of Belgorod.

Well-Being Due to What?

The land is beautiful around Belgorod: gently undulating hills on which teal-green patches of woods are interspersed with vast and groomed Kelly-green fields. The soil is so greasily black that even during the dry summer of 1995, when people had been hankering for rain for about a month, it seemed moist when looked at from a distance. Villages are large with white stone-made buildings awash in flowers. In those central villages where seats of rural Soviets are located one can find a capacious club, a day-care center, and sporting facilities.

The countryside of the province prospered in the late 1980s especially, when agricultural investments began to yield good returns. At that time rural amenities were given a boost. Also provincial leaders took advantage of good connections in Moscow (in particular, Ponomariov, the First Secretary of the Belgorod *Obkom*, used to be a confidant of Yegor Ligachev, one of the most influential people in the Politbureau), which helped to get extra budgetary funds for the area's rural infrastructure. By the late 1980s, the province was agriculturally one of the very best in Russia.

In distinction from other premier producers of food, the province has large metallurgy enterprises, which suffered little during the early 1990s.

Whereas in 1990, 41% of the gross economic output of the province was contributed by the food processing industry, mining and metallurgy furnishing only 17%, four years later the enterprises had switched positions: in 1994, iron ore and metal accounted for 55% of the industrial output and 94% of the total province-based exports from Russia. As a result, the province's tax-base appears to be stable, as Belgorod returns to the state treasury much more than it receives through federally sponsored programs. As for the provincial budget, 60% is provided by enterprises located in Stary Oskol and Gubkin, the two major centers of the Kursk Magnetic Anomaly (a rich deposit of iron ore). Also the province of Belgorod has taken full advantage of the December 1993 Presidential Decree giving provinces carte blanche in introducing local taxes. All this has helped to keep Belgorod economically healthy.

How is this money spent? In 1994, as much as 20% of the provincial budgetary spending was directed to subsidize retail prices. This is ten times Russia's average and even higher than in the provinces of Voronezh and Ulyanovsk, notorious for their pro-communist administrations. In terms of direct budgetary spending on agriculture, the province of Belgorod is second only to that of Voronezh in all of Russia.[2] In other words, the province is committed to pumping money from other branches of its economic activity to agriculture. Yuri Savchenko, the leader of the province's administration, believes that all socialized farms warrant support, both profitable and unprofitable (there were 166 of the latter out of the total of 360 in 1994). Agriculture faces real problems. In 1994 its output was 43% lower than in 1990. And yet the problems are at least in part smoothed over both by favorable nature and by the peculiar economic policy of the provincial administration. Let us take a closer look at it.

To Put Under Control

This sub-heading captures the gist of the economic policy with respect to farming. Measures of this policy include: subsidizing animal husbandry through centralized supply of discounted mixed fodder and vitamin supplements; centralized purchasing and distribution of saplings; setting uniform farm-gate prices for livestock produce; etc. Instead of loans, socialized farms receive partial, budget-based advance payment for produce. For example, by June 1995, 30% of the projected future output had already been paid for. Also there is a system of so-called investment projects, which in essence boil down to discounted loans for fertilizers, herbicides, and fuel. Farms conclude agreements with specialized Belgorod institutions required to supply farms with fertilizers, etc. It is, however, the administration that pays for such supplies, while farms, in their turn, repay with their produce.

A glance at the above policy measures suggests that the administration is leery of losing tight control over farming. While some of the measures seem reasonable, as both farms receive what they need and the administration is assured that money is not squandered, it is only the weak that need such strong direction. Those who are not weak are faced with a fait accompli: if you are averse to concluding an agreement with the administration regarding centralized supply of working capital, you will have to procure such funds on your own, which is risky given the price disparity between agricultural and industrial produce; if you are willing to conclude such an agreement with the province, you are "enslaved" by the supplier. If farms were allowed to borrow cash from banks, they might try to find cheaper fertilizers and fuel. That cash-based transactions are now routinely cheaper in Russia is a result of inflation and delayed payment for produce; as a result, any agreement on advance supply of working capital takes not only market prices into account but also projected rates of inflation and risk factors. It is because of that buying for cash is always cheaper. Alas, socialized farms have little choice, as they are usually cash-deficient. To arguments in favor of advanced bank loans, those in the administration would reply that the money loaned would disappear like water through sand or get bogged down in some intermediaries. Unfortunately this may be true. But, on the other hand, not everything is visible from an office in the city of Belgorod, and treating farms as callow babies is hardly compatible with basic free-market philosophy.

One of the authors attended a provincial seminar in Shebekino at which the leader of the administration repeatedly instructed farm chairmen and specialists when and how to mow and demanded that the district-based farming bureaus keep checking to ensure that "assignments" were fulfilled. It was so reminiscent of the worst communist-style practices that in our subsequent talks with the bosses of those bureaus we asked about their vision of their institutions' role under the new economic conditions. Are bosses needed at all, and how can they help farms? Interestingly enough, they offered no ideas at all about providing help, as district farming bureaus are no longer in charge of funds; nor can they set up a marketing service for farm units: staffed with former party operatives they do not know how. However, they take their control functions with the utmost seriousness, believing that farm chairmen are not mature enough to be fully responsible. So bitterly familiar resolutions like *Razobratsia v Administratsii* (To Investigate in the Administration); *Vziat' pod Kontrol'* (To Put Under Control); *Sostavit' Plan-Grafik* (To Set a Plan-Schedule) can be seen in the reports of provincial and district farming bureaus.

One third of the total number of cattle in the province are kept on large mechanized animal farms. The administration takes it as its prime task to ensure their economic survival. However, about 50% of livestock kept on more traditional farms incur huge financial losses. The administration believes that one of the ways to resolve the problem is to increase the share of personal auxiliary farming on which 20% of the province's cattle are now kept. The intention is to drive this share up to 50%. In order to make that happen, budget spending is devoted to such things as purchasing PAF milk to ensure that it can always be sold. Those who keep cattle and swine on their subsidiary plots are credited with additional years of service, which is essential for future retirement. In the 1960s and 1970s people who had no official job but working their subsidiary plots were considered social parasites and could be persecuted; in the 1980s the law was rarely enforced, but years of service were not credited either. So the attitude of those at the helm of local power today is a step forward. However, the desire to regiment and micromanage shows through this attitude as well. How are people lured into doing what the provincial power brokers want them to? Under Russian conditions -- by offering them money for housing, which is always in short supply. A special foundation was set up in the province to support individual housing construction in the countryside. It is the administration's favored child. This foundation concludes agreements with rural residents giving them a 1% interest loan (a huge benefit under the current level of inflation) to be paid off over ten years not in money but in kind, that is, by meat produced on subsidiary plots. One can pay off the loan by other farming produce in quantities price-equivalent to a certain amount of meat, which is thus taken as the yardstick. The activity of the foundation is not limited to loans. It also has a network of foodstores, refrigerators, and warehouses; along with cash-based retail items, it sells discounted food certificates to urbanites (attracting them to the foundation's retail facilities), and finances cooperative construction businesses. The foundation even issues its own newspaper with the characteristic title *Dobrostroi* (a Russian neologism for benefaction). While all this seems to be a highly commendable initiative, one aspect of it causes concern: total regimentation. Everyone who has received a loan from the foundation -- already over 3000 people -- works exclusively for the foundation. Those who would dare sell their produce elsewhere would be penalized. Actual use of loans is also under foundation control, which keeps a black-list of people who neglect their subsidiary plots. So while rendering real help, the foundation also acts like a socialist octopus that strives to put everything under its control.

Internal Differences

Since natural conditions are fairly homogenous all across the province, it is socio-economic factors that account for the differences in agricultural output. The western part of the province leads in output per unit of land. The district of Belgorod stands out in terms of the value of fixed assets per unit of land, as do the districts located along the major highways linking the city of Belgorod with Shebekino and Gubkin. The highest grain yields are in the district of Belgorod: in 1986-90 they were on average 34 centners/hectare (while the average yields for the province were 29 centners/hectare; both figures are very impressive for Russia); in 1994, yields averaged 38 centners/hectare. The lowest yields are in the province's east (Fig. 9.2); but here farming is less capital and labor-intensive: for example, the value of fixed assets per unit of land is four times higher in the district of Belgorod than in that of Krasnoye, and the employment per 100 hectares of arable land ranges from 12 persons in the district of Belgorod to seven in districts remote from the provincial center.

There is a substantial spatial differentiation in terms of the ratio of socialized and personal auxiliary farming (Fig. 9.3). Whereas in the province's periphery there are 45-60 cows in PAF per 100 households, in the district of Belgorod there are only 18. PAF is a very substantial source of income not only for country folk but for residents of small towns as well. Local experts believe that in 1995 rural households earned about 75% of their income in PAF, while small town residents earned about 50%.

To find out which factors influence farming efficiency the most under the current system of management (that is, large, stable socialized farms kept under rigid administrative control), we used as an indicator the sum of a district's ranks, in ascending order, in the production costs of milk, meat, grain, and sugar beets. It turned out that production costs are lowest in the districts both most and least endowed by machines and technology. It looks as if economy of scale shows up only when a certain threshold in the value of fixed assets is achieved, while operating at a level short of this threshold may be as efficient as simply relying on the natural fertility of the land. A more detailed analysis comparing socialized farms showed a similar result: the lowest production costs appeared in the largest (over 6000 hectares of arable land) and in the smallest (below 2000 hectares) farms. This is, of course, a very generalized conclusion as many factors, including the personality of a leader, affects each specific farm. Most local experts believe that the optimal size of a farm in the province is about 3000-3500 hectares. The same experts get more emphatic when it comes to the optimal size of one field. They are convinced that the current range of 100-250 hectares is way above the ecologically efficient norm for an undulating lowland prone to occasional droughts. Around fifty hectares seem to work better, they believe; and if woodland belts were more widely used to separate fields, they would facilitate preservation of moisture and diminish the damage caused by hot dry winds.

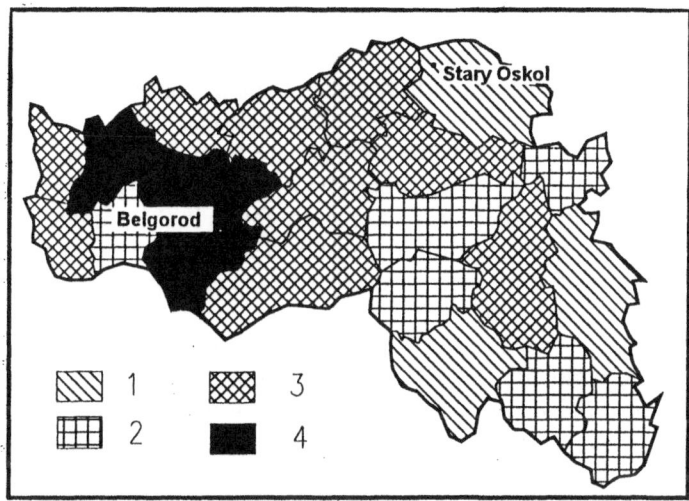

FIGURE 9.2 Grain Yields in Centners per Hactare (1994) in the Province of Belgorod. 1: Below 25; 2: 25-30; 3: 30-35; 4: 35-38.

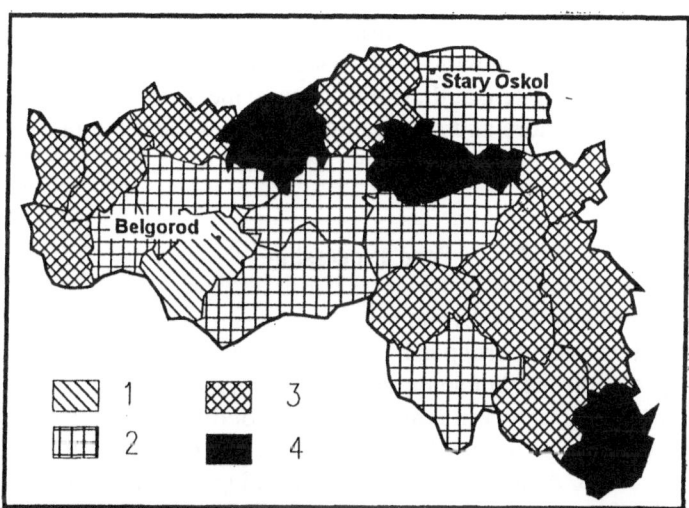

FIGURE 9.3 Cattle Kept in Personal Auxiliary Farming as a Percentage of the Total Number of Cattle in the Province of Belgorod (1994). 1: 1-18%; 2: 30-45%; 3: 45-60%; 4: 60-65%.

Which Reforms Are Better?

Inconsistent Russian reforms have affected this province as well. However, independent private farmers are not very welcome here. Their number peaked in 1994 when there were 2763 private farms in the province; only one year later this number was reduced by 1000. For every thousand people engaged in agriculture there were 14 private farms in the province in 1994, against the Russian average of 30 and up to 60 in some provinces. Most private farmers in the province gravitate toward exurbia in the districts of Belgorod, Shebekino, and Stary Oskol.

As for the large socialized farms, their leaders are obviously in favor of all federal and regional forms of financial support, and in terms of internal incentives for efficient work they opt for cost-accounting practices of the 1980s. The provincial administration believes those practices have been a success in contrast to the reform of the 1990s, which they feel was concocted at the federal level, using one template and without taking local conditions into account. Internal farm-governance practices are as follows: every technologically specific sub-collective of a farm (dairy operators, land-tilling teams, etc.) is assigned an upper limit of spending; should they spend less, 50% of the savings becomes their earnings; if they overspend, they do so at their own cost. In the district of Ivnya at "Niva," formerly a collective farm and now a joint-stock company, 70% of shareholders (formerly, just members) approved this system at their 1995 general meeting. Where the system works, earnings get more and more uneven. So does the economic standing of an entire farm.

Food Processing

This segregation into haves and have-nots is even more noticeable in food processing. Here output dropped by 25% in 1994/95 alone. Some production units are in dire straits, but some are well-to-do. An example of the latter is a milk-processing factory in Ivnya, northwest of the city of Belgorod, near the border with the province of Kursk. The factory has been profitable for most of the 1990s. And not as a result of robbing the dairy farms; on the contrary, the factory spends part of its profit by paying more than the Belgorod-set low purchasing price of milk and by supplying farms with cooling containers. Because of its favorable treatment of milk producers, the factory has been able to expand its market, accepting milk even from the neighboring province of Kursk. The creativity of the factory's director is the main cause of its success. Butter, sour cream, and cottage cheese produced here are sold far beyond the province including to the cities of Togliatti,

Ufa, and even Moscow, where a domestic supplier has reportedly been unable to penetrate because of imports.

A comparison of the two large sugar-processing factories located in the district of Shebekino, *Rzhevsky* (over 600 employees) and *Novotavolzhsky* (over 800 employees), is interesting in this regard. Whereas in the former, average monthly earnings in 1994 were 330,000 roubles (about $75), in the latter they were only 94,000 roubles ($21). *Novotavolzhsky* simply comes to a halt when the local sugar beet supply stops; *Rzhevsky*, on the other hand, is still working even though the local raw material supply lasts only 2-3 months. The reasons are the same: the director is resourceful. He found connections and is now processing raw cane sugar from Brazil. Also local suppliers enjoy good treatment from the factory: in contrast to most other producers of sugar in the region, which appropriate 35-40% of the sugar beet-based output (while giving the rest to the producers of the sugar beets), *Rzhevsky* appropriates only 30%. So farms are willing to deal with the factory, which is thus able to broaden its local supply base.

However, some earlier problems linger and some new ones have emerged. For example, after the privatization of food processing factories transformed them into unlimited joint-stock companies, a controlling share of the capital stock ended up in the hands of the factories' administrations. Since this enabled them to dictate conditions to primary producers, the province resorted to issuing a new round of stock. As a result, 51% of stocks are now held by farms. This was considered fair but such fairness turns out badly for food processors; now they cannot direct whatever percentage of their profit they deem necessary to the renewal of equipment because socialized farms want shareholders to be paid their dividends today. What is left for the factory is not sufficient; so the new system does not allow factories to think about tomorrow, which is regrettable since the production of sugar in Russia faces many other problems as well.

White Sugar on the Black Earth

The province of Belgorod produces about one quarter of the sugar in the Central Chernozem region and 11% of that in Russia as a whole. In 1990-1994 the area sown with sugar beets decreased from 151,000 to 113,000 hectares in the province, and the output dropped from 4.2 to 1.5 million tons. In the 1980s, sugar beets used to be highly profitable. But with production being capital-, energy-, and labor-intensive, the recent price disparity cuts back on profits.

It has been estimated that in the 1990s that sugar beets secure a good profit margin only under yields exceeding 250 centners per hectare. To reach this level of productivity requires 200 centners of mineral fertilizers per hectare plus herbicides. Through concessionary loans secured by the provincial administration, one half of the required fertilizers, one-third of the herbicides, and all the seeds have been delivered. But this is not enough. Labor is deficient as well. In the 1960s as many 44,000 women worked fields sown with sugar beets; now only 4,000 do. So the Belgorod farms recruit people from other areas, primarily from western Ukraine -- about 40,000. About 3,000 urbanites also help. In return for thinning out a one-hectare field of sugar beets one may earn two centners of grain and 50 kilograms of sugar. Manual labor is also widely used to clean soil from beets and to collect green tops left on the field after harvesting. New technologies do exist that allow accurate sowing of a pre-set number of plants thus eliminating manual thinning. However, this approach requires special seed drills and enhanced application of herbicides.

As was already pointed out, sugar-related problems are not limited only to cultivation. Russia consumes seven million tons of sugar. It can produce only 2.5 million tons from domestically grown beets. So the remaining portion can be derived only from imports. About 1.5 million tons will be delivered by the Ukraine, which will thus pay a part of its huge debt to Russia. For the remainder it would make perfect sense to buy not refined but raw sugar. Of the 300 sugar refineries in the former USSR, 100 are on Russian soil, and their combined capacity is more than enough to meet Russia's demand. Unfortunately most of these factories operate only 2-3 months a year and are idle once they use up their supplies of sugar beets. For example, in the province of Belgorod, only three refineries of eleven worked in June 1995.

The system behind the importation of raw sugar is now in disarray and at the mercy of highly inconsistent decision making. Raw sugar is being imported by private businesses. This has turned out to be lucrative because the state has rewarded domestic suppliers of imported raw sugar with oil-export licenses. Raw sugar, traditionally purchased from Cuba, is now being supplied by Brazil and some other countries. Since go-betweens want to ensure a good profit margin, raw sugar prices have soared. Additionally a 20% custom tariff on raw sugar has been introduced, which eventually made acquisition by processing factories prohibitively expensive. Note that the production costs of refined sugar are 60-70% dependent upon the cost of raw sugar. All told, the price of the final product appears to be about $500 per one metric ton. At the same time foreign producers (in the Ukraine and Latin America) offer sugar priced at $260-350 per ton, custom fees included. This is the principal reason why areas under sugar shrink and domestic sugar refineries are idle. But Russian sugar refineries are also worried that current low prices on imported raw sugar will disappear once domestic competition ceases to exist. What is the solution? Russian sugar producers believe that no more than 100,000 tons of the seductively cheap raw sugar has to be purchased abroad to enliven languishing refineries. Once refineries get to work they will generate profit and the whole technological chain will get back on track with farms becoming interested in expanding the area under beets. After all, the province of Belgorod alone could increase its total output of refined sugar from 400,000 tons a year to 1,000,000 tons, half from beets and half from imported raw sugar. But would things happen as planned? And will, say, issuing loans for raw sugar imports help *all* the refineries work efficiently? For some refineries, won't giving them loans be like beating a dead horse or, as Russians would say, like *attaching a poultice to the dead?* So the usual equity-efficiency dilemma, which in Russia so far has mostly favored equity, looms large.

Which Way to Choose?

The above situation is not sugar-specific. The same vicious circle applies to many production sub-spheres: domestic produce appears too expensive because of low efficiency and high production costs; prices that suit producers are not affordable to people; desirous to help people, the government suppresses price hikes; but that boosts the notorious price disparity between agricultural and industrial goods, and thus deals the final blow to impoverished farming. The circle is completed when the government pumps money from non-agricultural sectors of the economy into farming, thus attempting to put everything under its tight control, which suits the weak but strangles the initiative of the strong.

However, it is becoming increasingly difficult to fend off the free market. If in the past there were many different ways for farms to become diverse in terms of efficiency, there are more now. So the equity-efficiency dilemma will not go away and a conscious decision will soon have to be made whether to continue propping up helpless economic units or to support only the strong. The former strategy may well survive. According to a summer 1995 poll of 520 representative adults conducted by the Analytic Center of the Administration, 40% of the adults in the Belgorod province have no political preferences; of the remaining two-thirds, 16.5% support the Communist Party; 9.8% side with the "Liberal Democrats" headed by Zhirinovski; the pro-communist Agrarian Party enjoys the sympathy of 7.1%; the centrist block, "Russia is our Home," headed by Chernomyrdin has the support of 6.7% potential voters, while Gaidar's "Russia's Choice" carried 2.7%.[3] In the countryside, the Agrarian and the Communist Party advocating preservation of collective farming garnered the strongest support.

People Reach for Belgorod

The province has been losing population through mobility for a long time. During the 1960s, for example, out-migration reduced population by 18%; in 1970-1990 by 38%. Out-migration used to be mostly at the expense of the rural population, because the number of urbanites in the province was growing: in the 1960s by 18%, and in the following twenty years the urban population doubled, even though net migration for the province as a whole was negative up until the mid-1980s. In the second half of the 1980s, this all began to change. At that time the rural infrastructure was given a boost, which has made it now well above that in most of its neighbors. As a result, in the beginning of the 1990s the situation changed drastically. In 1990, the excess of incoming migrants over outgoing (net migration) was 1200; in 1992, net migration was already 9000, and in 1994, 29,000. The province has become relatively attractive, although, as in Russia as a whole, net migration has soared not so much because of the increasing number of newcomers as because of the dwindling number of out-migrants. But even this is changing: whereas in 1994, 199 newcomers came to the province of Belgorod per each 10,000 of its residents (including 181 per 10,000 villagers), in 1995, 230 came. This is four times Russia's average. Even Krasnodar and Stavropol, traditionally magnets for Russian migrants, receive fewer migrants now (respectively 184 and 158 persons per 10,000 in 1994).

Yet, because of the lingering effect of protracted rural out-migration, the rate of natural increase of the rural population in the province is negative: in 1994, the excess of deaths over births amounted to 8,300. However, the

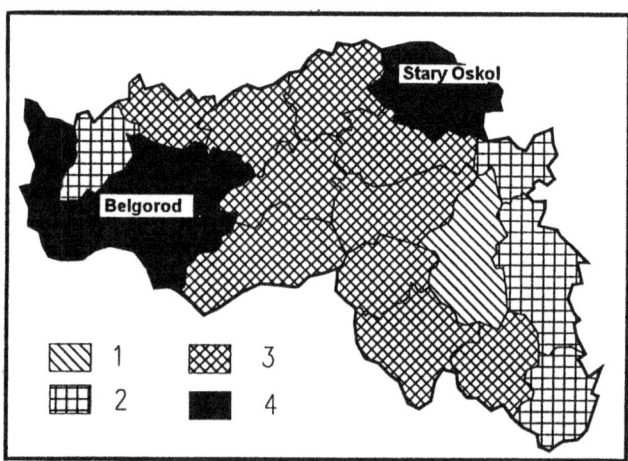

FIGURE 9.4 Average Rural Net Migration per 10,000 Rural Population in the Province of Belgorod (1992-1994). 1 - Below 30; 2 - 30 -50; 3 - 50-70; 4 - Over 70.

newcomers numerically outweigh this negative rate and the on-going migrations of the locals to cities (only in 1992 did the reverse flow gain the upper hand).

The majority of newcomers are from the ex-Soviet republics, especially from the Ukraine (specifically from the province of Kharkov and from the Crimea). The republics of Central Asia and the Transcaucasus contribute 50% of the entire inflow, which is mostly (85%) comprised of ethnic Russians. Thus only a minority of the newcomers come from Russia itself and they mostly head for cities. In contrast to them, ethnic Russians from republics go where they are assigned. Many of them replenish the labor pools of socialized farms. But farms may well be a temporary stop for them. Not only are most of them urbanites by birth, 22% have college degrees (compared with 9% in the province) and another 29% have some college education (vs. 18% in the province). No wonder many of the newcomers prefer the immediate vicinity of Belgorod and of Stary Oskol (Fig. 9.4).

Notes

1. Treivish, Andrei. "Blagodaria Anomalii." *Vash Vybor* 1995, No 2, p. 7.

2. Lavrov, Alexey. "Finansovyi Oazis ili Tikhii Omut." *Vash Vybor* 1995, No 2, p. 10.

3. Martynov, Mikhail, Zhurbin, Igor, and Nikitina, Valeria. "Bolshaya Igra so Zritelem." *Vash Vybor* 1995, No 2, p. 47.

10

The Non-Chernozem Zone

The name *Niechernozemye* or *Niechernozemnaya zona* is attached to about two-thirds of European Russia's total land area. According to the well-known Russian geographer Boris Rodoman, using this name is the same as calling England "non-France" or Russia as a whole a "non-banana republic."[1]

Yet there is some sense to the name. In the mid-1960s, when Soviet authorities became worried that agriculture might become the first stumbling block of their centrally planned economy and that something crucial needed to be done about it, research concerning the inter-regional disparity and the regional specificity of agrarian issues was encouraged. Resulting studies clearly showed that these issues did vary spatially, and when different region-specific investment programs were instituted in the 1970s, the most ambitious, expensive, and widely publicized was devoted to the "further development" of non-Chernozem agriculture in European Russia. Twenty-nine sub-divisions of European Russia were targeted in the following macro-economic regions: the Center, the Northwest, the North, the Volgo-Viatka, part of the Urals, and the free-standing province of Kaliningrad. According to the joint Party and Government ruling of 20 March 1974, the volume of agricultural investment in the "Zone" thus circumscribed was supposed to be drastically increased. This decision implied that the government had over-looked the Zone for quite some time because of its preoccupation with industry and with newly colonized agricultural areas in the east. As the thrust to the east had by then come to a halt, it was considered appropriate to attend to the rear. This "rear," which encompassed the most urbanized part of Russia and the cradle of its statehood, also required enhanced attention because the soil conditions here are inferior. Since, as was shown in Chapter 4, by the 1960s, agricultural productivity was correlated with the natural fertility of soils, it was considered appropriate to compensate this inferiority with heightened investment. Hence the emphasis on the adjective "non-Chernozem" for the Zone.

Initially the bulk of this investment was directed to the production sphere: equipment, construction of mechanized farm buildings, fertilizers, land reclamation, etc. Later in the '80s, investment in the social infrastructure was also given a boost. Whereas in 1961-1965, 7.1 billion roubles were directed to all these purposes, in 1971-1975, 19.5 billion were; in 1976-1980, 31.2 billion; and in 1981-1985, 39.3 billion roubles (bear in mind that roubles had not yet been devalued). After the initial 1974 ruling concerning the Non-Chernozem Zone, the Party and the Soviet Government revisited the issue two more times, in 1981 and in 1986, adopting special resolutions that targeted the ailing *Niechernozemye.* Since the mid-1970s, the area was accorded a special entry in official and classified data-books, a practice largely abandoned in the 1990s.

While the scale of changes that the Zone has experienced owing to its being elevated to the level of a prime political issue should not be underestimated, the returns on investment have left much to be desired, with the exception of the provinces of Moscow and Leningrad. In most other provinces not even the environmentally prescribed "norm" for grain yields (see Chapter 4) was achieved.

Is Nature Indeed Severe?

When Russians say, as they often do, that European Russia is an area of insecure crop harvesting, they usually mean the arid southern part of the Volga region and the Non-Chernozem Zone.

We will confine our analysis of the latter to a general overview of the area covered by twenty-three (Fig. 10.1) of the twenty-nine sub-divisions usually included in the Zone, because in most of Russia's north, agricultural land accounts for no more than 1% of the total land area, and the province of Kaliningrad warrants a separate analysis in view of its highly specific geographic location.

Most non-Chernozem provinces belong in the natural vegetation regions of either southern taiga or mixed forest, while the Zone's southernmost fringe stretches into forest-steppe. The climate of the Zone is *humid continental* with its annual temperature range increasing eastward. Assuming that an average mean daily temperature of 10 degrees Celsius marks the start of the growing season for most food plants harvested in the Zone, the annual sum of growing degree-days in the area ranges between 1600 and 2200. Only in the extreme south does it reach 2400. The average annual precipitation is 600-800 mm. Agricultural regions in Scandinavia and especially in Canada exist under similar conditions. Natural factors limiting agricultural productivity in the Zone include: occasional very harsh winters; frosts as late as early June and as early as September; uneven summer

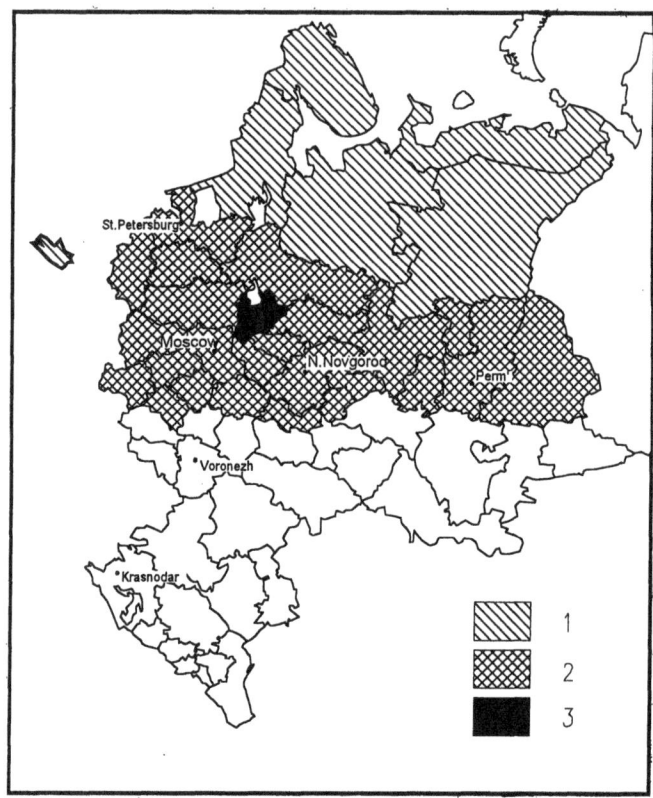

FIGURE 10.1 The Non-Chernozem Zone of (European) Russia. Shaded area is the officially designated area of the Zone, including: 1 - Provinces outside study area; 2 - Study area; 3 - Case-study area (the province of Yaroslavl) to be discussed in Chapter 11.

precipitation; an abundance of waterlogged soils; their low humus content; and their predisposition to acidification and turning into bogs. Fragmentation of fields interspersed with woodlands is also a factor.

Vasily Kliuchevsky associated specific features of the Russian national character with conditions in the Zone, particularly the area near the confluence of the Volga and Oka Rivers. "Nature," he wrote, "accords only a short time for working the land, which is why a Russian became used to exerting himself during a brief period when he was not allowed a single misstep; but following this, he would indulge in a lengthy, forced autumn and winter idleness. No other people in Europe are suited to such an exertion, but likewise nowhere in Europe can one find such a lack of even, moderate, and measured work habits as in Great Russia."[2] That in the Russian peasant mind no direct link existed between the amount of labor expended and its result, Kliuchevski also associated with the weather. Nature, he said, often mocked the most careful calculations, hence the peasants' stunning power of observation, their unpretentiousness, and, most importantly, the famous Russian reliance on luck or on the off-chance (*avos'*).

Thus erratic weather, a relative scarcity of the heat needed for the maturing of some crops (*Niechernozemye* is sometimes called a country of evergreen tomatoes), and a low soil fertility requiring a lot of fertilizers became commonplace descriptions of the Zone.

In terms of land use, the share given over to agriculture varies between 10% in some districts of Leningrad and in the Novgorod provinces, to 70% in the south. Note that the geographical distribution of this indicator, and that of its correlate, the percentage of arable land in the total land used in farming (Fig. 10.2), are closely related to the layout of natural vegetation regions; so is the land's bio-climatic potential when expressed as a plausible crop yield (usually grains) under the existing combinations of heat and moisture.

However, since yields in most of the Zone are short even of this bio-climatic potential (see Figure 4.8), a habitual argument that nature is to blame for hindering agricultural development appears dubious. The most powerful factors restricting productivity rather seem to be the Zone's rarified social space (discussed in Chapter 1) and the extraordinarily negative impact of collectivization on the area's fragmented and highly location-specific land use (Chapter 3) and on its dispersed rural settlements (Chapter 5). As for the first of these factors, Table 10.1 shows that within the European part of the former USSR, the most substantial differences in the habitability of the geographical space are between areas in the east and those in the west, with the Non-Chernozem Zone and the Volga region being on average the least habitable. Both the latter regions are characterized by a combination of low rural population density, low density of roads, low proportion of densely packed areas, and considerable inter-urban distances. These conditions typify what we mean when we say that agriculture and all other rural activities in the Zone unfold in a "rarified social space."

Given the sheer size of the area under consideration (shaded on Figure 10.1 are 1.5 million square kilometers), development indicators inside the Zone vary. The provinces of Riazan and Kaluga are the most plowed up, while the provinces of Kostroma and Leningrad are the most forested. The density of paved roadways is above the Russian average only because of the vast uninhabited spaces of Siberia. However, regarding this indicator, only the province of Moscow is on a par with the western republics of the former USSR. But contrasts in rural population density are even broader: in the province of Moscow it is 3-4 times higher than in adjoining provinces (Table 10.2).

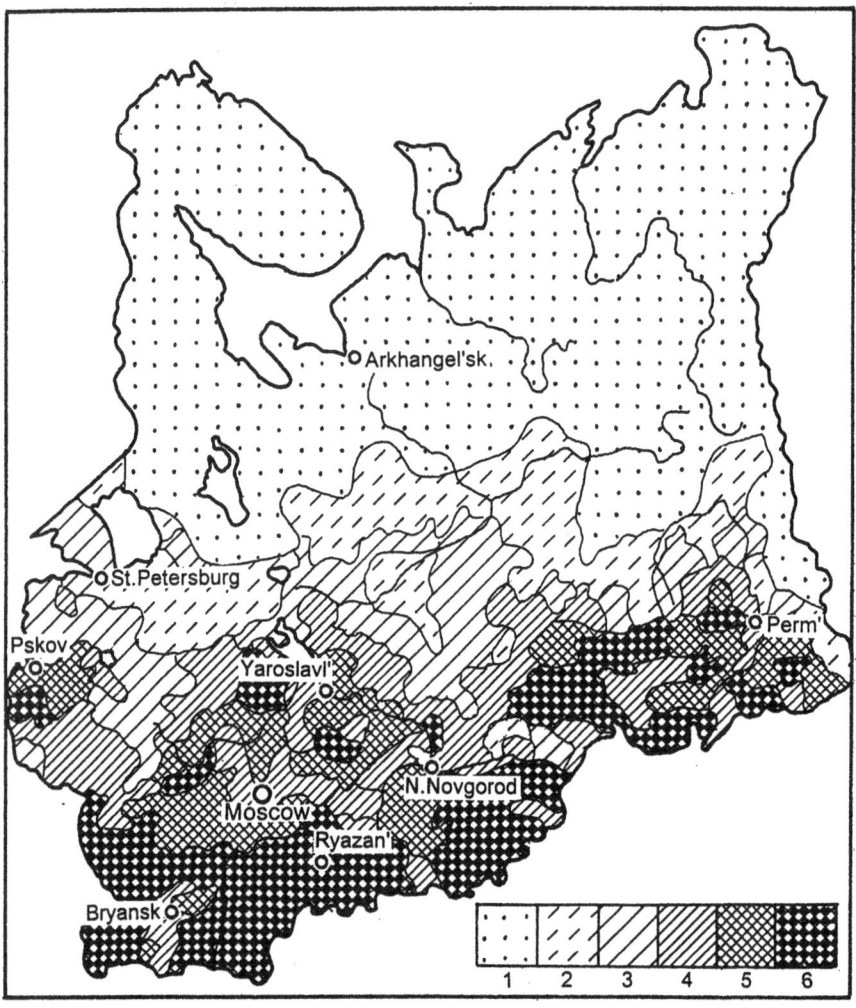

FIGURE 10.2 Agricultural Land as a Percentage of the Total Land in the Non-Chernozem Zone of Russia (1990). 1: Below 3%; 2: 3-10%; 3: 10-20%; 4: 20-35%; 5: 35-50%; 6: Over 50%.

Perhaps the most characteristic features of the Zone, in fact collateral to its rarified social space, are a heightened level of spatial concentration of population and the polarization of inhabited space involving breaches in the spatial continuity of its settlement. In 1959-1989, population was declining in 75% of the Zone's total land area, which is to say that even moderate urban growth in many districts did not outweigh their rural population's decline. As was shown in Chapter 5, all the demographically depressed areas

TABLE 10.1 Habitability of the Non-Chernozem Zone Versus Other European Regions of the Former USSR (early 1990s).

| | Population per sq km | | | Paved | % Share of Base | Average Inter-Urban Distance |
	Total	Rural	Railway[a]	Road[a]	Structure[b]	in km
Non-Chernozem Zone (as shown in Figure 10.1)						
	44	8	19	65	17	59
Central Chernozem						
Region	47	18	27	116	26	58
Volga	31	8	18	53	15	74
Northern						
Caucasus	39	22	14	115	18	65
Ukraine	86	28	37	260	32	38
Belarus	49	16	27	223	28	46
Baltics	45	14	31	321	35	31

[a]Length in km per 1000 sq km.
[b]See explanation in Table 1.4, p. 30.
Sources: Demographic Yearbook of the Russian Federation. Moscow, Goskomstat 1994; *Ekonomika Sodruzhestva Niezavisimykh Gosudarstv v 1994 godu.* Moscow, Goskomstat 1995; *Narodnoye Khoziaistvo SSSR v 1990 godu.* Moscow, Goskomstat 1991:620-622; *Territorialnaya Struktura Khoziaistva Staroosvoyennykh Rayonov.* Moscow, Nauka 1995:18.

are located away from large cities. Not only have the core-periphery population density gradients been becoming steeper in recent decades but so too have those of agricultural output per unit of land. Examples of such gradients are shown in Figures 10.3-10.5. For example, it can be determined based on Figure 10.4 that with each subsequent 10 kilometers distance from Moscow the output per 100 hectares dropped by 25,000 roubles in the mid-1980s. Taking into account that such a gradient was on a par with the average output per 100 hectares in many Russian provinces (for example in the province of Kirov it was 26,900 roubles in the mid-1980s), the gradient appears steep indeed. (The early 1990s data confirms that gradients reflected in both figures have even increased somewhat.) Given this, the economic depression between Moscow and Saint Petersburg that is shown in Figure 10.5 does not come as a surprise. It is not that the urban network in general between Russia's two largest cities is sparse, but that *attractive* urban centers able to invigorate their immediate surroundings are scant, and this scantiness has hindered agricultural development more than anything else.

TABLE 10.2 Land Use and Development Characteristics of Selected Provinces in Central Russia (circa 1990).

Province	% of Arable Land	% of Hayfields and Pastures	% of Forests	Rural Population Density	% of Urban Population	Paved Road[a]
Pskov	17	11	39	5	65	124
Vladimir	24	12	51	11	80	151
Kaluga	33	12	44	10	74	143
Kostroma	12	5	72	5	66	62
Moscow	27	10	41	29	80	249
Riazan	46	17	26	11	67	123
Yaroslavl	22	9	44	8	81	106
Nizhni Novgorod	19	6	47	11	78	106

[a]in km per 1000 sq km.

Sources: Demographic Yearbook of the Russian Federation. Moscow, Goskomstat 1994; *Zemelnyy Fond RSFSR na I Noyabria 1989 goda.* Moscow, Goskomstat 1990.

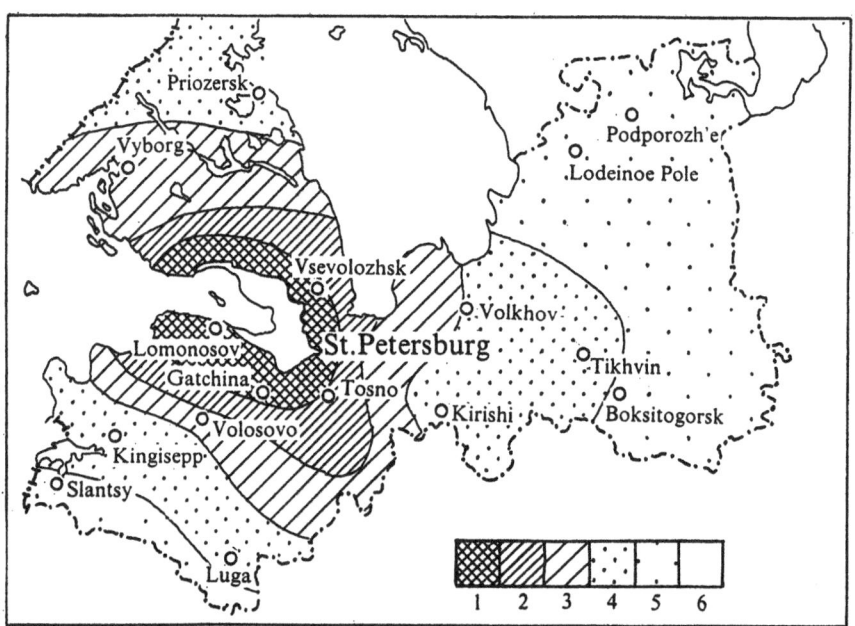

FIGURE 10.3 Gross Agricultural Output in the Province of Leningrad in Thousands of Roubles per 100 ha of Agricultural Land (1985). *Source:* Ioffe, G.V. *Selskoye Khoziaistvo Niechernozemya.* Moscow, Nauka 1990:74. 1: Over 600; 2: 300-600; 3: 200-300; 4: 100-200; 5: Less than 100; 6: City of Saint Petersburg.

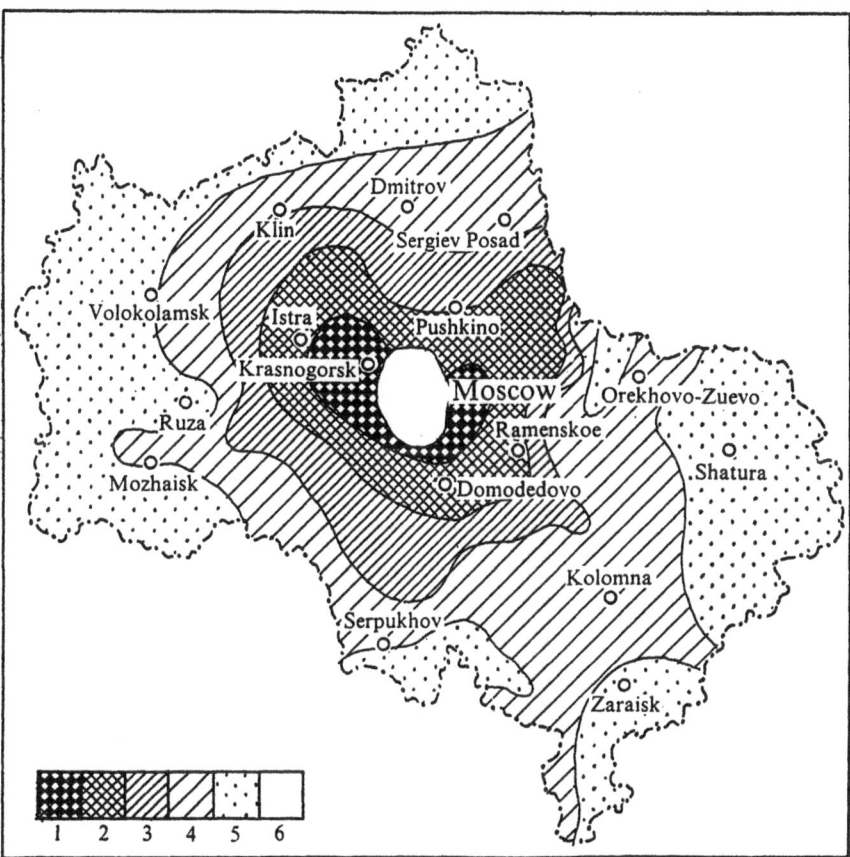

FIGURE 10.4 Gross Agricultural Output in the Province of Moscow in Thousands of Roubles per 100 ha of Agricultural Land (1986). *Source:* Ioffe, G.V. *Selskoye Khoziaistvo Niechernozemya: Territorialnye Problemy.* Moscow, Nauka 1990:71. 1: Over 600; 2: 300-600; 3: 200-300; 4: 100-200; 5: Less than 100; 6: City of Moscow.

The Industrial Center Early in This Century

The Industrial Center is at the heart of *Niechernozemye.* Many of its current features can be traced back to the past. It appears, for example, that the more reform-minded attitudes of its population (compared with those of the rest of Russia) are related to the population's higher mobility and taste for innovations, both of which were exhibited long ago.

The region has always been unevenly settled. During the initial Slavic colonization this unevenness was prompted by natural conditions. Settlers

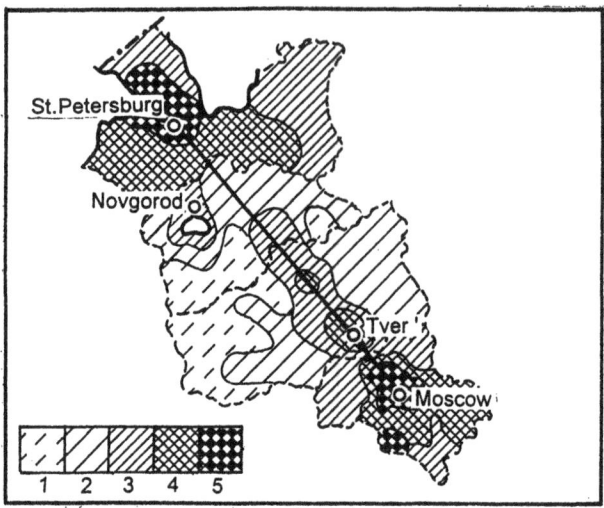

FIGURE 10.5 Agricultural Land Use Intensity Between Moscow and Saint Petersburg. 1-5: Ascending order of intensity (based on the sum of standardized indicators
of invested capital, number of people engaged in agriculture, and fertilizer application per 100 ha; the spacing of intensity gradations fits one standard deviation).

favored the elevated and dry area at the confluence of the Volga and Oka
Rivers, from where they subsequently moved into the depths of the country.
In 1724, the provinces of Tver and Yaroslavl had 5-10 people per square
viersta (1.12 square km), the province of Moscow and the northern part of
Kaluga had 25-30 persons per square *viersta,* and the provinces of Kostroma
and Nizhni Novgorod, less than 5.[3]

By the very end of the 19th century, population density in the province
of Moscow was already 83 persons per square *viersta*, in the province of
Yaroslavl, it was 34; and in that of Nizhni Novgorod, 35.[4] Agricultural land
in the Industrial Center was by that time somewhat more spacious than now.
About one-half of each peasant household's communal allotment was arable
land, while forest and hayfields each accounted for one-quarter. Forests
dominated private landholdings on which hayfields and plow-land occupied
15% and 13% respectively.[5] Peasants made up 88% of the total population,
though the percentage of actual rural population was lower, 83%. In the
province of Moscow, the mismatch between these two statistics was even
more striking: 45% of the province's population lived in urban areas while
only 22% qualified as non-peasants.[6] This difference shows that many people did not actually leave their peasant estate behind but maintained their
links with their village even while working for industrial factories. The highest level of attachment to land occurred in the southern and western

provinces of the Center. In the province of Kaluga, for instance, urbanites were 8% of the total population while non-peasants comprised 11% of it, which means that quite a few non-peasants lived in the countryside. In the northeast the situation was markedly different. Around Yaroslavl and Kostroma, urbanites outnumbered non-peasants because, as around Moscow and Vladimir, temporary labor migration was pervasive and peasants used to abandon their land more easily.

What influenced the countryside of the Industrial Center the most was the vigorous development of cities, manufacturing, and trade. In the mid-1800s Moscow accounted for one-half of Russia's wholesale and retail trade revenues. The development of large-scale manufacturing in the Center preceded that in other regions. So goods, capital, and people flocked there. By the beginning of the 20th century the region had 65% of all the spindles in the country and over 70% of all its looms.[7] About one-third of the output of Russia's machine-building and metal-working industries was being produced by the Industrial Center. In addition to the province of Moscow, the eastern and northeastern parts of the region stood out in terms of industrial development.

Progress in manufacturing, however, was accompanied by the decline of grain farming and by a growing emphasis on feeding cities. So potato and dairy farming increased at the expense of grain. Thus the division of labor set as early as the 1700s was gradually making the Industrial Center a grain-deficient region with the exception of its southernmost part. The most important commercial crop in the outlying areas of the Center was flax.

The proximity of the Center's countryside to large urban cores facilitated the accelerated diffusion of various innovations. Already by the beginning of the 20th century, the three-field cropping system was replaced by a more complicated crop rotation involving fodder herbs and root plants. Horticulture and vegetable growing expanded rapidly. The cradle of Russian vegetable farming was the province of Yaroslavl, especially the district of Rostov-the-Great, from which Peter the Great used to send peasants to Holland to master the trade. Chicory, mint, green peas, onions, and garlic did not only command markets in Moscow and Saint Petersburg but were also being exported.

Moscow and Saint Petersburg, like two giant pumps, sucked country people off the countryside. The decline of grain farming, the expansion of initially temporary labor migration and of handicrafts and cottage industry in small towns -- all these facilitated a spatially polarized development pattern decades before the commencement of actual rural depopulation.

Landlords were losing their grip on land much faster than in the Chernozem provinces. Privately owned land accounted for 40% of the Industrial Center's total land area in the beginning of this century. But only a half of the private land was owned by the gentry. The dominant form of peasant-landlord economic relationship was a quit rent. The average size of a noble-man's landholding was about 500 hectares. Merchants bought up more and more estates; on average they controlled one-fourth of all the private land in the Industrial Center in the 1890s. Their share was much higher in the most industrialized areas of the Center. Peasants owned 24% of the private land.[8] About half of all land was in peasant communal use being redeemed from landlords up until the cancellation of redemption payments in 1905. The average parcel of land immediately attached a household's residence was 7-10 hectares for a household of 6-7 members. Together with distant strips of plow-land in household use, the size of an allotment amounted to 40 hectares.[9] Given repartitioning (see Chapter 2), only the land around the house was perceived as privately owned -- a forerunner of personal auxiliary farming in the years to come.

Clearly there were considerable land-use-related differences between the Industrial Center and the Chernozem Center early in this century. In the former, landlords played a smaller role; quit-rent prevailed over corvee, making peasants more interested in working their own parcels of land; merchant land-ownership was more widespread; and, last but not least, household agriculture was very important. Thus, on the one hand, peasants were lured by cities and labor migration, and more ready to quit the land,[10] and, on the other hand, they attached more economic significance to land of their own.

As we saw in Chapter 6, even 60 years of Soviet-style collective farming had not whittled away those differences: it was in the Non-Chernozem Zone that the crisis of the 1990s triggered the greatest expansion of personal auxiliary farming.

With the higher mobility of peasantry and their less conservative attitudes, the productivity of agricultural land was no lower, and often was higher, than on the better endowed soils in the Central Chernozem region. However, the superiority of non-Chernozem farming was subsequently undermined by collectivization (Chapter 3).

"Intensification" and Spatial Pattern of Agriculture

In the late 1980s one of this book's authors undertook an in-depth research of the agricultural output in 28 sub-divisions of the Non-Chernozem Zone (the province of Murmansk was excluded from the sample) based on a factor analysis of 75 indicators.[11] The research tasks were to uncover the most significant factors of spatial variance and of the dynamics of agriculture in the Zone. By the late 1980s, agriculture already appeared to be a catastrophic drain on the budget; it was becoming clear that the monumental investments during the whole decade had not resulted in commensurate returns, and were unlikely to do so in the future. The economic situation was patently out of control, not so much administratively but rather in some more fundamental ways including both the inherent flaws of the economic system and the likelihood that it would focus on inappropriate economic indicators. Needless to say, the second problem was easier to discuss openly inside the country (even under *Perestroika*, agricultural authorities were the least prone to any reform-minded discourse). The ensuing restrictions impacted the whole design of the research undertaken under the auspices of the Soviet Academy of Sciences. Yet it allowed for quite a few conclusions and observations that do not appear outdated even today.

One of the most significant inter-provincial differences in non-Chernozem agriculture is the degree of emphasis on animal husbandry. The province of Leningrad and the Udmurt republic lead the rest of the Zone in this category; the province of Moscow and the northernmost agrarian enclaves (in Karelia and Komi) are also in the leading group. Aside from the southernmost provinces of the zone (Orel, Tula, Riazan, and Mordva) that historically emphasized crop farming more than livestock production, some of the most demographically depressed provinces (Pskov, Smolensk, Tver, Kostroma, and Kirov) are also at the bottom of the list. Whereas in the former group of provinces grain crops prevail, in the latter fodder crops do.

However, perhaps the most significant element of agricultural variation inside the Non-Chernozem Zone is output per unit of land. As Figure 10.6 shows, the provinces of Leningrad, Moscow, and Karelia (where farming is conducted mostly in the south) are absolute leaders. The monetary value of output per 100 hectares of agricultural land in the three leading provinces is 4.2 times higher than in the final three (Kirov, Smolensk, and Perm). A few questions suggest themselves in relation to these figures: 1) how does the variance of the late 1980s compare with that 20 years earlier, that is, prior to the commencement of the investment program? 2) to what extent can the different variances be attributed to the sheer size of the inputs (fixed assets and labor) and to what extent to the variance in returns (labor productivity and capital productivity expressed as a ratio of output to fixed assets)? 3) what is the most significant factor of agricultural growth? and 4) what is the effect of the sheer size of agricultural land?

The analysis showed that while the output-based ranking of non-Chernozem provinces had changed only slightly over two decades, the range of variance increased 4.5 times. The provinces of Yaroslavl, Orel, and Briansk now rank lower than they did twenty years ago, while the republics of Komi and Mari rank higher. The growth of contrasts in output refers to the output's spatial concentration. This is especially noticeable in the Industrial Center (11 provinces) and in the Northwest (three provinces). While in the former the proportion of output produced in the province of Moscow is on the rise, in the latter the percentage share of Leningrad is rising. For exam-

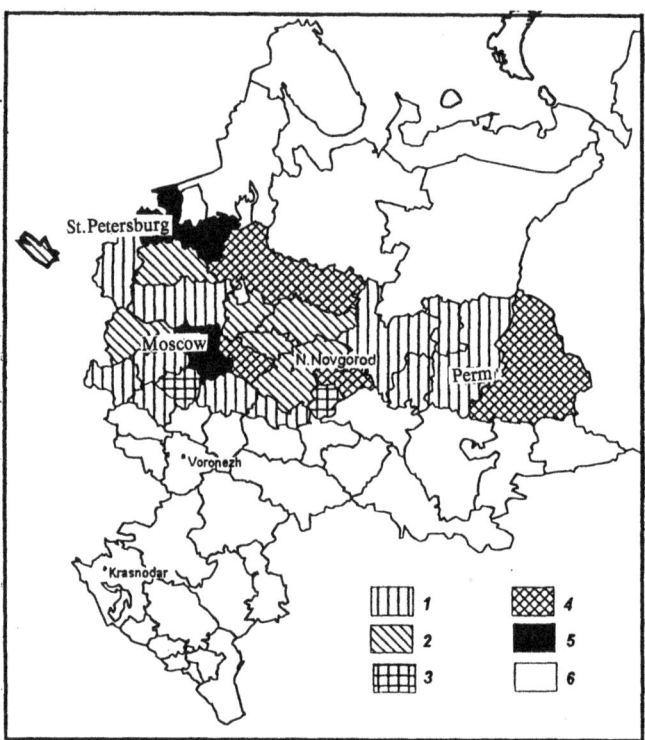

FIGURE 10.6 Agricultural Output per Unit of Land. 1-5: Ascending order of productivity (leaders exceeding the least productive by a factor of four); 6: Areas beyond consideration.

ple, whereas in the early sixties only 13.4% of the Industrial Center's meat was produced around Moscow, in the mid-eighties 24% was; for milk the proportions were 16% and 25% respectively. The dominance of the Leningrad province in the Northwest is even more conspicuous: in the mid-eighties it produced 56% of the milk of the region (compared to 32% in the early 1960s) and 62% of the meat (vs. 35% in the '70s).

The answer to the *second* question parallels our discussion of agricultural productivity in the European USSR as a whole (Chapter 4). In the Non-Chernozem Zone the sheer *sizes* of invested capital (fixed assets) and labor are much more important predictors of agricultural output than labor and capital *productivities*. That is, the latter two factors do not vary nearly as much as the former two do. And yet spatial differences in agriculture's efficiency do exist even under the same system of management. Using the formula shown underneath Table 4.8 (in Chapter 4) one can estimate the contribution of the *size* of inputs per unit of land versus the *efficiency* of formula shown underneath Table 4.8 (in Chapter 4) one can estimate the contribution of the *size* of inputs per unit of land versus the *efficiency* of their utilization in the inter-provincial output differential. For example, in the province of Yaroslavl, in the mid-eighties, agricultural output per 100 hectares was 52,000 roubles lower than in the province of Moscow; it appears that 78% of this output differential is due to lower inputs in the province of Yaroslavl (which translates into lower land-use intensity), while the remaining 22% is due to lower returns per unit of inputs. An analysis of the entire 28-by-28 matrix of such differences leads to the conclusion that in the Non-Chernozem Zone higher efficiency is correlated with higher inputs of capital and labor, that is to say, a certain economy of scale shows up when there is a heightened concentration of farming inputs. On the other hand, the on-going investment has been accompanied by declining efficiency. This conclusion can be drawn if one changes the format of the statistical analysis based on the formula in Table 4.8 (in Chapter 4) from a cross-section format to a time series, that is, if one switches from an analysis of inter-provincial differentials at a certain point in time to an analysis of one and the same province's performance at different points in time. It is thus possible to relate the investment-based effect of increased labor productivity to the "price" at which this effect has been achieved. Assessments of that kind were in high demand in the former USSR in the late 1970s and in the 1980s because of the "intensification" campaign. The meaning of that term, as explained in thousands of popular brochures, was to shift from a sheer amassing of resources to a more efficient use of them. Needless to say, the Non-Chernozem Zone was considered one of the most critical probing grounds of the ability to "intensify" the economy in this sense, and this ability was being called in question by the analysts. Specifically the time-series-based calculations showed that while between 1966-70 and 1971-75 (two consecutive five-year investment programs), 91.8% of the surplus output was due to increased capital productivity, during the following period, between 1971-75 and 1976-80, this percentage share dropped to 50.8%, and during the next period, between 1976-80 and 1981-85, it further dropped to as low as 37.4%.[12]

The answer to the *third* question posed above (concerning the most critical factor of economic growth in agriculture) was even more discouraging for the ideology of central planning. It appeared that growth in output could be sustained only in those provinces in which the agricultural labor force was declining the slowest (or was stable or even growing, as in the provinces of Moscow, Leningrad, and Kaliningrad). The correlation coefficient of the growth rate of output and of the number of persons engaged in farming was as high as 0.87 for the period between 1961-65 and 1981-85. By contrast, no meaningful correlation with the growth rate of fixed assets of agriculture was recorded. In all non-Chernozem provinces the ratio of output to fixed assets (capital productivity) was declining due to a growing capital/labor ratio; however, the decline was smallest in the provinces with relatively stable labor forces in agriculture, even though in such provinces the capital/labor ratio grew more rapidly than on average in the Zone because investors were increasingly cautious and favored provinces with a relatively stable farming population. Moreover, it was calculated that the net effect of each 10% reduction in employment in agriculture in the period from 1968-1970 to 1981-1985 was a 10.6% average loss of the total sum of gross output, whereas the net effect of a 10% increase in invested capital produced only a 2% increase in gross output.[13]

Thus what appeared to be the most important factor controlling agricultural growth was itself the least controllable by the Soviet system of management. While the system could and did regulate the spatial distribution of investment, it could not exert the same control over migration and the aging of the rural population. Hence the discouraging nature of the research results for the whole ideology of central planning. Although the efforts to diminish this discouragement included an increased proportion of investment directed to the social infrastructure and amenities in the countryside, these activities were like a drop in the bucket: the spatial pattern of population change described in Chapter 5 remained intact. At the province level, it was clear that only areas with the lowest percentage shares of farming in the employment composition were able to retain much of their agricultural labor pool -- suggesting the seeming paradox that farmers in Russia feel better when surrounded by people of other occupations than by comrades-in-arms.

As was demonstrated in Chapter 5 (see Table 5.4), two factors provided a meaningful statistical explanation of the agricultural population change in 1960-1986: *inertia,* meaning that the more significant the rural population decline had been in 1926-1959, the more agricultural employment declined during the following time period of 1960-1986; the second factor was the proportion of *large cities* in the total population. Because of this latter factor the historical pattern of rural population density was gradually being changed: whereas early in this century the rural population's pressure on the land was inversely proportional to the size of land parcels open for agriculture, now the availability of large cities is an equal factor in population pressure.

Cities also provide the answer to the *fourth* question posed above (concerning the effect of land size). As cities exerted their pull on rural population, more and more agricultural land was abandoned in outlying areas. Rural exodus has been the greatest catalyst in this regard. Various techniques used to assess the effect of land size on production (production functions, correlation analysis, and location quotients) all result in the same conclusion: agricultural land looks more and more like an over-abundant and

TABLE 10.3 The Dynamics of Agricultural Land, the Dynamics of Rural Population in Selected Provinces of Central Russia, and Rural Amenities.

Province	Agricultural Land in 1989 as a % of 1960	Arable Land in 1989 as a % of 1960	Rural Population in 1959 as a % of 1939	Rural Population in 1989 as a % of 1959	% of Rural Dwellings with Plumbing in 1993
Pskov	78	89	51	45	27
Vladimir	96	98	64	56	62
Kaluga	99	93	62	56	44
Kostroma	82	82	66	46	24
Moscow	90	98	83	66	90
Riazan	97	95	59	45	58
Yaroslavl	75	95	58	46	46
Nizhni Novgorod	95	106	72	49	59
Russia (Total)	95	100	78	70	37

Sources: *Zemelnyi Fond RSFSR 1961 and 1990; Demographic Yearbook of the Russian Federation.* Moscow, Goskomstat 1994.

therefore superfluous resource. Stated differently: other things being equal, the larger a piece of agricultural land in a province is, the less it produces.

While Table 10.3 shows correctly the pre-eminence of the urbanized provinces in terms of rural infrastructure, it definitely underestimates the degree of contraction of agricultural land that had occurred since 1960. We came to understand the reason for this discrepancy only during numerous field and business trips to the non-Chernozem provinces. For a long time the size of land per collective and state farm was an important variable in investment decisions made in provincial capitals and in Moscow. Consequently farm chairmen and directors did whatever they could to prevent parcels of land from being officially written off even though they might have long since stopped being farmed. Aerial photographs occasionally revealed a stunning mismatch between the actual and reported land in farming operations.

The crisis of the 1990s impacted non-Chernozem agriculture (Table 10.4). Yet, as was shown in Chapter 6, the decline of output in the Zone was not as drastic as elsewhere and it was to some extent offset by the expanded productivity of subsidiary plots.

The following chapter provides a more geographically detailed account of the current situation in agriculture in a fairly typical non-Chernozem province.

TABLE 10.4 Productivity and Fixed Assets of Agriculture in Non-Chernozem Provinces by 1990 and at the Height of Crisis in 1993-1994.

Region/Province	1986-90 Grain Yields in Centners per ha	1993 Grain Yields in centners per ha	1990 Milk Yield per Cow in kg	1994 Milk Yield per Cow in kg	Amount of Cattle per 100 ha	1980 Agricultural Fixed Assets in Thousand Roubles per 100 ha	1989 Agricultural Fixed Assets in Thousand Roubles per 100 ha	Mineral Fertilizers in kg of Active Ingredient per ha in 1990	Mineral Fertilizers in kg of Active Ingredient per ha in 1994
Northwest	9.2	9.1	3098	2386	47	115	233	162	63
Leningrad	16.3	14.7	4089	3086	76	293	522	192	100
Pskov	8.7	9.1	2259	1772	36	63	134	176	46
Industrial Center	15.4	16.0	2812	2237	42	83	154	161	77
Vladimir	16.3	14.7	2880	2313	46	98	190	200	85
Kaluga	13.7	13.4	2527	2084	42	74	155	162	77
Kostroma	9.3	9.7	2294	2065	33	70	140	120	45
Moscow	25.8	20.2	3922	2867	71	225	398	266	122
Riazan	16.3	17.4	2881	2186	38	67	116	193	76
Yaroslavl	10.2	8.2	2253	2023	45	127	192	143	35
Volgo-Viatka	13.4	15.8	2797	2584	38	68	130	142	74
Nizhni Novgorod	14.4	16.3	2860	2532	43	106	139	174	77
Chuvash Republic	18.8	22.1	2860	2994	51	142	179	143	79
Russia Total	15.3	16.0	2781	2249	27	52	90	88	46

Source: *Narkhoz RSFSR 1991.* Moscow, Goskomstat 1991; *Ekonomicheskoye Polozheniye Regionov Rossiiskoi Federatsii.* Moscow 1995.

Notes

1. Rodoman Boris. "Peizazh Rossii." Moscow: ROU 1994, p. 15.

2. Kliuchevsky, Vasily. *Russkaya Istoriya. Polnyi Kurs Lektsii v Triokh Knigakh.* Moscow: Mysl 1993:57.

3. Semionov, V.P. (ed.). *Rossiya. Polnoye Geograficheskoye Opisaniye Nashego Otiechestva.* Volume 1. Saint Petersburg: Devrien 1899, p. 87.

4. *Rossiya. Entsyklopedicheski Slovar.* Saint Petersburg: Brokhaus i Efron 1898, pp. 106-115.

5. Semionov, V.P. (ed.). *Rossiya,* p. 93.

6. *Rossiya. Entsyklopedicheski,* p. 115.

7. *Ekonomicheskaya Geografiya SSSR.* Moscow: Prosveshcheniye 1966, p. 268.

8. Semionov, V.P. *Rossiya. Polnoye Geograficheskoye,* p. 123.

9. Ibid., p. 144.

10. Pallot, J. "Peasant Manufacturing in Nineteenth-Century Moscow Province." In Pallot, J. and Shaw, D. J. B. *Landscape and Settlement in Romanov Russia 1613-1917.* Oxford: Clarendon Press 1990, pp. 216-240.

11. Ioffe, G.V. *Selskoye Khoziaistvo Niechernozemya: Territorialnye Problemy.* Moscow: Nauka 1990, pp.42-67.

12. Ibid., p. 50.

13. Ibid., p. 57.

11

The Province of Yaroslavl

The province of Yaroslavl (Fig. 11.1) is one of the most urbanized in Central Russia: 81% of its 1994 population are urban; however, 43% of all provincial residents live in just one city, Yaroslavl (631,000). Rybinsk is the second-ranked urban center (249,000); together both cities have 60% of the province's entire population. However, in view of the city of Yaroslavl's central location, the province is not as markedly bi-central, in the sense of its rural hinterland's diverging social gravitation, as, say, the province of Pskov.

Our experience with Yaroslavl has been long-term. Our first field observations took place in 1974 and the most recent in July 1995 when we interviewed local officials in the exurban vicinity of Yaroslavl and local officials in one of the most remote areas, the Nekouz district, whose center is 150 km from the city of Yaroslavl. Covering this distance in a Volga car was an experience valuable in itself. Upon leaving the provincial center we observed a landscape fairly typical for a large Russian city's countryside: vast fields, girded by thin woodland belts, neat-looking dwellings belonging both to urbanites and to country folk. As we drove, the woodlands were visibly expanding and the fields shrinking, turning into islands in a forest vastness. Close to Nekouz, abandoned fields turned up already overgrown by bushes and trees. The patches of land still under cultivation drew closer to villages, while the remaining open spaces looked dishevelled and neglected, as if their proprietor was long past his prime and could not keep this much land in check any more.

The Case of the May Day Cooperative

Under Soviets many socialized farms were named after May 1, the so-called International Day of Working Class Solidarity. For this reason, and because May in general is regarded as the springtime of one's life, youth, and prime, anything that bears its name but looks shabby and run down smacks of acute irony.

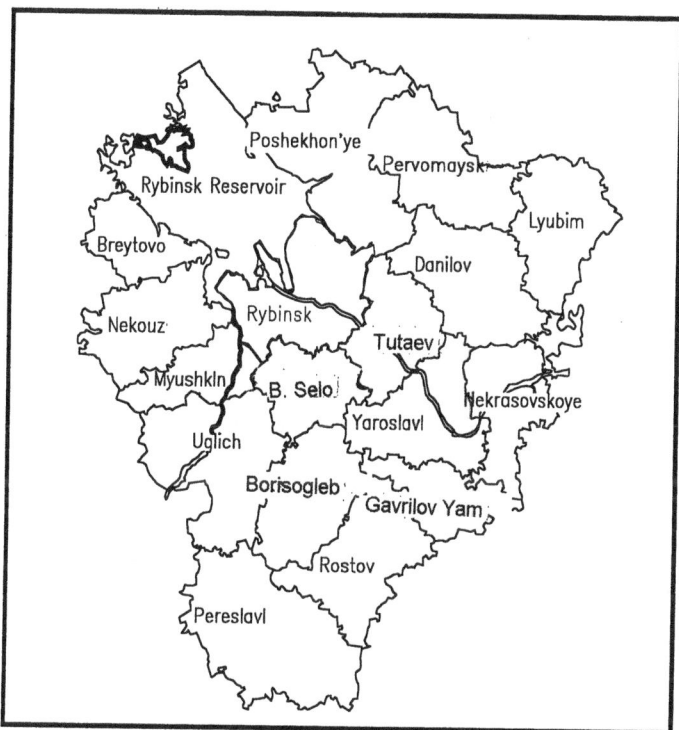

FIGURE 11.1 Districts of the Province of Yaroslavl.

The "First of May" cooperative is located ten kilometers away from the formerly small (3,800 residents) town, now turned village, Nekouz. The cooperative has 3000 hectares of agricultural land with fifteen villages attached to it. In the 1970s it was still a large state farm composed of two subdivisions. Gradually the whole business fell into decay, and by the 1990s it was already one of the worst production units in one of the province's least agriculturally efficient peripheral districts. Suffice it to say that milk yields in the cooperative are a slight 1000 kg per cow and grain yields are 5-9 centners per hectare. As of now 70 workers are employed by the "First of May," while the number of shareholders is over 200, which means that the majority of them are retirees. Major specialists in charge of the cooperative (the agronomist, the animal technician, and the bookkeeper) are fresh, seem to have good educational backgrounds, and are not afraid of hard work. But the kind of problems they face would discourage even a zealot.

By the time of our arrival (14 July 1995) the 200 cows on the farm (down from over 500 in the 1980s) had not been milked for three days. Three days earlier employees had received their long-delayed pay. Since that time the dairymaids had been dead drunk and had not shown up for work. There was nobody around to replace them, a situation highly typical of a depopulated outlying area. Actually, some retirees would have volunteered, but only for instant cash. However, cash was unavailable: even staffers' scheduled pay was delayed and thus was being devalued by on-going inflation. The stock-piled animal feed was being stolen by shareholders for the benefit of their own cattle. The cooperative attempted to assign part of its calves to households (together with a part of the necessary feed), to get them back at maturity when the cooperative would pay for labor -- to no avail: such is the level of apathy that nobody is willing to help out, even in summer. As a result calves are being slaughtered, at a huge disadvantage for the farm.

Not that people are so lazy. Machine-operators can work fervently on occasion. But almost everybody has fits of hard drinking, and one had better stay away from them during these times. Those of a different mold are long gone. Only alcoholics and retirees remain, and a few others attached to the area due to uncontrollable circumstances. Still, up until 1995 the cooperative used to make ends meet, but only because of subsidies. In the 1990s, as a result of the crisis in cities people began to take to villages, but the local resettlement "boom" soon subsided without tangibly benefitting the farm: the few newcomers either do not work in agriculture or go to seed through drink after having joined the local environment.

So they continue to languish, trashing their cooperative yet unable to do without it. Who else indeed would sell meat at a discount? Who else would lend a hand in one's own vegetable garden? Who would plow it up? Who would help with animal feed?

And should help not be *given*, people will *take* what they need, since whatever-is-the-*kolkhoz's*-is-mine psychology has proved to be tenacious. There are indeed many things available nowhere else except from the farm cooperative one belongs to. Nobody in the "First of May" has volunteered to become an independent farmer. People are already basically working for themselves anyway. Every third cow in the whole district of Nekouz is private. And yet everybody keeps an eye on the cooperative's profits. When it actually succeeds at something, discussions on how to distribute the gain never end. Some even propose to distribute it according to the number of mouths in a household. Ideas of that kind nip in the bud any residual initiative to go private or to organize in small collectives. Because ten people, say, would work while the remaining 60 would see their profits and swell with envy. And that would not be the worst case scenario: the 60 may set the property of those ten on fire.

In such an environment a growing number of people opt for unemployment benefits. They resign from their cooperative and live off the dole and PAF (private auxiliary farming). In the district of Nekouz, 11% of the people of working age are registered unemployed, of which over half resigned from their jobs of their own accord. Moreover, employment officers have begun to deliver benefits themselves so one need not travel to receive payments. If the welfare benefit is supplemented by what a cow or two can give, one can easily make ends meet.

Not everything, though, is so hopeless in outlying areas. In the nearby village Danilovo is the cooperative *Lnovod* (Flax-grower). It is in much better shape. There is even a bus between it and the district center twice a day. Aside from flax itself, the cooperative runs a dairy farm producing sufficient animal feed of its own; recently mineral fertilizers were purchased with borrowed money. People say that this relative well-being is due to a dynamic chairman. Indeed, almost all the indicators of output are twice as high as in the "First of May." It looks as if *Lnovod* will make it through reforms unless it is stifled by non-payments for its produce, a scourge afflicting Russia's economic life. However, in many respects life on this cooperative is not different from that on its neighbor. People drink a lot and mostly survive on their PAF, not on earnings from the enterprise.

The situation in exurban areas around the city of Yaroslavl could not differ more greatly from the above. A typical example would be the partnership *Kurba,* a former state farm on 4000 hectares. While it too experienced a downward trend, the enterprise's entire setting is different and so are the repercussions of the setting. For example, a cow herd was reduced because the arable land was redistributed in favor of PAF so that each family has one hectare at its disposal; even so the reduction was partly made up by the growing number of cows in PAF. What is more important, the number of employees was reduced, from 530 people in 1990 to about 400 in 1995, mostly because drunkards, pilferers, and truants were fired. And yet there are many fewer rural unemployed here than in the reportedly labor-deficient outlying areas.

Our brief and unsystematic survey of the situation in the field suggests the following conclusions.

- Despite attempts to emasculate or hinder reform in the countryside, the business environment that former collective and state farms face has changed. Before, the principal driving force of productive work was district and provincial Party bosses: *oblast* operatives pressured *rayon* officials, who in their turn put pressure upon farm chairmen and directors, who issued commands to the rank-and-file, who could either comply fully or partially, or exercise their right to flee the countryside, which many did. Now this purely executive system is evolving into a more complex economic system in which such factors as money earnings, sales, and production costs are gaining weight. However, it is still far from a genuine market-driven system, if only because payments for produce, distribution of discounted loans, and many other activities depend on administrators. These are no longer Party bosses but provincial executives; since the district executives no longer wield real power, the chain of command has gotten shorter on both ends.

- The degree to which this economic-administrative system adopts market elements depends upon the political preferences of the variously ranked bosses. When a critical mass of executives dedicated to central planning (or to reform) forms, the result is regional variation of the kind we wrote about in Chapter 6, wherein such phenomena as the Red Belt of European Russia's mid-south were described. As for the variation within provinces, it is usually linked not so much to the political preferences of the bosses as to location relative to a provincial center. The principal difference between peripheral (outlying) and exurban (centrally located) socialized farms is that the former have in fact long since ceased to be economic entities in the true sense of the phrase. A notice of dismissal would not work here because people are in large measure already confined to their subsidiary plots, and there is nobody available to take over their obligations on the farm. Economic sanctions would also not work because payments arrive after a long delay and, given their size, do not matter much anyway. Such farms actually work for themselves, supporting their own members, rather than being assets to the Russian economy, even though the farm members may not quite understand or appreciate that fact. On the other hand, the exurban farms like *Kurba* are more similar to regular production units using hired labor who can be affected by regular economic incentives and administrative sanctions. In fact, the 1994 grain yield in *Kurba,* which used no mineral fertilizers that year, was 26 centners per hectare, which is five times that of the "First of May," while the milk yield per cow was 2600 kg, three times the "First of May" level. The differences in natural conditions are negligible.

- Finally, the above observations call into question the traditional Russian view that urbanization has caused the shortage of agricultural labor. The number of unemployed in the periphery and the number of people employed outside agriculture in a city's countryside testify to the opposite.

Spatial Pattern of Rural Population Change

In 1959-1989, the rural population of the province of Yaroslavl almost halved (582 versus 270 thousand people), the heaviest demographic losses occurring in the northern outlying districts and in the district of Niekrasovo situated between the cities of Yaroslavl and Kostroma (Fig. 11.2). The district of Pereslavl adjoining the province of Moscow also lost over half of

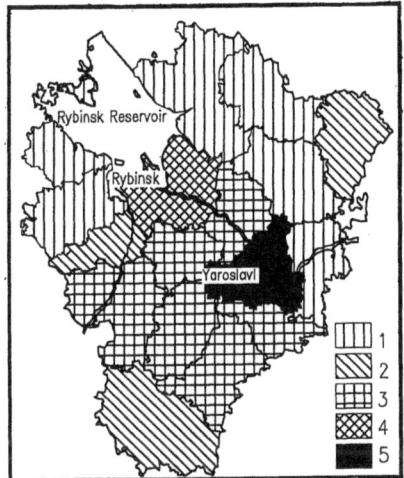 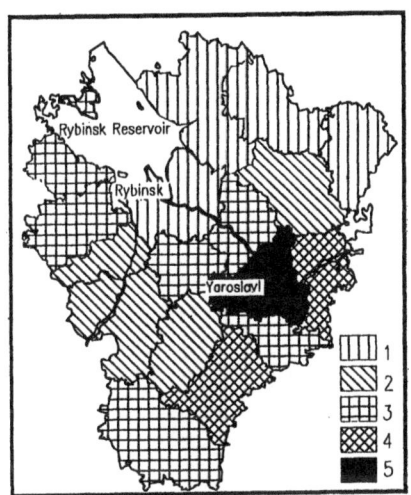

FIGURE 11.2 Rural Population in the Province of Yaroslavl in 1989 as a Percentage of 1959. 1: 80; 2: 69; 3: 46-51; 4: 41-45; 5: 35-40. *Source:* see endnote 1.[1]

FIGURE 11.3 Rural Population Density per Square Kilometer in the Province of Yaroslavl (1995). 1: 4-5; 2: 6-7; 3: 8-9; 4: 10-11; 5: 12-27.

its rural population. On the other hand, the two districts headed by the two largest cities of the province, Yaroslavl and Rybinsk, have become only 20% and 30% less populous. Whereas the population density in the district of Yaroslavl is 27 persons per sq km, it is merely 8-11 persons per sq km in neighboring districts, while in the northeast it is only 4-5 persons (Fig. 11.3). One-fifth of the entire rural population of the province now lives on 5% of the land around the city of Yaroslavl.

Rural population dynamics have been influenced in part by rural amenities. For example, Figure 11.4 shows the proportion of the rural population served by piped water. Only in the Yaroslavl and Rybinsk districts do over half of the country folk have this rudimentary service at their disposal; in the northern districts only 10-15% of the rural population enjoy piped water.

In the 1990s, the tide of rural population change reversed. The rural population, in thousands, over the last half century looks like this: 1939: 995; 1959: 582; 1970: 419; 1979: 316; 1989: 270; 1993: 274; and 1994: 282.[2] The rate of the rural population's natural increase in the province of Yaroslavl has been negative since the late 1960s. In 1992, the gap between deaths and births began to grow explosively, and by 1995 the rate of natural increase in the districts ranged between minus 10 and minus 25 persons per 1000 population. So on the surface, depopulation ceased, but as defined by vital rates it only deepened. Migration, therefore, was a factor in reversing the tide of

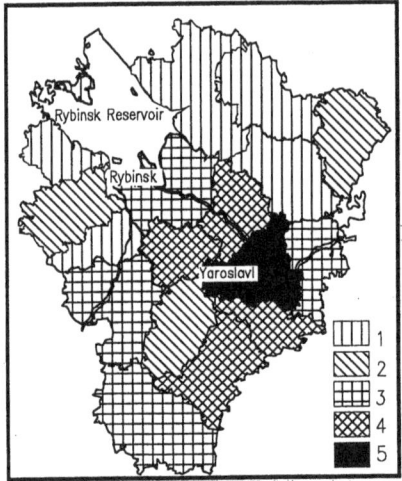 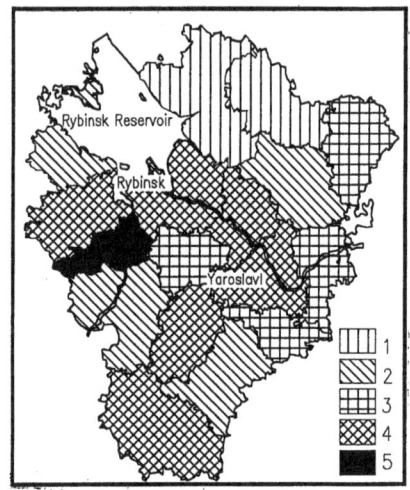

FIGURE 11.4   Percentage of Rural
Dwellings with Pumped Water in the
Province of Yaroslavl (1994). 1: 6-15; 2:
15-30; 3: 30-35; 4: 36-50; 5: 50-62.

FIGURE 11.5 Rural Net Migration per
10,000 Rural Population in the Province
of Yaroslavl (1994). 1: (-3)-0; 2: 0-2; 3:
3-5; 4: 6-8; 5: 10 .

population dynamics after 1989: whereas in 1979-1989 the net migration
amounted to minus 71 persons per 10,000 rural population, in 1989-1993 it
was plus 17. At the same time the net migration inflow to the provincial
urban areas shrank from 43 to 32 per 10,000 population.[3] However, as the
above figures may indicate, migration could not change the rural population
dynamics alone, so the settlement re-classification contributed to that: in
other words, change of status. Whereas up until the 1990s settlements
changed from rural to urban, in the 1990s, the opposite became true:
because of population decline in the previous years many small towns
(*poselki gorodskogo typa*) have been recently renamed villages.

Migration to the countryside was at its highest in 1991-1992. At that time
it more than canceled out the excess of deaths over births in many districts.
In relative terms, the semi-periphery, that is, districts not quite outlying but
not exurban either (Bolshoye Selo, Rostov, and Uglich) and the eastern
periphery led the way, with a net inflow of 10-14 persons per 10,000 residents. Most of the newcomers were first generation urbanites escaping the
empty shelves of urban food-stores and the threat of hunger. However,
when by 1993 it became clear that it was possible to survive in cities and
towns, the inflow to the countryside lessened to 5-7 persons per 10,000 residents. Simultaneously the composition of newcomers changed: the proportion of migrants from within the province grew smaller while the proportion
of resettlers and refugees from outside the province, including those from

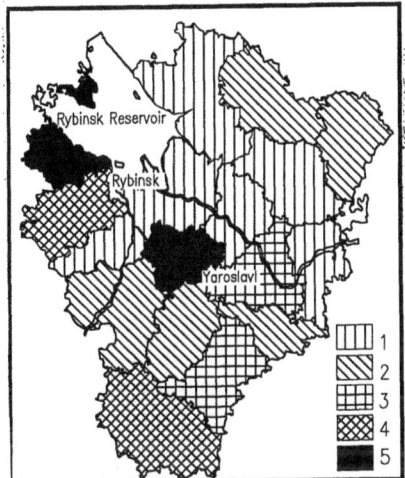 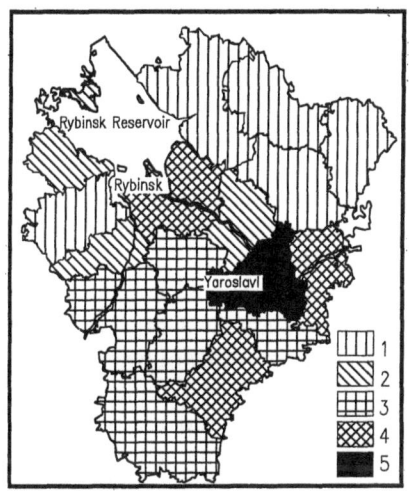

FIGURE 11.6 Rural Population in the Province of Yaroslavl in 1995 as a Percentage of 1989. 1: 90-95; 2: 96-100; 3: 100-110; 4: 110-130; 5: 130-180.

FIGURE 11.7 Grain Yields in the Province of Yaroslavl in Centners per Hectare. 1: 8-9; 2: 10-11; 3: 11-13; 4: 13-15; 5: 16--20.

other post-Soviet republics (44% of all "outsiders" in 1993), increased. Ethnic Russians from Central Asia account for 47% of the migrants from the republics. Rural districts receive between one-half and three-quarters of all "outsiders" coming to the province. Most of them are coming to the two southern districts, Rostov and Pereslavl, and to other districts stretching along the Volga River (Fig. 11.5).

Settlement re-classification affected districts of the province to a variable degree. For example, in the Nekouz and Pereslavl districts rural population increased 30% due to re-classification only, while the district of Bolshoye Selo almost doubled its rural population due to the same reason. The combined influence of all the growth factors in Figure 11.6 shows that in the 1990s, rural population grew dramatically only in the districts which experienced settlement re-classification. So in real terms it grew only in the districts of Yaroslavl and Rostov. In all other districts it continued to decline, though not as fast as before.

The Contraction of the Activity Space

The public sector's agriculture is unprofitable everywhere in the province except the districts of Yaroslavl, Rybinsk, and Tutayev. Animal husbandry is in especially dire straits. In 1994 the ratio of revenues minus production costs to production costs (so-called *rentabelnost*) in animal husbandry ranged from -0.07 to -0.40 across the 17 districts of the province.

The most striking aspect of the geography of crop harvesting is the supremacy of the Yaroslavl district: it tops the remaining 16 districts in the yields of all grains (Fig. 11.7), potatoes, and root plants. For example, in 1986-90, yields of winter wheat in the district of Yaroslavl exceeded those in peripheral districts by a factor of 2.5. Even in the areas adjoining the capital district of the province, grain yields were in some cases substantially lower than in it.

However, by and large the geography of agricultural output is characterized by relatively gentle core-periphery gradients, meaning that the output declines smoothly outward from the city of Yaroslavl and, to a less extent, from the city of Rybinsk. Figure 11.8 captures these gradients based on data from the mid-1980s. Today the situation is roughly the same, except that the gradients have become steeper. By 1995, for example, the gap in the grain yields between the district of Yaroslavl and those of the northeast periphery of the province had reached the 4:1 level, whereas in the mid-1980s it was only 2.5:1. Such a drop-off in output per unit of land is equivalent to the spatial concentration of farming output or, in other words, to the contraction of the activity space. A study conducted by one of the authors in the late 1980s revealed that the process in question had already by then reached its mature stage. For example, whereas in 1968-1970, the districts of Yaroslavl and Rybinsk accounted for 16.7% of all of socialized agriculture's output in the province, by 1981-1983 these districts' share had grown to 26%.[4] For comparison, in 1994-1995 the same share was at the level of 40%. As for the percentage of agricultural land in the same areas, it was only 11% in the mid-1980s and 13% today.

According to the provincial statistics of the late 1950s, the variation in the milk output per district did not hinge so much on the milk yield per cow as on the number of cows. At that time the ratio of the yields of the lowest and the highest districts was 1658 kg (in Breitovo) to 2361 (in Tutaev), whereas the ratio of the lowest and highest number of cows per 100 hectares of agricultural rate was 4.3 (Prechistoye) to 9.1 (Yaroslavl). In other words, the yields differential was, roughly speaking, 2:3 while the livestock density differential was 1:2. By the mid-1980s the situation had been reversed: while the variation in the livestock density had been substantially levelled (it ranged between 13.8 and 19.8 cows per 100 hectares), the gap in the yields per cow had widened to 1:2.[5] As far as today's milk yields are concerned (Fig. 11.9), we are witnessing the same gap as in the mid-1980s. However, the gap between exurban and peripheral districts in terms of the sheer number of cattle has grown once again; the core-periphery output gradient has therefore steepened because it is now being sustained by both components of the output (yields *and* number of cattle) rather than by just one.

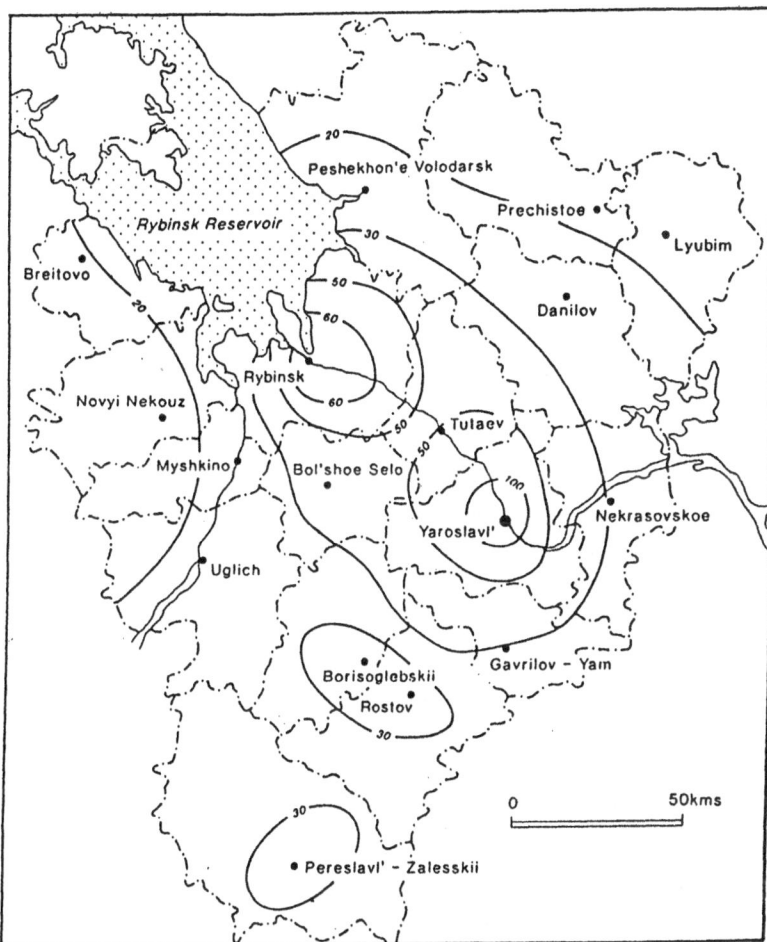

FIGURE 11.8 Gross Agricultural Output in the Province of Yaroslavl in Thousands of Roubles per 100 ha of Agricultural Land (1986). *Source:* Ioffe, G.V. *Selskoye Khoziaistro Nechernozemya.* Moscow, Nauka, 1990:15.

Bio-climatic conditions of the province favor the southern districts, and thus offer little if any explanation for the above trend. In fact, the districts of Yaroslav and Nekouz have very similar natural conditions.

Other production factors differ from place to place more dramatically. For example, the doses of mineral fertilizers per hectare varies (among districts) by up to a factor of three and so does the monetary value of fixed assets ("capital") per unit of land. Needless to say, the district of Yaroslavl tops the list. However, because the peripheral districts are most affected by out-migration, there is no appreciable variation in the ratio of capital to

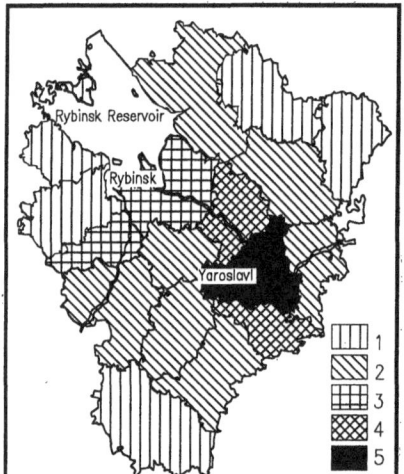 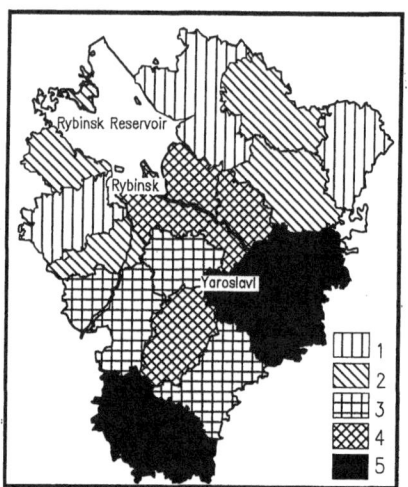

FIGURE 11.9 Milk Yields per Cow in Kilograms in the Province of Yaroslavl (1994). 1: Below 1600; 2: 1600-1800; 3: 1800-2000; 4: 2000-2200; 5: 2900.

FIGURE 11.10 Cattle in Personal Auxiliary Farming in 1995 in the Province of Yaroslavl as a a Percentage of 1990. 1: 122-130; 2: 130-135; 3: 135-140; 4: 140-150; 5: 150-171.

labor among the districts of the province. There has been, however, an appreciable variation in returns to investment: the districts with the most ample labor supply (in this case exurban districts) showed higher returns already in the 1970s. Since then these exurban districts have been the preferential sites for agricultural investment. Labor supply has told upon the production of flax, long a traditional staple of the province. Flax offers good returns, but is labor intensive. So while in 1958, as many as 57,000 hectares were devoted to flax in the province, by 1981 the cropping area had been reduced to 30,000 hectares and is now barely 10,000 or 2% of the cropland.

It is clear from the above that relative location vis-a-vis major urban centers matters the most for the geography of agricultural output. It matters the most for the specialization in specific kinds of produce as well. Only in grains is the playing field more or less level: grains' share of the area under crops ranges only between 35 and 45% among the districts. On the other hand, the districts of Yaroslavl, Rybinsk, and Tutayev lead the others in terms of the percentage share of meat in output. In the six centrally located districts between Rostov and Rybinsk, 70% of all meat is produced, whereas these districts' share in the rural population of the province is only 42%.

The ratio of socialized versus personal auxiliary farming (PAF) also depends upon relative location. Figure 11.10 shows the speediest growth in the

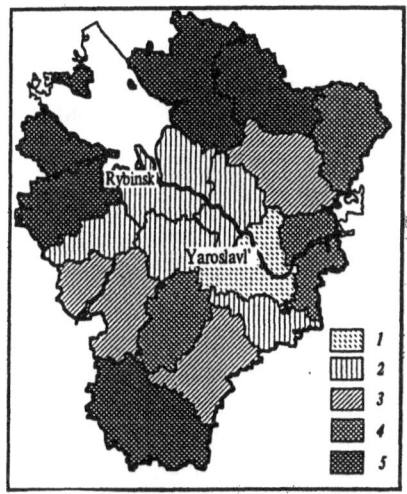

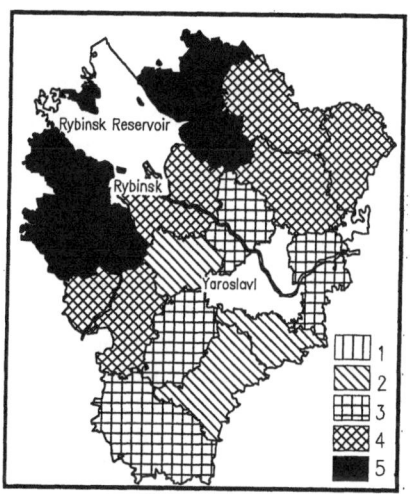

FIGURE 11.11 Cattle in All Kinds of Farms in the Province of Yaroslavl in 1995 as a Percentage of That in 1990. 1: 97%; 2: 81-90%; 3: 75-80%; 4: 70-75%; 5: 60-70%.

FIGURE 11.12 Percentage of Cows in Personal Auxiliary Farming in the Province of Yaroslavl (1994). 1: 11%; 2: 15-18%; 3: 19-22%; 4: 23-25%; 5: 27-34%.

number of cattle kept in PAF in exurban districts and in the district of Pereslavl adjoining the province of Moscow, and Figure 11.11 reveals that the reduction of cattle total numbers has been the greatest in most peripheral districts. Yet PAF's *share* in the number of cattle is at its highest (almost one-third) in the peripheral northern districts (Fig. 11.12). Recently the same situation held true for pig-breeding. But by 1994 it had changed. Pig-breeding began to markedly gravitate to exurban areas where it is easy to deliver pork to the customer. The pig thus ceased being essential to subsistence farmers (though cows still are) and became simply a means of earning money. In the areas most accessible to major cities and towns, from 50-90% of all pigs are kept in PAF, proving that relative location not only differentiates the evolution of socialized farms but the evolution of PAF units as well: only in the most accessible areas does PAF bear a commercial character; elsewhere it is a form of subsistence farming.

All in all, the activity space of provincial agriculture has been contracting like drying rawhide, with exurban spaces, the narrow belt stretching along the Volga, and the wider belt along the transportation axis between Moscow and Yaroslavl being the only ones that retain vitality in contrast to the decay in the vast northern periphery.

"As Compared with 1913"

This subheading was a catch-phrase of Soviet agitprop up until the late 1970s, since 1913 was considered a reference point for many Soviet achievements. In later years this phrase assumed a rather anecdotal meaning.

To be sure, 1913 is no worse a reference point than any other year, especially since the resultant comparisons do not always present the current situation as an achievement. An excellent 1915 book by Gurevich[6] is based on statistics circa 1913. At that time 85% of the provincial population was rural, and rural population density was 28 persons per sq km versus seven today. The area under crops was smaller: 500,000 hectares versus 718,000 in 1991. Grains accounted for 70% (versus 36% today) of the cropping area, but it was mostly rye (40%) and oats (25%), not wheat, an intruder from the south which prevails today and shows meager yields by current standards. Interestingly, the 1913 grain yields were almost as high as today, 10 centners per hectare on average. At that time this represented decent productivity, especially given the province's geographical location (in terms of bio-climatic potential it is more or less analogous to central Ontario).

By 1913, the agriculture of the province already bore a quite commercial character. The main commercial crops were potatoes and flax. Potatoes accounted for 10% of the whole area under crops and yielded 414,000 tons a year; today the yield is about 313,000 tons. The per capita production (for the whole province) was 319 kg in 1913 versus 212 kg today. Flax occupied 14% of the cropping area in 1913; now it is 2.5%.

At that time the richest areas of commercial agriculture were the bands stretching along the major railway lines. Two major bands had rural population densities and land values higher than elsewhere: one along the Trans-Siberian line which flows north-south within the province and another, the Rybinsk-Tver line, flowing west. The former band was mostly potato-oriented, while the latter emphasized flax. Flax farming in the district (then *uyezd*, now *rayon*) of Myshkin has by now almost withered away, though the culture of flax diffused southward (Uglich) and northeastward (Poshekhonye). Today the districts of Uglich, Danilov, and Poshekhonye produce 50% of the flax in the province. Today, as in 1913, flax and potatoes are mutually exclusive, largely because of the high labor intensity of each.

It can be pointed out that of the two regions of pronounced commercial farming early in this century, the flax-growing northwest and the potato-growing center/southeast, only the latter has survived. Other areas lagged behind those two even then. Allowing many fields to be fallow and cultivating mostly rye, their farming was in large measure of a subsistence type; they stood out for their poverty and the low number of cattle in peasant possession. Today the situation of those districts relative to the Rostov-Tutaev-Rybinsk axis is little changed. The inertia of spatial development appears to have persisted through all the trials of the last 80 plus years: the communist revolution, collectivization, world wars, rural depopulation, and the collapse of communism. And even though in the meantime the productivity of agriculture grew by fits and starts in the early post-war years and in the 1970s, its level today is not far apart from 1913 despite revolutionary changes in farming technology.

What Have We Arrived at by 1995?

The issue of the on-going reform in agriculture is addressed in Chapters 6 and 7. There is nothing peculiar about the province of Yaroslavl in this regard, as it is neither in the vanguard of reform, nor a part of the "red belt" of Russia flouting all government reforms and grass-root initiatives.

As in the nation as a whole, the change of status of collective and state farms is more indicative of the general attitude toward reform ("joint-stock company" appears to be a code for a non-committal social endorsement of the very idea of market economy) than of any profound systemic shifts in farming management. The highest frequency of status change has been in the district of Yaroslavl, some adjoining districts, and in the district of Pereslavl which neighbors the province of Moscow. In contrast, in the north-western and northeastern periphery most collective and state farms have retained their status. The districts of Pereslavl, Bolshoye Selo, Gavrilov-Yam, and Rostov are the leaders in terms of private farmers' percentage share of cattle (1.7-4% versus 1.3% in the province as a whole; in the district of Yaroslavl, 0.6%).

But while private farming, or whatever it is, is languishing, PAF is on the rise; by 1994 its share in the provincial output of meat had risen to 38%, in milk to 36%, in potatoes to 76%, and in vegetables 90% (Table 11.1).

TABLE 11.1 Percentage Shares of Farms of Different Categories in the Agricultural Output of the Province of Yaroslavl in 1994.

Produce	Collective and State Farms	Collective and State Farms That Changed Their Status	Private Farmers	Personal Auxiliary Farming
Meat	18	42	0.2	38
Milk	25	39	1.4	34
Grains	34	64	1.0	0
Potatoes	8	15	1.2	76
Vegetables	No data	No data	No data	90

Source: Unpublished data of the Provincial Bureau of Statistics.

The decline of agricultural output in the socialized sector of the province had commenced already in the 1980s; in the 1990s its pace has accelerated. This was especially evident in animal husbandry when it was deprived of imported grain and some habitual types of subsidies. As a result, in the 1990s the number of cattle in the province in all farm categories (socialized farms, PAF, and, since 1991, private farms) dropped by 23%, although in the 1980s in many districts the number of cattle had already been reduced by 10-20%. As in the 1980s, the district of Yaroslavl was all but spared this fate, retaining the number of cattle it had had in the late 1970s. The overall downsizing trend is being inhibited by the growth of PAF; however, the socialized farms in the outlying districts have seen their cattle reduced by one half, while those in the central districts have experienced below 20% reduction. In crop harvesting the downward trend has not been as pronounced as in animal husbandry, with the exception of flax. In the 1990s the area under crops has been restructured in response to the worsening supply of animal feed. Specifically, the area under grain crops was reduced and that under green and succulent feed expanded.

Despite all the measures that have been taken, the economic survival of socialized farms is a haunting problem; in two-thirds of them costs exceed revenues and without subsidies many would drown immediately. The district of Nekouz tops the list with 78% of its farms being unprofitable; in contrast, in the district of Yaroslavl only 8% of the socialized farms cannot make ends meet on their own. With subsidies, however, the picture is not so grim; in 1994 the ratio *revenues-minus-costs to costs* in 1994 ranged between -7% and 20%.

As discussed in Chapter 6, the volume and mechanism of subsidizing changed after 1991. Whereas today subsidies amount to about 50% of the total output of socialized farms, in 1991 this percentage varied between 70% and 200% among the districts of the province. Whereas in the 1980s, subsidies were mostly channeled through farm-specific price mark-ups set at various levels to rescue sinking farms, today price mark-ups are flat and thus may not rescue the weakest. Additionally farms receive partial budgetary compensation of their fuel and fertilizer purchase costs, and this form of support is not tailored to the economic situation of a farm either. So in this regard, the underpinnings of the pre-1991 cradle-to-grave welfare economy have been abandoned. Not totally, however, because there are forms of indirect budgetary support, such as writing off debts, which do aid the weakest; and, on the other hand, when the state procrastinates, as it often does, in remitting payment for purchased produce, the strongest farms are hit the hardest because of inflation.

Nevertheless, the economic differentiation of farms goes on, and there is no special policy of supporting those that have successfully adjusted to the new economic conditions.

Regionalization

The preceding discussion points to the existence of different types of agriculture, including spatially variable combinations of different farm types, of levels of output per unit of land, and of relationships with non-agricultural land use.

- In the central belt (Rostov-Rybinsk) the output per unit of land is the highest under average natural conditions. Labor and capital (fixed assets) per unit of land are also higher than elsewhere. Labor fluidity is high given the many non-agricultural employment options, but the region remains the most attractive to those living in outlying districts and to migrants from outside the province. The socialized farms in the region are large, employing between 170 and 270 people. Most socialized farms have changed their status to joint-stock companies and associations. The proportion of unprofitable farms is minimal. Although the socialized farms have retained almost their pre-crisis level of output, PAF is growing vigorously and is heavily commercialized.
- The extreme south (the district of Pereslavl) is agriculturally the best endowed by nature but has been very much impacted by rural out-migration because of its location next door to the province of Moscow. This location does not allow this region to be classified as peripheral; "semi-peripheral" would be a more fitting term. The level of output per unit of land is lower here than in the previous region. The extreme south has proven very attractive to migrants from outside the province, to private farmers, and especially to seekers of summer recreation homes.
- In the west (the districts of Bolshoe Selo, Uglich, and Borisogleb) the output per unit of land is below-average to average under natural conditions similar to those in the central belt. The west, too, is "semi-periphery." In the early 1990s it experienced a short-lived in-migration boom. Although the boom has subsided by now, the region remains attractive to migrants from without the province.
- The Trans-Volga northeast is a classic periphery; in addition it has the worst natural conditions for agriculture in the province. Pockets of arable land are dispersed and oftentimes have limited accessibility. Its rural population is thin and aging most rapidly; roads are the worst in the province. In the 1980s small collective farms (employing less than 100) abounded here; in the 1990s most of them have retained their status. Unlike other regions, rural out-migration has not been reversed here in the 1990s. The output of socialized farms is low and in large measure they function as umbrellas for PAF, which is crucial for people's survival in the region.

- The northwest (the districts of Nekouz, Myshkin, and Breitovo) shares the social conditions of the previous region (it is periphery as well) but is better endowed by nature. The output per unit of land is, nevertheless, the lowest in the province. In the past it was one of the two most prosperous agricultural regions of the province, thriving on flax. Any attempts to restore flax-growing have been unsuccessful. Just as in the previous region, socialized farms mostly ensure the survival of PAF. However, in contrast to the northeast, the region has been attractive for migrants.

The Local Scale

We have selected four rural districts, representing four out of five regions distinguished above, for the more detailed analysis of the evolving spatial pattern of agriculture. We used a rural Soviet as the information unit to ensure a spatial match between economic and socio-demographic indicators. In many cases the land area under supervision of a rural Soviet includes one socialized farm; in some cases, however, there are up to three farms per one Soviet. In such cases we averaged economic indicators for farms grouped in one Soviet.

The following economic indicators were used: grain yields, milk yield per cow, monetary output per unit of land, production costs, labor input per unit of produce, and the ratio of revenues minus production costs to production costs (see Figure 11.13). The following social indicators were used: population density, percentage share of the elderly in population, 1970-1990 percentage change in population, and 1990-1994 percentage change in population.

Each information unit was ranked on each indicator and five gradations ranging from the "best" to the "worst" were identified. Average ranks for the group of economic indicators and that of social indicators were calculated.

It should be noted that three-quarters of all the information units analyzed reveal a perfect or almost perfect match between their rankings on the social and on the economic scale: the better the social situation is, the better the economy is. In most of the remaining units economic indicators are superior with respect to social indicators, which may reflect a higher flexibility in the economic sphere. A mismatch of this type is most frequently encountered in the district of Yaroslavl. In the more outlying districts social and economic spheres correlate very closely.

The accessibility of a district center is not a valid predictor of the economic situation inside the outlying districts; it matters only in the district of Yaroslavl.

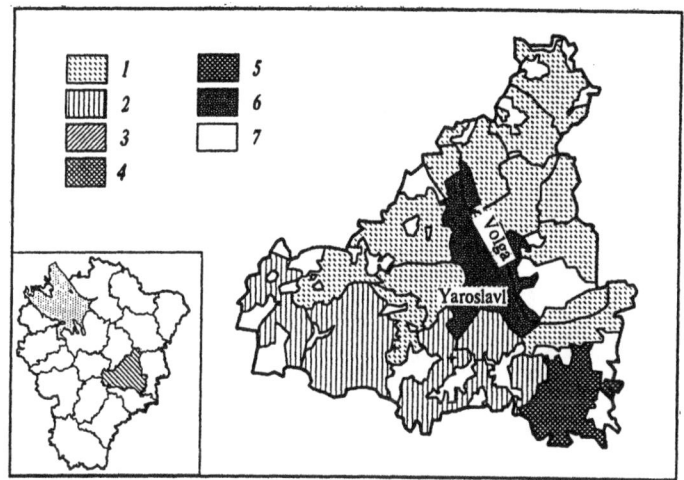

Figure 11.13 Profitability of Socialized Farms in the District of Yaroslavl: Percentage Ratio of Revenues-Minus-Costs to Costs (1994). 1: 0%-157%; 2: minus 20% - 0%; 3: minus 35% - minus 20%; 4: minus 50% - minus 35%; 5: minus 85% - minus 50%; 6-8: Non-agricultural land, including 6: Built-up areas, 7: Forests; and 8: Lake.

Whereas in the district of Yaroslavl, the size of farms is positively related to their economic performance (as larger farms with 300-500 employees fare better), the situation is opposite in the periphery, where smaller farms (less than 100 employees) seem to have a higher potential for survival.

The following six chapters address problematic and other issues that we have not given sufficient attention thus far in this book, but without which any account of the current situation in the russian countryside would be far from complete.

Notes

1. Figures 11.1-11.7 and 11.9-11.12 are based on current data gathered by the authors from the Yaroslavl Bureau of Provincial Statistics.
2. *Demographic Yearbook of the Russian Federation.* Moscow: Goskomstat 1994, p. 19.
3. Ibid., p. 406.
4. Ioffe, G.V. *Selskoie Khoziaistvo Niechemozemia: Territorialnye Problemy.* Moscow: Nauka 1990, p. 80.
5. Ibid., p. 80.
6. Gurevich, M.B. *Differentsiatsia Typov Khoziaistva Yaroslavskoi Gubernii.* Yaroslavl: Statisticheskoie Otdelenie Yaroslavskoi Gubernii 1915.

12

Polarization of the Rural Activity Space

Based on our account of the evolution and current situation of the Russian countryside, both Chernozem and non-, polarization appears to be pervasive, geographically the most pronounced trend of its development. Exurban and outlying areas are growing increasingly different; the transitional belt between them, the semi-periphery, continues to share some characteristics of both even as the gap grows more pronounced.

The crisis of the 1990s affected these segments of the countryside differently. Economically weak collective and state farms in the periphery were impacted first; by 1994 the wave of crisis had reached strong socialized farms of the exurbia, and by 1995 many technologically advanced poultry and pig-breeding farms located close to large cities came to a halt. At the same time the outlying areas were completing a switch to subsistence farming that survived as a result of a symbiotic relationship between socialized farms and PAF.

The extent to which the exurban rings of heightened agricultural land use intensity and core-periphery contrasts are pronounced depends upon two major factors: the population's pressure on the land and the size of an urban center. The former factor is colonization-related, and the historical east-west and the largely natural north-south gradients of population density play the greatest role in this regard. In sparsely settled areas, however, the polarization is reaching a higher level than in the densely settled ones. The size of the city contributes to the polarization as well, with less accessible segments of the countryside falling prey to seemingly irretrievable rural depopulation.

These underpinnings of polarization are visible in Figures 12.1 and 12.2. Both are fragments of the same land-use intensity map, compiled by Tatyana Nefedova, based upon summing up standardized estimates of several production inputs (fixed assets, labor, fertilizers, and livestock); with the

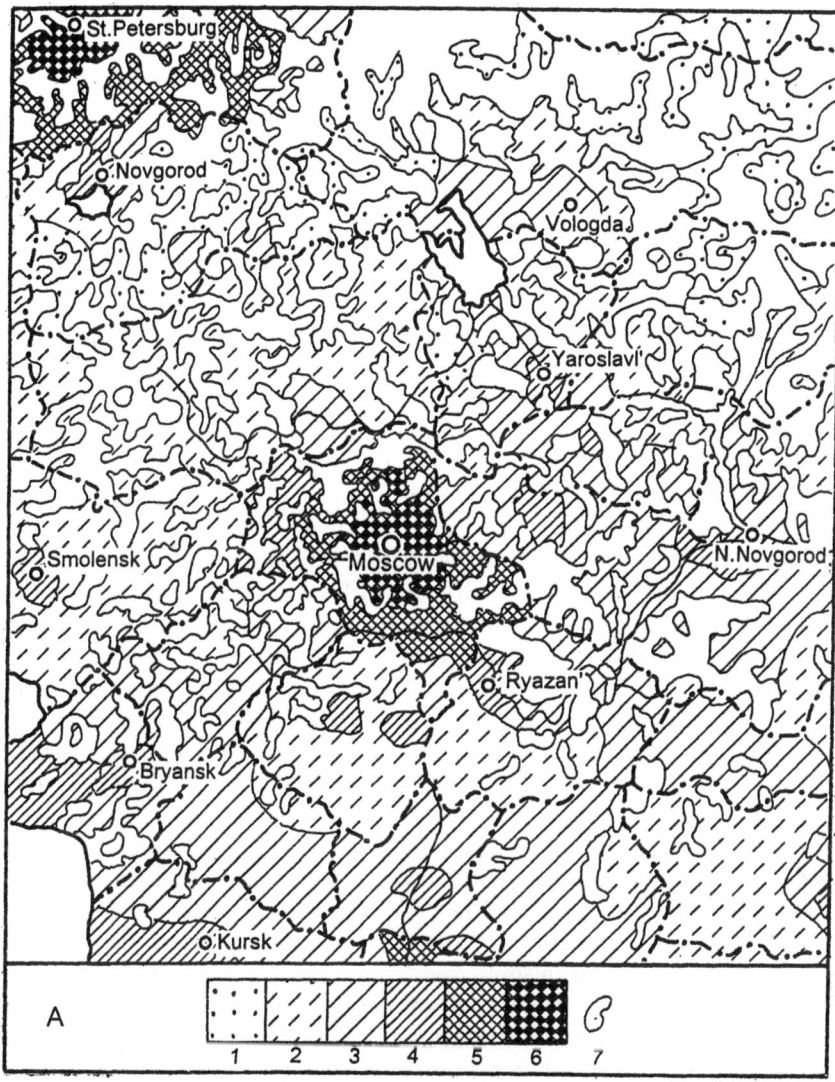

FIGURE 12.1 Agricultural Land Use Intensity in Central Russia. 1-6: Ascending order of intensity assessed as the sum of the standardized estimates of the following: monetary value of fixed assets per unit of land; application of mineral fertilizers per unit of land; grain yields; amount of livestock per unit of land; milk yields per cow; and share of arable land in agricultural land; (the spacing of the gradations 1-6 is equal to one standard deviation); 7: forests and water reservoirs.

spacing of the sums' gradations based on one standard deviation from the average. In particular, Figure 12.2 suggests that while in European Russia's south, spatial contrasts in land-use intensity are high, they hinge upon the core-periphery pattern only in part and can be attributed largely to the natural setting (e.g., river valleys as in Kuban' or aridity, as in the southeast of Saratov, etc.). In contrast, in the north-central part of European Russia (Fig. 12.1), spatial contrasts are of a markedly core-periphery brand. Central Russia, positioned at the junction between the historically sparse and dense belts of rural settlement, follows the pattern of the north with extensive polarization in large measure due to well-established urban cores and their lasting demand for labor. This influence was considered in Chapter 5.

In central Russia it makes a huge difference what kind of urban center "presides" over the neighboring countryside. An urban concentration of around 50,000 people brings to its nearby countryside a heightened agricultural land-use intensity, including output per unit of land. Larger cities produce more land-use intensity gradients that are steeper and/or stretch over larger chunks of the countryside. For the delimitation of three spatial zones, *exurbia, semi-periphery,* and *periphery,* in our study of non-Chernozem Russia we used the following specific indicators: the output per unit of land, the density and dynamics of rural population, the ratio of agricultural output of the public sector to that of PAF, and indicators of rural infrastructure (percentage share of houses with plumbing; paved roadways density, etc.). Each of these indicators typically is a descending function of accessibility to a major urban center. So on each province's map based upon constituent districts' averages, the spatial gradients invariably showed up. The delimitation was subsequently reinforced by the location quotients, i.e., ratios of population percentage share to the percentage share of a specific type of produce (usually milk or meat). In virtually no case did the demarcation of exurbia pose serious problems: even though there is no single threshold criterion for all the provinces studied, the drop-offs between the exurbia and the rest of the provinces are clear-cut. It was the periphery whose delimitation was in some cases more difficult to ascertain. Needless to say, the size and configuration of the exurban ring or zone, in which agricultural output per unit of land normally peaks, depend not only upon the size of the closest urban center or centers, but also upon the centers' location relative to each other. For example, in the province of Pskov, where there are two not very large urban centers, Pskov and Velikiye Luki, there are two separate exurban agricultural zones, of which the former includes three rural districts and the latter only one. That this is really the case is verified by our spatial analysis using the above indicators. The situation is different in the province of Yaroslavl with the two *large* urban centers, Yaroslavl and Rybinsk, located *not far from each other.* This configuration

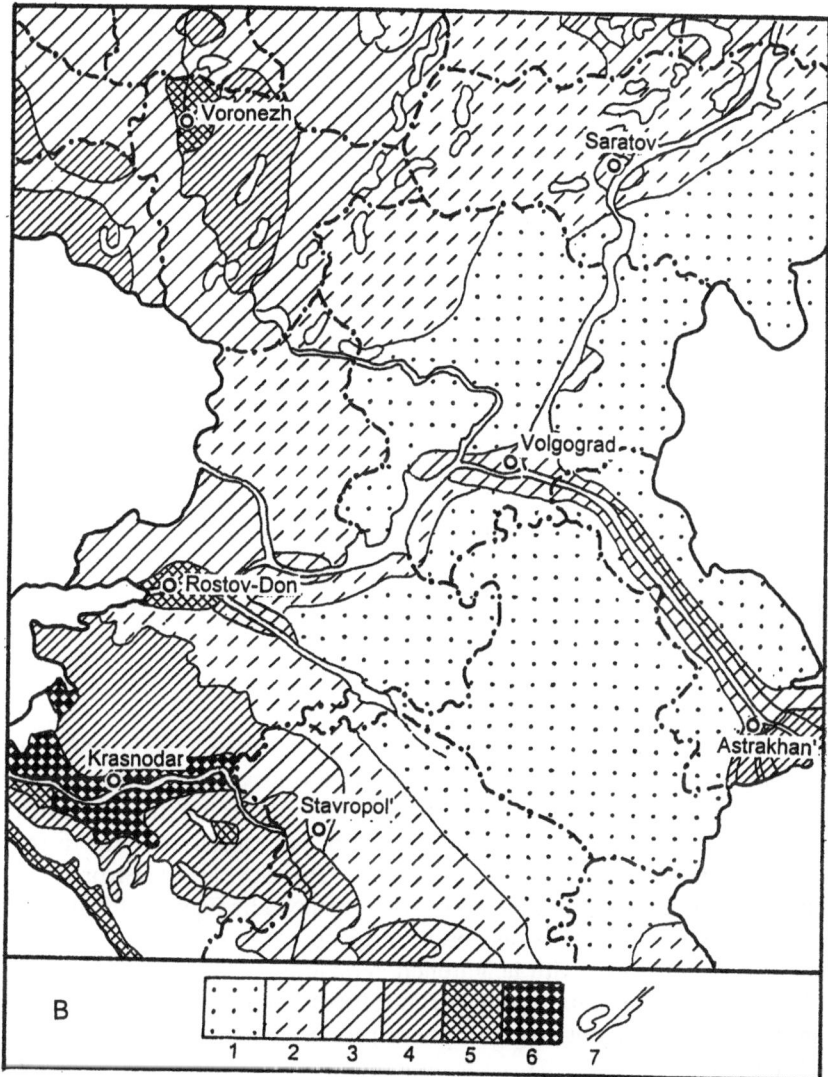

FIGURE 12.2 Agricultural Land Use Intensity in Southern Russia. 1-6: Ascending order of intensity assessed as the sum of the standardized estimates of the following: monetary value of fixed assets per unit of land; application of mineral fertilizers per unit of land; grain yields; amount of livestock per unit of land; milk yields per cow; and share of arable land in agricultural land; (the spacing of the gradations 1-6 is equal to one standard deviation); 7: forests and water reservoirs.

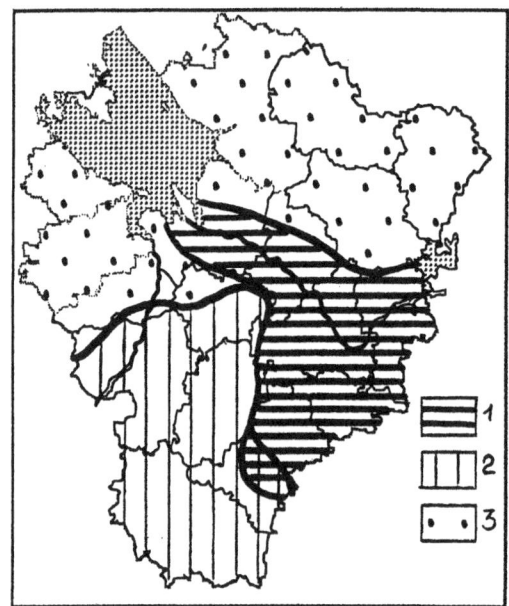

FIGURE 12.3 The Trichotomy of the Inter-Urban Space in the Province of Yaro-slalv. 1: Exurbia; 2: Semi-Periphery; 3: Periphery.

led to the formation of a large exurban agricultural zone including not only the four rural districts (those of Yaroslav, Rybinsk, Gavrilov-Yam, and Tutayev) situated between the two major urban centers but also the district of Rostov along the railway and roadway leading to Moscow and the district of Niekrasovo located between the cities of Yaroslavl and Kostroma (see Fig. 11.1). The example of the delimitation is shown by Figure 12.3. In most provinces, however, there is only one conspicuous urban center and, conse-quently, only one area with heightened agricultural output.

The division of the countryside of central Russian provinces into exurban, semi-peripheral, and outlying (peripheral) zones is characterized in Tables 12.1 and 12.2.

The exurban agricultural zone of Saint Petersburg includes six districts which emphasize vegetables and house some of the largest stock-raising farms in Russia. It must be pointed out that while the countryside of that province perfectly matches the others (featured by Tables 12.1 and 12.2) in one respect, i.e., the spatial polarization with outwardly or centrifugally descending land-use intensity, it is different from them in terms of the absolute level of this intensity, especially in the exurban zone. In those terms, the province of Leningrad is one step ahead of the others in that its semi-periphery is on a par with the exurbias of the three other provinces.

TABLE 12.1 Size and Percentage Shares of Three Types of Countryside.

Territory	Land Area in 1000 sq km	% of Land Area	% of Rural Popu- lation	% of Cattle	% of Pigs	% of Potato Output	% of Meat Output
Pskov	55	100	100	100	100	100	100
exurbia	9	16	30	21	22	26	28
semi- periphery	25	46	37	49	44	36	39
periphery	21	38	33	33	34	38	33
Kostroma	60	100	100	100	100	100	100
exurbia	6	10	27	30	47	85	51
semi- periphery	8	44	35	38	33	13	31
periphery	26	46	27	32	20	2	18
Yaroslavl	37	100	100	100	100	100	100
exurbia	13	35	48	42	78	76	68
semi- periphery	10	25	24	24	13	18	12
periphery	14	40	28	34	9	6	20
Leningrad	84	100	100	100	100	100	100
exurbia	17	20	53	42	86	60	62
semi- periphery	47	44	40	53	13	36	34
periphery	20	36	7	5	1	4	4
Average Non-Chernozem *Province*	64	100	100	100	100	100	100
exurbia	13	20	40	30	60	55	50
semi- periphery	23	36	33	45	25	30	30
periphery	28	44	27	25	15	15	20

Source: Authors' calculations based on unpublished provincial statistics.

As Tables 12.1 and 12.2 show, the spatial contrasts in the most urbanized province of Leningrad are the most striking. They are the least pronounced in the least urbanized province of Pskov. Yet there are commonalities as well. Covering between one-tenth and one-third of the overall land areas, the exurban rings carry about half of the rural populations, about one-third of the cattle, and the bulk of the pigs. The spatial proportions shape up differently in the periphery: occupying half of the countryside, it houses

TABLE 12.2 Selected Characteristics of Exurbia, Semi-Periphery, and Periphery Relative to Provincial Averages (averages for each province are assumed to be 1.0).

Province	Rural Population Density	Grain Yields per ha	Milk Yield per Cow	Total Farm Output per Unit of Land	Private Farms per 1000 Engaged in Farming	Share of PAF in Farm Output
Pskov						
exurbia	2.0	1.2	1.1	1.4	1.2	0.9
semi-periphery	0.8	1.0	0.9	0.9	1.3	1.0
periphery	0.8	0.8	1.0	1.0	0.7	1.2
Kostroma						
exurbia	2.8	1.6	1.2	2.5	0.7	0.6
semi-periphery	1.0	0.9	0.9	0.8	1.8	1.5
periphery	0.7	0.7	0.8	0.5	0.8	2.5
Yaroslavl						
exurbia	1.4	1.4	1.2	1.5	1.0	0.8
semi-periphery	0.9	1.0	1.0	0.9	1.5	1.2
periphery	0.6	0.7	0.8	0.5	0.8	1.6
Leningrad						
exurbia	2.6	1.0	1.1	1.7	0.7	0.7
semi-periphery	1.1	1.0	0.9	0.7	1.3	1.3
periphery	0.3	0.5	0.4	0.2	1.4	2.4
Average Non-Chernozem province						
exurbia	2.3	1.5	1.2	2.0	1.0	0.7
semi-periphery	1.0	1.0	1.0	0.8	1.3	1.2
periphery	0.6	0.7	0.7	0.6	0.8	2.0

Source: Authors' calculations based on unpublished provincial statistics.

between one-tenth and one-third of the rural population, and its contribution to the agricultural output ranges between one-twentieth and one-third. The percentage share of PAF in the agricultural output is three times as high in the periphery as in exurbia.

It seems obvious that any reasonable regional policy has to take those differences into account.

The biggest unknown is the type of change that needs to be advocated for the periphery. Rural depopulation has dealt it a most severe blow, which is why agriculture cannot be sustained as a form of land use everywhere in the periphery. In this regard spatial continuity has long been breached. In fact, land abandonment and spontaneous reforestation of formerly arable land and especially of meadows and pastures has been under way since the early 1960s. The periphery is ideal for nature reserves, for forest energy plantations, and perhaps for other experimental forms of land use in open spaces. Within the pockets of surviving agriculture, leasing land to exurban farms could be promoted since the latter are deficient in animal feed in part because they cannot afford to assign a lot of land to it. However, such links can be forged only when local roads are substantially improved. PAF in the periphery is not likely to survive without budgetary aid. At this point such aid is rendered indirectly by keeping ephemeral collective and state farms from closure. Most of the latter have ceased to exist as commercial enterprises anyway; but their official closure would deprive PAF of implements, fertilizers, maintenance, etc., and thus would deal it a mortal blow.

Exurban areas require a totally different approach. Large socialized farms here still feed urban areas. Under current conditions many such farms make ends meet only when propped up by subsidies. But these subsidies ought to support the strongest and promote their takeover of weaker farms. Zoning ordinances ought to be introduced to restrict selling valuable agricultural exurban lands to the "new rich," who build castle-like cottages on them.

The semi-periphery forms a special, transitional case. Whereas the periphery contains archaic but robust PAF, and exurbia boasts of viable socialist agriculture, the semi-periphery is a no-man's land. Depopulation has affected the semi-periphery to almost the same extent it has the most outlying districts, leaving myriads of abandoned villages. On the other hand, unlike in exurbia and the periphery, neither viable collective farms nor robust subsistence economy (for those who choose to stay) are available. So the niche is now filled by two kinds of developments: summertime recreational land-use (*dacha*s or recreational second dwellings) and, to a smaller extent, private family farming. Both developments would in fact gravitate toward exurbia if land were available but since in most cases it is not, the semi-periphery, usually located 1-2 hours from the provincial centers, becomes the only option. People now populating the semi-periphery tend to be the most open-minded: they are relatively less addicted to archaic communal bonds, or to reliance on strong leadership and socialist enterprises. *Dacha*-owners, private farmers, and forced migrants at the crossroads of their lives may constitute the human capital with which the long-term prospects of the Russian countryside at large rest. However, the infrastructure in the semi-periphery can hardly be upgraded solely on the basis of private investment, and its current condition is not conducive to economic development.

Thus the peculiarity of the situation is that at this formative stage of the market economy in Russia, there is an unavoidably crucial role for the government in creating extrinsic conditions for economic growth.

13

Urbanites in the Countryside

A representative 1993 survey conducted in various cities of Russia by the All-Russian Center of Public Opinion Polls revealed that three-quarters of the respondents or their close relatives have a parcel of land which they use to grow agricultural produce. In large cities, the number was 60%. And two-thirds of the "have-nots" would like to obtain such a parcel.[1]

Such a pull towards land and the desire to engage in farming along with one's principal job go back to the recent agrarian past of the nation, and to the peasant roots of many urbanites. However, such factors as poverty, long-term food shortages, urban overcrowding, and very meager personal living spaces, especially in large cities, contribute to this thirst for land as well. For all these reasons urbanites seek a second dwelling in the countryside, usually a low-quality dwelling surrounded by a small parcel of land. According to the 1993 poll, only 3% of the residents of large cities (population over 200,000) and 10% of those in medium and small cities (below 50,000 residents) have parcels sized 0.2 hectares and over. Between one-quarter and one-third of the landholdings are less than 0.05 hectares.

Currently the country "estates" of Russian urbanites fall into *four* major categories. *Dachas,* the oldest category, appeared long before the revolution of 1917 as recreation sites of white-collar people, their version of the traditional estates of the landed gentry. A *dacha* could be privately owned or rented and was located, as a rule, close to a city -- a major difference from traditional estates.

In the 1920s, government-owned and "departmental" (*viedomstvennye*) *dachas* appeared to serve three major segments of the new Soviet elite: the Party, economic management, and the intelligentsia. The expansion of private recreation sites, but without formal ownership of land, began in the 1930s and continued for about 20 years after the war. Since that time *dachas* have largely remained a perk of elite social groups. Land parcels for private recreational construction were being distributed through places of work, and *dachniks* were required to be members of a cooperative set up on an institutional basis (e.g., a *dacha* cooperative of aircraft specialists) which concluded long-term land-use agreements with local authorities on behalf of its members and provided some maintenance and supply services for *dacha*-owners. The demand for *dachas* grew quickly and could not be met. So in the 1950s and '60s renting parts of *dachas* and of houses in nearby villages was becoming widespread, an activity usually confined to a 10-50 km radius from cities.

Initially, *dachas* were a phenomenon exclusively of the vicinities of Moscow and Leningrad, with ownership of a *dacha* being a mark of wealth. However, these wooden two-storey recreational dwellings, with 50-80 square meters of living space, were gradually simplified bearing more and more similarity to ordinary village homes. Rural infrastructure long stayed inferior (lack of plumbing and piped water and primitive heating devices) even in the vicinity of Moscow, which made it impossible to use a *dacha* all-year round.

Dachas represent the largest-sized parcels of land available to urbanites; they range from 0.12-0.5 hectares. Initially agricultural activity on the land was not intensive, with forest and meadows accounting for the bulk of land parcels. Many *dacha* holdings girding Russian cities have retained this picturesque idleness to this day.

In the 1960s, during what Soviets remember as a second, Khrushchev-led collectivization (see Chapter 3), *dacha*-owners fell into public disfavor as new *kulaks*. Many of them even hurried to sell their second dwellings (at that time land itself had no price, but it affected the price of dwellings mostly through location). While anger against subsidiary plots in villages arose because they reportedly produced and sold too much (see Chapter 3), similar anger against *dacha*-owners was spurred by the fact that land at *dachas* was exempt from farming, as if publicly-run agriculture would have necessarily put it to efficient use. An apparent lack of logic never thwarted any domestic political campaign, including this one, which appealed to rank-and-file Russians' traditional attachment to equality. However, when the campaign subsided, though it affected subsidiary plots in villages more than urbanite-held *dachas,* owners of *dachas* decided to pay more attention to farming -- just in case. The growing deficit of food items accentuated this trend. Some *dacha*-owners began to sell fruits, berries, and vegetables at peasant markets (so-called *kolkhoznye rynki*) in nearby cities.

Cooperative orchards and vegetable gardens began to be set up in the 1960s when it became clear that demand for second dwellings could not be met by *dachas.* By the 1970s, about 80% of today's cooperatives or comradeships were already established. In 1980 in the province of Moscow alone, 189 comradeships of this kind existed along with 256 *dacha* cooperatives. Together they accounted for 50% of the recreational land-use in the province of Moscow. Overall, in Russia, by 1990, five million people owned parcels run by orchard-and-vegetable comradeships. They represent the most widespread type of country "estates" held by urbanites. This type is rooted in two phenomena. On the one hand, it is a reduced *dacha,* as typically the size of the landholding is 0.06 hectares at the most. On the other hand, historically it was a means to eke out a living during war-time dislocations: it was during the war of 1941-45 that urbanites for the first time were allotted small parcels of land (to grow potatoes and vegetables). In the sixties and seventies they were intensively used to grow produce and the agricultural use of land in comradeships is now much higher than on *dachas.* In most cases inferior land was allotted for this type: used-up quarries, marshes, strips immediately underneath high voltage transmission lines, and other areas inappropriate for collective and state farms. In many cases parcel-owners created topsoil themselves at the expense of enormous investments of time and labor. Initially dwelling construction on these parcels was either legally banned or confined to primitive shacks. However, the statutory transition from a one- to a two-day weekend in 1967-69 resulted in the initial restrictions being loosened and led to setting up of tiny houses, unsuitable for living but adequate for spending the night. Such parcel-and-dwelling clusters surrounded by ominous fences are the most aesthetically offensive segments of Russian exurbia.

Village homes, the third type of urbanite recreational estate, began to develop in the 1970s. Dissatisfied with overcrowded garden parcels and sundry associated restrictions, urbanites were desirous of a freer vacation life in a traditional village. Although plenty of village houses had been and were being abandoned by their owners due to out-migration and aging-related mortality, local authorities initially did all they could to prevent urbanites from obtaining those dwellings. The reason: it was thought that urban vacationers would ultimately corrupt villages and wean them of any remnants of a zealous commitment to working land. For a long time, until the late 1980s, central authorities acted on this attitude. As a result, the inevitable "colonization" of many traditional villages by urbanites initially assumed a semi-criminal character: to obtain a house, an urbanite concluded an inofficial agreement with a village *babushka* or her urban relatives, paying her or them a small sum of money for an abandoned house. Another widespread way to circumvent short-sighted laws was to find a figurehead buyer among country folk. In the seventies a dilapidated but still livable village house could be bought for what amounted to two or three moderate monthly salaries. In due course the abandoned houses in all villages more or less accessible to a city had been bought up by the early 1980s and the radius within which this type of second dwelling could be found began to expand further. Only in 1989 did the government issue a ruling officially allowing urbanites to buy a house in a village and use the land attached to it. Prices skyrocketed and yet the number of those willing to buy a village house continued to increase, even though by the early 1990s, virtually all vacant village houses within the corporate limits of the Moscow and Leningrad provinces had already been sold. So those Muscovites and Peterburgers still wanting to buy village property could do it only outside their respective provinces, which translates into 3- to 5-hour car trips in one direction. Taking into account that the rate of private vehicle ownership in Russia was 84 per 1000 population in 1994 and only 59 in 1990,[2] one could not escape being amazed at the sight of thousands making onerous trips *na pere-kladnykh* (literally "by post-chaise"), that is, by several consecutive means of public transportation, usually overcrowded commuter trains and/or buses. Needless to say such a distant second dwelling is more conducive to spending several weeks a year in rural privacy combined with hunting, fishing, and gathering than to regular weekend trips, let alone to the daily summertime commuting *dachniks* are used to.

Second dwellings in villages are regular wooden village huts slightly or substantially repaired. The farming component of such recreation sites is well under that of *dachas* and especially that of vegetable cooperatives. However, this statement applies only to second, third, and later generation urbanites. Those who were born in villages, then abandoned them leaving family behind, and who now return to the countryside in a second dwelling, are more active in their farming pursuits. Yeltsin's Decree of March 7, 1996 (see Chapters 6 and 7) attempted to expand the legal rights of urbanites who have inherited village houses. These heirs can now claim possession not only of 0.15 hectares around the house but the adjoining parcel as a whole, which frequently exceeds that threshold.

To this point there have been very few systematic surveys of urbanites in village settings. The best known to us was conducted in the summer of 1985 by Fingerov in the Firovo district of the province of Tver. This is a peripheral district located midway between Moscow and Saint Petersburg. Throughout the nine years preceding the survey the negative rate of natural increase accounted for two-thirds of the 20% total population decline in the district, so the ratio of urbanites to permanent residents in 1985 was approximately 1:1 and ranged from 1:3 to 3:1 in different villages. For many villages in the district serving the needs of urbanites at leisure has become the main source of income. In these cases the regular seasonal incursion of urbanites is the only thing that provides, occasionally, a demographically normal base for a local population pyramid: it is only in summer that kids in a village can outnumber the elderly. Numerous services rendered to vacationers by rural folk include keeping an eye on the house all year round and on private vehicles in summer, selling milk and vegetable produce raised on subsidiary plots, baby-sitting, leasing boats and boat engines, pointing out places rich in wild berries and mushrooms, etc. In their turn urbanites spread innovations like new breeds of strawberry, raspberry, and vegetable seeds, and different household devices, and render various services that are occupation-specific (especially health care and the repair of electric appliances and equipment).[3]

The fourth type, *cottages,* which showed up only in the 1990s, fits the Western notion of a single-family home most closely. Initially they were wooden, surrounded by a plot of 0.08-0.1 hectares, located in close proximity to a city, and not exceeding $10,000 in price. But with more and more "new rich" turning up, the size and price began to grow fast. Cottages became stone and brick, two-to-four-storeys, and looking like stylized medieval castles. The price of such monstrosities spilled over $1,000,000. They have mushroomed recently not far from large cities either inside existing *dacha*-settlements or on the fringes of farming fields. The demand for such cottages, however, was quickly met and has now subsided. The newest development is two-storey cottages grouped in clusters with their own semi-autonomous infrastructure. These townhouse-like cottages cost about $100,000. In contrast to the other three dwelling types, in many cases these are not just recreational; they are suitable for permanent residence. The demand for such cottages is definitely higher than for "castles" but is also very limited given that a system of mortgage loans has not yet emerged in Russia. So the demand is generated more by banks, large enterprises, and ministries than by private residents.

The Hinterland of Moscow

Using the province of Moscow as an example, let us now take a look at how urbanite expansion into the countryside unfolds and how a land market takes shape.

Moscow is not only a national capital, a city with a population of nine million, but also a principal research, cultural, and financial center, the un-disputed national leader in all sorts of innovations. Moscow has been immensely attractive because of its heightened living standard and enor-mously broad spectrum of opportunities: everybody seems to want to move into Moscow but nobody wants to move out. The city and the province of Moscow together account for 10% of Russia's total population.

In 1995, the average per capita income of Muscovites exceeded that of Russia as a whole by a factor of three (even though the poorest strata is also quite numerous in Moscow). Income, together with the sheer size of the population, created colossal pressure on the surrounding land. This pressure revealed itself in the following:

- A long-term and on-going expansion of industrial and residential land-uses, a sprawl leading to merger with nearby towns; urban land-uses account for 15% of the whole land area inside the province's border;

- Development of a robust agrarian infrastructure for large and relatively well-to-do socialized farms (prior to 1990 most of them were state farms) and with the output per unit of land being a descending function of distance from the city of Moscow (see Figure 10.3);
- Creation of a complex network of public recreation sites (recreation homes, sanatoria, "pioneer" camps for kids, etc.) with the density of this network also descending as the distance from Moscow grows: in the mid-1980s, for example, there were 20 public recreation sites per 100 sq km at a distance of 15-20 km from Moscow but only 2-3 sites per 100 sq km at a distance of 80 km[4];
- Generation of a colossal demand for short-term individual recreation facilities; in the 1980s, the number of people spending one to two summer days a week outside Moscow and other urban settlements of the province amounted to 2.2 million; these included tourists as well but most were those leaving town for their second dwelling in the countryside. The density of such private and semi-private recreation sites also changes with distance from Moscow. In the city's immediate vicinity the density is low because so-called dormitory settlements prevail, all of whose residents work in Moscow. The number of people taking a short-term rest per unit of land peaks within a 20-60 km radius from Moscow and it gradually descends outside these limits. Whereas in 1979 the percentage share of the rural settlements in the province of Moscow, in which a major part of the home owners was engaged in agriculture (not counting PAF), was 75%, ten years later, in 1989, it was only 34%.[5] This stunning change is almost exclusively due to urbanites' buying up all vacant village houses. Such houses are located mostly in the periphery of the province (Fig. 13.1).

There are only two kinds of villages urbanites are reluctant to approach in pursuit of second dwellings: large farm villages where the management offices of socialized farms are located; and small villages with extremely low accessibility, in terms of transportation routes, to Moscow. So the bulk of the most attractive villages are located in an intermediate accessibility zone, not very close to major thoroughfares but not too far from them either. By 1994 about one-half of all village dwellings in the peripheral districts of the province of Moscow had already been bought by urbanites. And once the supply was exhausted, Muscovites began to reach out to neighboring provinces. According to Liudmila Filippovich, the percentage share of urbanite-(mostly Muscovite-)owned rural houses in some of the adjoining provinces, particularly Tver and Kaluga, is even higher than in the province of Moscow itself.[6] This is because the proportion of houses abandoned by original owners due to rural depopulation is much lower in the province of Moscow than in those around it. Although there are especially many Muscovite second-

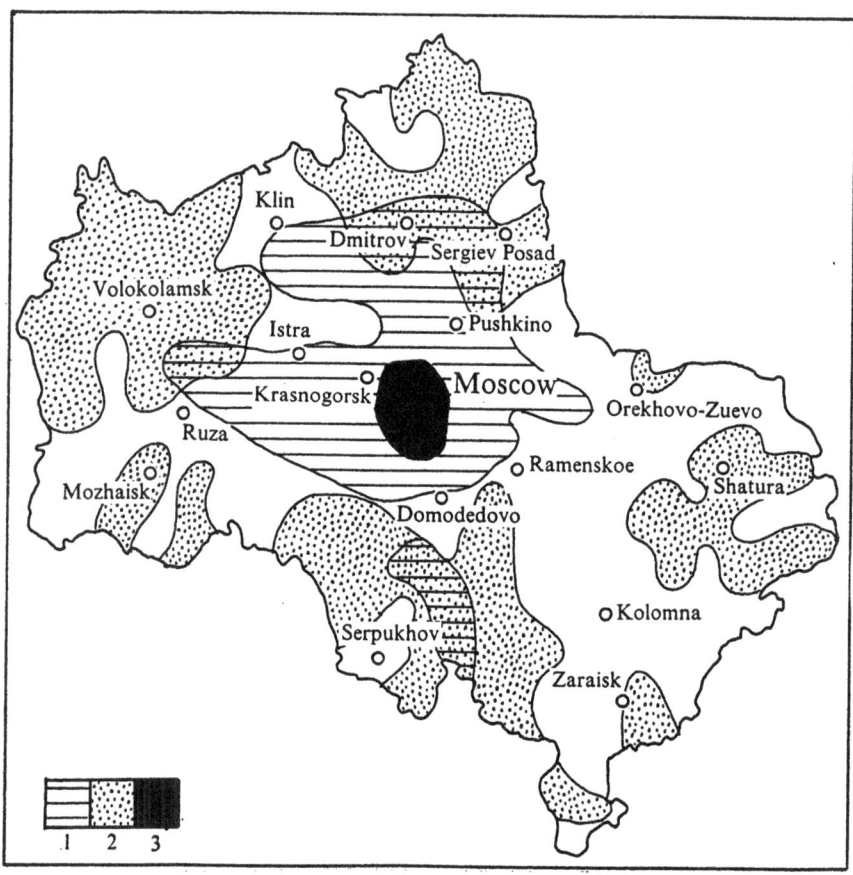

FIGURE 13.1 Urbanites' Second Dwellings in the Villages of the Province of Moscow. 1: Most attractive areas for summer recreation; 2: Areas in which most abandoned villages are located with all dwellings bought up by urbanites. *Source: Vash Vybor,* 1994, 3:24. Reprinted by permission.

homeowners in the districts immediately adjacent to the Moscow province's boundaries, the recreational expansion of Muscovites has embraced areas even farther away. The recreational hinterlands of Moscow and Saint Petersburg have now extended to the south of the Novgorod and Pskov provinces. It is here that these hinterlands overlap. Urban vacationers have thus filled the void or social desert that emerged as a result of the same urban areas' long-term sucking away of the permanent rural population. After reaching its peak in 1990-91 the number of purchase deals on rural houses recently subsided somewhat.

As of now, the most vigorous process of urban intrusion into the countryside is associated with cottage construction. In 1992 a Presidential decree was issued concerning assignment of land "for low-storey and cottage construction." It was expected that by the end of this century over 140,000 single-family homes would emerge in the province of Moscow alone. Land was allotted mostly from agricultural holdings, as a result of which in the districts immediately outside the city limits of Moscow, cropland reduced by 20-50% in just two or three years. However, the ambitious construction projections are unlikely to be fulfilled. The most significant reasons: shortage of consumers with appropriate buying power and a system of mortgage loans still in its embryonic stage. In 1994-95, for example, only 80% of completed cottages were sold, even though in 1995, the scope of cottage-related construction activity had already been reduced by 40% compared with that in 1994.[7]

Polarized Landscape Around Moscow

Rural polarization has been one of the leitmotivs of this book; the phenomenon was taken up in our discussion of the Non-Chernozem Zone (Chapter 10), because it is there that core-periphery contrasts are most pronounced, and then summarized in Chapter 12. But any focus on *Podmoskoviye* (the area around Moscow), which is located inside that Zone, cannot escape revisiting the subject yet once again.

Northeast of Moscow lie industrial towns; all the space between them is filled with *dachas* and vegetable cooperatives and over half of all the houses in remaining traditional villages have been bought up by urbanites. West of Moscow are situated the most elite recreational sites. Southeast and east of the city, along with *dachas* and orchards, are the largest socialized farms. The agricultural component of land use increases with the distance from Moscow. Forest clearing restrictions in the province of Moscow have been in force since the reign of Catherine the Second (1762-1796); as a result woodlands account for 44% of the total land area, higher than almost anywhere else in Central Russia. Unfortunately maintenance of many forests leaves much to be desired.

Multi-functional land-use, and its vivid core-periphery spatial arrangement with an almost symmetrical system of a few ring roads and many evenly spaced radial spokes of transport thoroughfares, have long influenced researchers' thinking about the ideal models of rural land use in the Moscow hinterland. The most popular and well-known model of this kind was elaborated by Boris Rodoman (Fig. 13.2). It is based upon two ideas that became interlocked in the context of *Podmoskoviye*: a strict functional zoning ordinance and separating the network of human activities as far as

FIGURE 13.2 Rodoman's Polarized Landscape. 1: Urban architectural landmarks; 2: Urban industrial and residential areas with public utilities and transportation routes; 3: High-intensity farming; 4: Natural meadows and pastures, forestry, hunting grounds, and exurban state parks; 5: Nature reserves; 6: Recreational dwellings and tourist trails. *Source*: *Resursy, Srieda, Rasseleniye*. Moscow, Nauka 1974:160.

possible from the network of nature reserves. Rodoman's model has long served as a kind of idealistic template, a reference line which real physical planning could only asymptotically approach.

Land Is Being Sold After All

De facto buying and selling of land became possible around Moscow even before the introduction of any kind of legislation on land deals. Responding to the spontaneous development of a land market, the Government of the Russian Federation ruled in November 1994 that the normative price of land would be the previously set land tax times 200. Setting the level of land tax has been a prerogative of provincial administrations.[8] The administration of the province of Moscow adopted a system of function-specific price co-efficients: for urban uses, for *dachas*, for subsidiary farming, etc. The average price of one hectare of land in the province in 1995 was 60-120 million roubles or \$12,000-\$25,000; however, it was \$1,000-\$12,000 if the land was in the countryside and was intended for personal auxiliary farming, vegetable cooperatives, or hayfields; for land intended for the same purposes within the corporate limits of urban settlements a much higher price was set; for *dachas* the normative price was \$20,000. These prices, however, apply to new land deals. Charging prices for land parcels already in actual use and not violating zonal ordinances has been banned by Yeltsin's decree of March 7, 1996. Prior to that decree *dacha*-owners were required to redeem all their land in excess of 0.12 hectares from the state.

In actuality, however, the above price norms are only used as reference points at best, since the land market is in fact being regulated only marginally. So prices are "geography"-specific, depending upon supply and demand. In 1992 newspapers for the first time began to publish ads with land sale offers. Based on a sample of 1000 ads Filippovich compiled a first-ever map of land prices in the province of Moscow in 1992. Later such maps were regularly released by local real estate firms. Figures 13.3 and 13.4 show the geography of land prices in early 1994 and in late 1995. It appears that land within a 40-60 km radius is in highest demand. In addition, prices are influenced by proximity to a railway station (indicating the low level of private ownership of cars), availability of a forest and/or river nearby, and the condition of the rural infrastructure, especially the availability of electricity. The most prestigious radial spoke is to the west; along it the *dachas* of well-known writers, actors, and politicians are located. In the district of Odintsovo (west of Moscow), for example, 0.01 hectares cost \$4000 (that is, \$100,000 per acre) by the end of 1995. Prices are relatively high south and north of Moscow, \$600-800 per 0.01 ha; though they dropped somewhat in 1994. Land is cheapest in the industrial southeast, especially outside the line Chekhov-Voskresensk-Orekhovo: by the end of 1995, prices here were \$90-\$100 per 0.01 ha. Figures 13.3 and 13.4 show that land price differentials are on the rise both in absolute terms and as spatial gradients. As of now the price drop-off between the most and the least prestigious parts of *Podmoskoviye* is by a factor of 80. Contrasts are high even within single districts because of variable ecological situations and pressure on land.

Overall, the demand for rural land generated by urbanites in the province of Moscow is close to saturation: the wealthy have already bought land and the poor do not have the means to do so. According to some experts, in 1995 land sale offers exceeded demand by a factor of 20.[9]

Muscovites in the Province of Pskov

Seventy percent of the summer vacationers in the district of Nevel, in the southern part of the province of Pskov, are permanent residents of Moscow

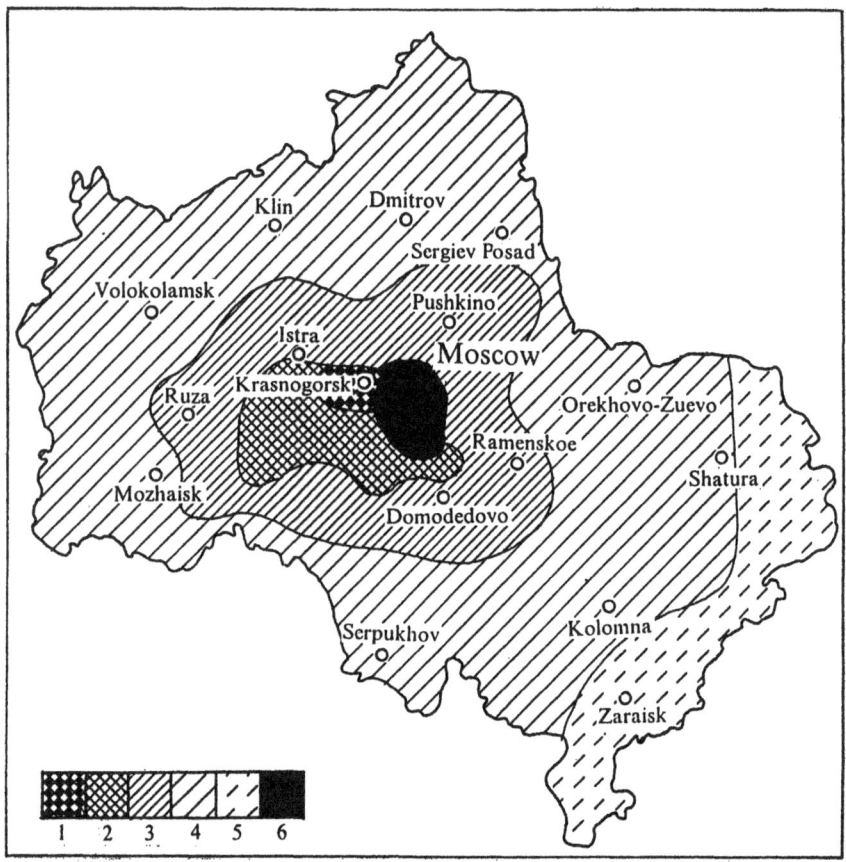

FIGURE 13.3 Price of 0.01 ha of Land for *Dachas* in the Province of Moscow in US Dollars, January-March 1994. 1: Over $5000; 2: $500-$1000; 3: $300-500; 4: $100-300; 5: Less than $100; 6: city of Moscow. *Source: Vash Vybor* 1994, 3:26. Reprinted by permission.

or Saint Petersburg, even though distances from these cities are 300 and 280 miles respectively. In summer, 7500 local dwellers are supplemented by 4000 urbanites.[10] Many villages with not a single surviving aboriginal inhabitant still show up on maps only because of recreational migrants. True, a reversed tide of rural migration (see Chapter 7) in the 1990s has been adding 150-200 people per year to the permanent residents of the Nevel district; however, the excess of deaths over births due to the very advanced aging of the local population outweighs the incoming flow, annually "subtracting" 200-250 people. But because of summer recreation and incoming migrants a land market is taking shape even in so peripheral a district like Nevel. The

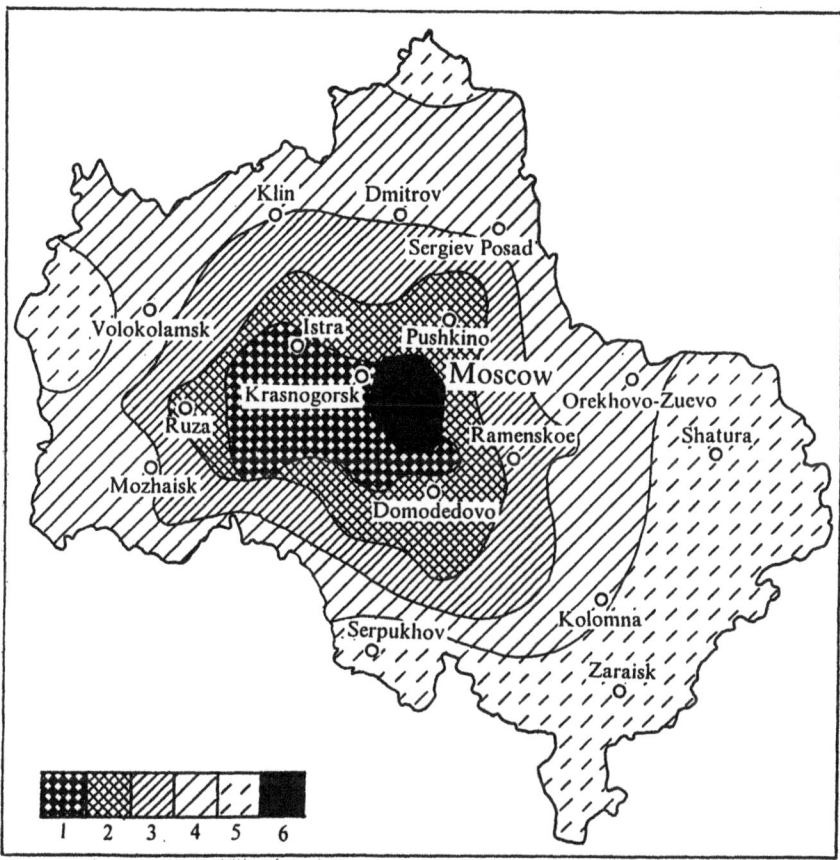

FIGURE 13.4 Price of 0.01 ha of Land for *Dachas* in the Province of Moscow in US Dollars, September 1995. 1: Over $5000; 2: $500-$1000; 3: $300-500; 4: $100-300; 5: Less than $100; 6: city of Moscow. *Source: Argumenty i Facty* 16-22 September 1995.

market is, of course, in its embryonic stage and is under the tight control of the local administration, which, therefore, is turning into a large landed estate-owner. Only the administration decides how much land to divert from decaying socialized agriculture, whom to assign it for recreational use, how much to sell it for, and where to direct migrants.

But even though nobody manages and releases information concerning these issues, a certain geography of land values is shaping up in Nevel. Land around the district town of Nevel is in highest demand: not only are farming yields higher here than elsewhere, and the rate of land abandonment lower, but also new settlers want to take root here, and the industrial enterprises of Nevel itself want to set up vegetable cooperatives nearby. Thus a miniature model of the Moscow region is being reproduced. Heightened demand leads to higher land values: the district administration, therefore, rejects requests for free allotments of nearby land even for vegetable cooperatives while it acquiesces to similar requests from the wealthy to assign land for cottages, although it is never known for sure who pays whom and how much. On the other hand, in peripheral rural soviets of the district any urbanite can get 0.15 hectares free in exchange for only a commitment to conduct farming on that land. Also land market pressure is correlated with population composition. For example, in the rural soviet of Plis, the closest to the district town, the percentage share of seasonal recreational migrants in the population (43%) is almost equal to permanent residents' share (45%; the rest are recent settlers); in the more distant Glubokoozersky rural Soviet, the permanent population accounts for 60% of the total summer population, vacationers for 20%, and there is a heightened proportion of new settlers (20%) directed here by the district administration.

We will take a closer look at the resettlement process in the following chapter.

Notes

1. Levinson, Alexander. "Goroda i Ogorody." *Izvestia,* 29 May 1993.

2. *Sravnitelnye Pokazateli Sotsialno-Ekonomicheskogo Polozheniya Regionov Rossiiskoi Federatsii.* Moscow: Goskomstat 1995, pp. 244-246.

3. Ioffe, G.V., and Fingerov, G.M. "Selskoye Khoziaistvo i Recreatsiya." In *Territorialnaya Organizatsiya Proizvoditelnykh Sil kak Factor Ekonomicheskogo Razvitiya* (edited by G.V. Ioffe). Moscow: IGAN SSSR 1987, pp. 122-138.

4. *Territorialnaya Organizatsiya Otdykha Naseleniya Moskvy i Moskovskoy Oblasti.* Moscow: Nauka 1986, p. 73.

5. Filippovich, Liudmila. "Gorozhanie za Gorodom." *Vash Vybor* 1994, No 3, p. 25.

6. Ibid., p. 24

7. Vakhitov, R. "Kottedzhnyi Rynok Ugnetion Rostom Sebestoimosti I Ischeznoveniem Sprosa." *Finansovye Izvestiya* 1 March 1996.

8. "Zemelnyie Spory." *Rossiiskaya Gazeta* September 8 1995.

9. Barmina I. "Sotka za Sotku." *Argumenty i Facty* 16-22 October 1995.

10. Filippovich, L., Pavlova, I. "Kto Budiet Zhit' v Opukhlikakh?" *Vash Vybor* 1995, No 5, pp. 22-24.

14

Re-Settlers: A New Diaspora?

Incoming migrants in the province of Pskov, the case study region for this chapter, outnumbered outgoing migrants in 1989-1993 by 34 persons per 10,000 population. In comparison, the average net migration for Russia's *oblast*-level sub-divisions was plus 14 in 1994.[1] Forty percent of all the Pskov newcomers are from the nearby Baltics, Saint Petersburg, and the Leningrad province.[2] And even though Pskov is no leader in terms of attractiveness to migrants, this province (*oblast*) is of interest to this study because it is fairly typical and also because it has experienced an abrupt change in its geo-political situation: the emergence of a national border. So traditional rural problems are enmeshed in some new processes.

Where Do People Go?

The industrial boom of the late 1800s sidestepped Pskov. Even during the splash of socialist industrialization the area remained overwhelmingly agrarian. However, during the last thirty years the proportion of agriculture in the total output of the economy dwindled from 57% to 19%.[3] Yet even now the overall value of fixed assets in agriculture exceeds that in industry.

Land used for agricultural purposes accounts for 30% of the total land area of 55,300,000 sq km. Fifty-eight percent of the working population of the countryside is engaged in agricultural production units; 15% more work in personal auxiliary farming only. Most of the retirees in the area (35% of the rural population!) also work subsidiary plots. Thus the overwhelming majority of country folk deal with farming in one way or another. In the 1990s, agriculture absorbed one-third of all the investment in the province. The economy of the whole province has been a recipient of federal aid, with the bulk of the subsidies poured into agriculture.

A productivity drop-off of the kind that exists between the province of Pskov and neighboring Estonia (a two-to-threefold difference in output per unit of land) has nothing to do with natural conditions. Pskov is no worse off in this regard. As pointed out in Chapter 4, an indicator of those conditions traditionally used in Russia is the yield of grains recorded at state testing stations in a year with statistically normal weather and under mainstream farming technology. For the province of Pskov this level ranges between 15 and 20 centners per hectare, with an average of 19; formerly Soviet Estonia was characterized by exactly same average. However, in the 1970s the actual grain yield in Pskov's collective and state farms was only 12 centners per hectare; in the 1980s, it was 9-11 centners/hectare, and in 1991-94, 8-11. Across the border in Estonia, grain yields in the 1970s and '80s ranged between 24 and 28 centners per hectare.[4] The application of mineral fertilizers per unit of arable land in the '80s in the province was 60% of the Estonian level. However, in this regard Pskov has been better off than most other Russian non-Chernozem provinces with the exception of Moscow and Leningrad. Pskov is also one of Russia's leaders in capital/labor ratio, that is, in the monetary value of agricultural fixed assets per one worker.

It is social conditions that are the most significant factor in the province's under-achievement. For example, there are 21 km of paved roads per 1000 square km in the province of Pskov versus 328 km in neighboring Estonia. In terms of flush toilets in rural residences, the province is 52nd among Russia's 89 civil divisions, and in terms of central heating it is 65th.

From Overpopulation to a Demographic Desert

In the past, the province of Pskov stood out for its high number of rural settlements, their smallness, and the short distances between them. The latter is no longer the case, as rural depopulation took a heavy toll on the countryside, and myriads of villages were abandoned. In 1900 the population of the province was 1,191,000; its density, 25 persons per sq km, exceeded European Russia's average (20 per sq km). Eighty-six percent of household heads earned their income from farming. Arable land accounted for 26% of the total land area (compared to 15% today). The 1861 Emancipation of Serfs Act ushered in the era of speedy transformations. Whereas in 1877, two-thirds of all land belonged to landed gentry, ten years later only one-half was in their possession, and in three more years as little as one-third. It was not individual farmers, however, who were largely taking over the land but rather peasant communes and comradeships. In the very beginning of this century there were only 282 private farms (*yedinolichniki*) in the whole province and 874 comradeships with an average of five-to-six households in each. The principal benefactor-crop for the province was flax, occupying one-third of Spring-sown cropland. The development of commercial agriculture and trade was especially facilitated by the construction

TABLE 14.1 Agriculture in the Province of Pskov in the 20th Century.

	1900	1925	1941	1961	1971	1981	1985	1992
Flax Output in thousand tons								
	41	ND[a]	25	28	30	9	8	3.4
Potato Output in thousand tons								
	ND	ND	1441	799	997	551	407	417
Number of Tractors in thousands								
	ND	ND	3.5	6.3	11.8	19.3	20.7	15.2
Grain Yield in centners per ha								
	8	8	9.4	5.2	12.7	11.5	12.3	10.7
Potato Yield in centners per ha								
	104	74	127	92	145	101	82	103
Flax Yield in centners per ha								
	3.5	2.7	1.6	2.4	3.8	1.2	1.8	1.2
Milk Yield per Cow								
	1800	800	940	1986	2522	2071	2140	1642

[a]ND -- no data.

Sources: Nefedova, Tatiana. "Byl i Lion i Khleba Vdovol." *Vash Vybor* 1993, 6:27; *Narkhoz RSFSR* 1962. Moscow, Gosstatizdat 1963.

of a railway line between Saint Petersburg and Warsaw and of the main flax roadway between Rybinsk and Riga. Pskov was the western outpost of a huge flax-growing region, so it collected not only home-grown flax but also flax from the southern part of the province of Novgorod, the western part of Tver, and from northern Byelorussia.

Pskov owed its flax-based fame to its agrarian overpopulation. Flax is very labor intensive, and at that time it was even more so because there were no harvesting and flax-pulling combines available. However paradoxical it may sound today, it was the very labor-intensiveness of flax, not just its profitability, that attracted peasants. However, it was not only flax that was produced in abundance in the province before the communist revolution. The turn of the century was one of the few heydays of the province's farming: Pskov was not only able to meet local demand for grain but had a surplus of it. As for the province's agricultural output, many farms of today would envy it (Table 14.1).

Historically one of the cores of ethnic Russian settlement and colonization, along with the adjoining Novgorod and Tver lands, the province now has a population density on a par with many areas in Siberia and in European Russia's north. The rural population density in the Pskov province now is only 5.3 people per square km, down from 25 in 1926. Even worse, no

TABLE 14.2 Population of the Province of Pskov.

	1926	1939	1959	1970	1979	1989	1993	1994
Total	1678	1551	953	875	850	846	840	837
Urban	128	200	258	373	470	534	542	542
Rural	1550	1351	695	502	380	312	298	295

Sources: Demographic Yearbook of the Russian Federation. Moscow, Goskomstat 1994:13-21; *Chislennost Naseleniya Rossiiskoi Federatsii po Gorodam, Rabochim Posiolkam i Raionam na 1 Oktiabria 1995.* Moscow, Goskomstat 1995.

other province of the Russian Federation has as large a negative rate of natural increase of its rural population: in 1993 the difference between the number of births and the number of deaths in the countryside was -20.8(!) per 1000 persons. Compared with -4.9 in the Russian countryside on average,[5] this is clearly a lethal gap: it could lead to total depopulation in less than 70 years.

The 1994 rural population of the province was 25% of what it was in 1897, and 19% of that in 1926 (Table 14.2). In comparison, the neighboring Leningrad province and the Baltics have retained 55-60% of their 1926 rural population levels (Figure 14.1).

Rural population dynamics have proved to be a stronger predictor of agricultural output than investment has. The districts of Pskov and Velikiye Luki (the two largest cities of the province) have lost fewer people through migration and negative natural increase than any other district in the last 20 years, and these two areas also top all the rest in agricultural output per unit of land. On the other hand, the districts in the province's middle, located at some distance from both centers, have lost half of their rural population since 1974 and are also languishing in farm production. And that is despite the fact that the distribution of agricultural investment in the province has been spatially equitable, unlike in many other non-Chernozem provinces where exurban areas have been favored by investors. Due to the peripheral position of the two largest urban centers of the province, it is its middle that appears "outlying," along with the north-east, and it is here that the reduction of arable land has been the largest. Aside from its sheer proximity to urban centers with their jobs and retail stores, the cities' countryside appears to be relatively attractive because of a higher quality of rural infrastructure. In the Pskov district, for example, about half of all rural residences enjoy flush toilets and central heating. In comparison, in the districts located in the middle of the province, only 20-40% of residences have these facilities, and in the north, only 10-20%.

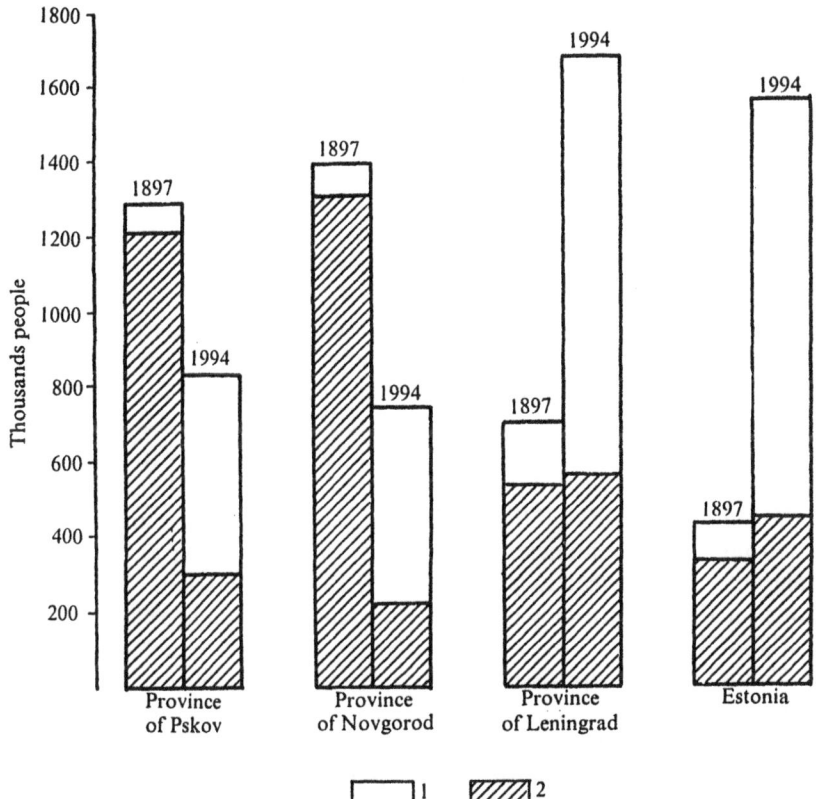

FIGURE 14.1 Population of the Province of Pskov and Its Neighboring Regions. 1: Urban population; 2: Rural population. Based on: *Rossiya. Entsyklopedicheskiy-Slovar.* Saint Petersburg, Brokhauz i Efron 1898:106-115; *Chislennost Naseleniya-Rossiiskoi Federatsii po Gorodam, Rabochim Posiolkam i Rayonam na 1.01.1994.* Moscow, Goskomstst 1994:3; *Statistika Aasti Raamat.* Tallinn 1994:5.

Who Are These Newcomers?

In the 1990s the province of Pskov, for the first time in decades, began to attract migrants. A government-sponsored "resettlement" program has been given high priority by the provincial administration because of the general consensus that the province is too thinly populated.

The program includes five components entitled Individual Migration, Communal Migration, Ethnic Germans of Russia, Northerners, and Job-

Creation for Re-Settlers. According to the planners' design, the principal emphasis was to be put on communal migration, that is re-allocation of close-knit groups of mainly ethnic Russians from the ex-Soviet republics. However, it is actually "individuals," that is, unrelated persons, who generate the bulk of the incoming flow. Though the communal resettlement segment of the program envisioned only about 2000 migrants a year in 1993-95, as many as 30,000 individual volunteers registered annually. By and large, the re-settlers must rely on themselves. The only asset that the province of Pskov supplies them with is land: two hectares per volunteer, and twenty hectares in case a newcomer wants to establish a private farm. Land is assigned free of charge.

About one-half of the actual resettlers are from other provinces of Russia. In 1991-93, the proportion from the Baltics increased fourfold and now accounts for 25% of the whole inflow; together with migrants from the non-Russian Slavic republics, the flow to the Pskov countryside accounts for over one-third of the migrants arriving in the province. Fifteen percent of the newcomers originated from southern republics. About half of all newcomers, regardless of their origin, are people with roots in the province returning to their homeland after many years of being elsewhere. The migrants are overwhelmingly Russian. When they are asked about the reasons for their return they point to worsening inter-ethnic relations and to social problems.

This all sounds impressive, but there is a rub: about 80% of the migrants from other ex-Soviet republics are *urbanites,* according to their last place of residence, and their percentage share is even higher among migrants from the Baltics and from the province of Leningrad (Table 14.3). It is they who are being offered land. Those who have no other resettlement option take it. During only nine months of 1993, about 7000 people arrived in the urban areas of the province and about 6000 in its countryside. The rural population was thus substantially replenished. Compared with this increase from outside the province, the net urban-to-rural migration *within* the province itself was minuscule: +754 persons in 1993. Overall, whereas in 1979-1989 the countryside experienced a net outflow of 67 people per 10,000 rural population, in 1989-1993, there was a net inflow of 34 people per 10,000.[6]

Every cloud has a silver lining; in this case, a crisis of truly national proportions turned the tide of migration. In the past many efforts to recruit re-settlers proved fruitless. Contract agents used to make forays into labor-surplus regions like the Northern Caucasus, selecting the most promising volunteers. Some of them got lost en route after receiving cash advances to cover moving expenses; others vanished later.

Everything changed in 1992 when a provincial Office of Employment in Pskov was besieged by refugees from all corners of the fallen empire. The flow soon subsided, since the province was unable to finance residences for

TABLE 14.3 Percentage Composition of 1992 Migrants into the Province of Pskov.

	Total	*From Saint Petersburg and the Leningrad Province*	*From Estonia*	*From Latvia*
Coming FROM the Countryside	23	11	9	23
Coming FROM Urban Areas	77	89	91	77
Total	100	100	100	100
Moving INTO the Countryside	47	52	61	65
Moving INTO Urban Areas	53	48	39	35
Total	100	100	100	100

Source: Stiepanov, V. "Smotritie Kto Priyekhal." *Vash Vybor* 1993:14.

all those interested. The resettlement allowance itself was highly insufficient: in 1992 it was 3500 roubles per household head plus 1700 roubles per household member. Compare this with the 10 million roubles it cost to construct a single cottage. The ability of newcomers to cover this cost by selling their former homes was and still is fairly limited, because both in the Baltics and in the southern republics demand for housing is short of supply, which translates into low prices. So individually managed resettlement is an option for either the relatively wealthy or for those still having relatives among country folk. The latter resettlers, of course, have a certain advantage: they are more prone to be absorbed among the locals, they do not hang out with other migrants, and do not set themselves against the locals.

Given the above conditions, the resettlement program aims at having the potential migrants themselves construct cottages. They work in stints with one brigade giving way to another. Small industrial enterprises are being incorporated into the physical plans of the newly-emerging settlements. Nine such settlements are already under construction, and twelve more are being projected. Where they are located is a crucial question in terms of their prospects for taking root and surviving. Unfortunately the answer is: the provincial administration has located them in outlying sites not conducive to new settlement. Their average distance from district centers is 35 km. They are being built by construction teams from Latvia and Murmansk composed of people still holding onto their current residences in their home areas. This is understandable as for the time being there is no other place for these people to live. But the local officials suspect that in the future there will still be no real replenishment of farming labor; what the province is likely to receive, in their view, is new *dachniks,* or owners of second, usually recreational, dwellings. And that, for local officials, is no good at all. However, in all fairness, the construction teams build residences for the locals as well. For example, the "Latvians" (ethnic Russians from Latvia) have recently bought from a collective farm in the Ostrov district eighteen shells, seven of which they have finished and sold back to the farm while the remaining ones they are finishing for themselves. Still, as our survey has shown, no less than 60% of the resettlers would like to take root either in the city's countryside or in the westernmost districts of the province, which they or their ancestors had left for the Baltics, and only an insignificant portion of these returnees have retained skills in working land.

The geographical distribution of the newcomers from different places and their starting positions in the province are different as well. For example, migrants from Saint Petersburg and the Leningrad province are generally young (over half of them are below 30 years of age). As many of them have failed to find their niche in or around the second Russian capital, they are returning home; they are mostly unmarried, active, well-educated, and tend to settle down in towns. Resettlers from the Baltics are generally older, about two-thirds of them are married couples over thirty years of age, they find it more difficult to adjust, and being urbanites they are often unable to obtain a residence in towns and thus settle in villages. Since many of them or their ancestors had been from the western part of the province, they prefer to move there. Thus, over one-third of the newcomers from Estonia settle in the districts of Piechory and Gdov along the border with Estonia, while 40% of the migrants from Latvia find their new homes in three districts along the Latvian border. Migrants from the southern republics tend to concentrate in the province's middle or in its south.[7] So groups of ethnic Russian migrants from different areas appear to set themselves apart from each other.

Adaptation to life in the new area occurs in three stages. The first stage is tainted with euphoria and exaggerated optimism as it seems to the migrant that the main difficulties have already been left behind. This stage, however, gives way to gloom and to a state of confusion and frustration. The third stage, stabilization, may or may not come at all. The more obstacles newcomers confront, the more they are prone to set themselves against the locals. As for the latter, antipathy toward newcomers may result from the new rules of behavior the migrants seem to be setting. This is just the case in the province of Pskov where new migrants generally have a more entrepreneurial bent compared to the locals and drink less.

As a result some features of diaspora are evident in the attitudes of these ethnic Russian returnees-newcomers: their close-knit communities impenetrable by the locals, a peculiar combination of pariah and superiority complexes, and an acute sense of purpose. Their way of living is different from that of the locals, and their labor productivity is much higher. However, it is exactly this set of features that has traditionally alienated the Russian peasant, just as he turned away from the classic diasporas created in Russia by Greeks, Armenians, and Jews. While urban areas of Russia are no longer totally hostile to them, a Russian village is still in the habit of rejecting "aliens" even if they are actually ethnic Russians. It would be regrettable if local authorities fail to tap into the energy of newcomers and not help turn it to the benefit of the Russian countryside.

Notes

1. *Demographic Yearbook of the Russian Federation.* Moscow: Goskomstat 1994, p. 407.

2. Stiepanov, V. "Smotritie Kto Priyekhal." *Vash Vybor* 1993, No 6, p. 14.

3. The industrial profile of the province includes construction materials, agricultural implements, food and flax processing. After the war some machine-building enterprises were put in operation, and most recently, electronics enterprises have been added. However, even now the structure of the labor force is early-industrial in contrast to most provinces of Central Russia where the structure is hyper-industrial (and, in the province of Moscow, transitional to post-industrial).

4. *Selskoye Khoziaistvo SSSR.* Moscow: Goskomstat 1988, p. 136.

5. *Demographic Yearbook,* p. 63.

6. Ibid., p. 406.

7. Stiepanov, V. "Smotritie Kto Priyekhal." *Vash Vybor* 1993, No. 6, p. 15.

15

Disappearing Crops:
A Case Study of Flax in the
Province of Kostroma

Russia used to be the world's premier producer of flax. The country's flax-growing band stretched from the provinces of Pskov and Novgorod eastward. Today this branch of agriculture is in a shambles. Even worse, it is rapidly disappearing although Russia still accounts for 43% of the world's flax output. However, of 2200 flax-growing socialized farms in 23 provinces of the Russian Federation, only 20% believe it is worth specializing in this product in the years to come.[1]

In characterizing the value of flax, figures like the following are quoted: the flax grown on one hectare can produce 2400 square meters of fabric, and 160 kg of linseed oil, and, from the waste products, construction slabs, cardboard, resins, and vinegar. All this is all possible if the yield is 10 centners per hectare (as it is in Holland, Belgium, and France) and if the processing technology is state-of-the-art.

In the province of Kostroma, one of Russia's traditional flax-growing regions, flax fibre yield is the '90s has been 2-4 centners/ha and the non-labor production costs per hectare, including application of four or five required fertilizers, are twice as high as for a field of grain. Labor input per hectare is also colossal: 100-150 man-hours, five times that per one grain-sown hectare. How can this demand for labor be met in Kostroma, one of Russia's most depopulated provinces? Especially considering that 80% of the labor is required at the time when potatoes are harvested. No wonder that the area under flax in Kostroma has been contracting. In the 1970s it was 70,000 hectares, in 1995 barely 6,000. The biggest slump occurred in the last two years. Before, the high price of flax made it profitable even with the seemingly intractable problems growing and harvesting entailed. But in 1992-93 the cost of producing flax straw grew twice as fast as its selling price. Flax thus became unprofitable in the province. However, when price hikes aimed at ensuring a reasonable profit margin occurred in the last two years, flax in Kostroma, and in Russia in general, lost its competitive edge: instead of purchasing 1 kg of domestic long-fibrous flax for $2.50, one could obtain the same amount in Belarus and the Ukraine for $1.50 which many textile factories now do.[2]

So how did it come to this?

A Bit of History

The 1912 book by Rybnikov[3] is an excellent source of historic data on Russian flax. In the 13th-14th centuries only Novgorod and Pskov grew flax commercially. By the 16th century the flax-producing area had greatly expanded eastward gravitating to the rivers flowing to the White Sea across which flax was shipped abroad. At that time domestic processing of flax resulted only in coarse fabrics; delicate fabrics were imported.

In the 18th century Peter the Great implanted state-owned flax-processing factories in the Central Industrial region: in the provinces of Moscow, Yaroslavl, Vladimir, and Kostroma. At the same time, some landlords developed processing on their estates that could produce delicate fabrics. As flax-manufacturing machines were introduced in the 19th century, the manual labor that was thus released provided extra employment that could be applied to the cultivation of flax. This facilitated the expansion of cropland under flax. Under the rural population density in the non-Chernozem areas, 30 persons per sq km, labor was over-abundant. So peasants were interested in flax and its high labor-intensity did not scare them away. On the contrary, it attracted them because it allowed maximum use of able-bodied persons and not just in the summer.

In the meantime cropland under flax began to shrink in Western Europe. Whereas in the 1860s, England held 229,000 flax-sown acres, in 1908 it had only 47,000. During the same time France reduced its cropland under flax from 264,000 acres to 64,000, and Germany from 473,000 to 93,000 acres. The main reasons for this reduction were competition from cotton, which was more suitable for machine-based processing, and competition from cheap Russian flax. By the end of the 19th century Russia held 81% of the world's total area under flax. However, because of low yields, the percentage share of output was much lower, 70%, and was dominated by low-grade fibre. Almost two-thirds of Russia's output of flax was exported: to Belgium, Britain, France, and Germany. Domestic factories processed only 72,072 tons, or slightly over one-sixth of the output in 1913.[4] By the end of the 19th century the cropland under flax in Russia peaked as did flax spinning; in 1880, 35% of all the fibre was processed in the province of Kostroma. However, this heyday did not last long because in Russia, not unlike Western Europe, the percentage share of flax fabric, among all fabrics produced, soon began to decline (to as low as 10% early in this century). The production costs of cotton fabrics were lower, waste products were not as abundant and were easier to deal with, and the technology was simpler. So flax prices dwindled and the output was scaled back drastically: whereas in 1913, Russia's output of flax fibre was 417,636 tons, in 1921 it was only 43,158 tons.[5] This figure, however, may have substantially underrated the domestic production because in the 1920s peasants were not willing to sell flax to the state which set a very low purchasing price. So the country could sell no more than 40,000 tons of flax fibre abroad while many more tons were in sacks in peasant households. That situation led to setting up institutional conglomerates in which flax-growing farms and processing factories were under one management. It also allowed the creation of large flax-growing farms and accelerated machine-based processing, which appreciably cut back on labor-intensity. But other problems emerged: the amount of useable fibre was significantly reduced. Whereas under manual scutching the percentage share of waste did not exceed 10% of the flax straw, the amount of waste produced by machines was 50-60%; also the quality of fibre worsened. Table 15.1 shows that while the flax-sown area had expanded before the war years of 1941-45, the output had actually dropped.

After the war two main tasks were set: to expand the network of primary flax-processing sites producing fibre from straw and to improve flax cultivation and processing technology. Whereas the first task was accomplished (currently there are 230 primary processors in Russia, though the bulk of them had come to a halt due to a lack of raw material), the second task was not, even though it was of vital significance for flax-growing regions, where the rural population density dwindled to 5-8 persons per sq km and the elderly came to prevail.

Although both in the 1980s and 90s there has been no shortage of government programs aimed at resolving lingering flax-related problems, problems still exist. Some promising signs showed up, however. For example, there has been a substantial growth in yields in the last two years attributable to heightened attention to the choice of growing fields (flax is very demanding in terms of soil preparation and should be the seventh or even the eighth crop in rotation.) Let us follow the whole production chain from field to retail store using Kostroma-grown flax as an example.

Russia Miniature

The province of Kostroma is located at the crossroads between three macro-economic regions of Russia: the Industrial Center, the North, and Volgo-Viatka. One could say that the province looks like a miniature model of Russia as a whole. As does Russia at large, it has a sizable southwest-

TABLE 15.1 Cultivating and Manufacturing Flax in Russia.

Year	Flax-sown Area in thousands of ha	Output of Fibre in thousands of tons	Flax Fibre Yield in centners per ha
1887	2200	567	2.6
1900	1600	467	3.0
1910	1100	410	3.8
1925-27	1264	272	2.0
1940	1500	239	1.6
1970	720	248	3.4
1981-85	550	152	2.9
1986-90	418	124	2.5
1991	328	102	3.1
1992	327	78	2.4
1994	134	55	4.1
1995	120	72	4.1

Sources: Nefedova, T. "Shagrenievaya Priazha." *Vash Vybor* 1994,2:16; *Finansovye Izvestia* 1996, 30 May: 4.

northeast climatic gradient and is colonized fairly unevenly. The city of Kostroma (304,000 residents in 1994), located in the southwestern corner, holds 37% of the total population, and together with the three nearby districts stretching along the Volga River, over 50% of it. Just as in the Russian heartland as a whole, which gravitates to meridional waterways leading "from the Varangians to the Greeks" (*iz variagov v grieki*), the west-leaning historical core of this province focuses on north-south trending rivers. Also the vast forested area to the northeast of the Volga is a minia-ture copy of Siberia, and just as in the "original," a latitudinal railway line crosses rivers. Even the structure of the economy is close to the overall Russian economy, although the proportion of the rural population in the province of Kostroma (34%) is higher than in the nation (27%).[6]

Kostroma has sundry local problems that are typical for Central Russia. However, one of the peculiarities of the province is that only 38% of the economically active rural population works in agriculture. In comparison, in the province of Pskov 58% do farming work, and in Russia as a whole, 53%. There are three pivotal economic sectors in the countryside of Kostroma: flax-textile, dairy, and forestry. All three are crisis-stricken in the 1990s so that the province's overall economic health is partially dependent on enter-prises not as aboriginal as the above three, that is, on power engineering, and, to some extent, on machine-building.

The proportion of arable land in the province ranges between 41% in the southwest and 4% in the north. The spatial layout of the rural population density is a close correlate of land under plow: it ranges from 26 people per sq km around the city of Kostroma and 21 in the district of Krasnoye Selo, to just 3 in the north. However, the extreme northeast includes some arable islands in the ocean of the forest vastness: the proportion of arable land here is 12-14%.

Districts close to the city of Kostroma have been given priority in terms of fertilizers and other investment, a factor that has influenced returns. For example, in 1986-90, the district of Kostroma led in grain yields (26 centners per hectare), with other southwestern districts close behind (20 centners-/hectare). At the same time, in the south and southeast grain yields ranged between 10 and 14 centners, and in the north they were merely 7 to 9 centners. In the 1990s only the southwest has retained decent grain yields. The output of other crops per unit of land correlates with that of grain.

However, heightened capital inputs by themselves have not led to higher returns. Labor supply, amenities, and other factors matter. For example, only the southwest has stable, 4-5 centners/hectare yields of flax fibers; in the majority of other districts yields range between 1 and 2 centners, although capital inputs per hectare of flax-sown land in the southeast exceed those in the northwest. As a result, flax turned out to be most profitable in the southwestern corner of the province where it covered 6-7% of arable land. Overall the output differences per 100 hectares between the southwest (around the city of Kostroma) and the east are on the order of ten and over (Fig. 15.1).

Crisis over the Whole Production Chain

The best conditions for flax straw are 18 degrees Celsius, no sharp ·temperature drop-offs, and a humidity of 50-60%. It is considered optimal in the province to harvest and spread out flax in August so that it can lie 16-20 days on the field. If flax is harvested in September, a longer lying period is required and quality usually goes down, with fibre being graded no higher than 11 out of 70 (the world's highest grade; 30 is the highest grade now available in Russia). Such fibre is used only for coarse fabrics.

If flax is harvested manually, 44 laborers per 100 ha are required for harvesting alone. But even when flax-pulling is mechanized, a lot of manual operations still remain: binding up sheafs, gathering them in shocks, and drying, threshing, sorting, and loading them. In order to mechanize most of this, one has to have no less than three combines per 100 hectares, two scutching swords, three blast engines, etc. Farms could not afford all this even in earlier days; they can afford even less today. And yet given the current catastrophic reduction in flax-sown area, manual harvesting, aided by local retirees, is not the most important problem. On the other hand, it

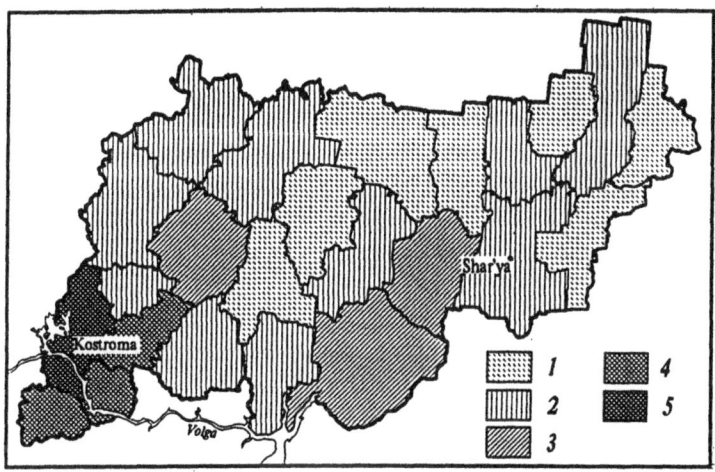

FIGURE 15.1 Gross Agricultural Output in Thousands of Roubles per 100 ha of Agricultural Land in the Province of Kostroma (1994). 1: 13-20; 2: 20-30; 3: 30-50; 4: 50-100; 5: 164.

has been calculated that flax could be profitable under capital-intensive technologies on large, 200-300 ha, fields: production costs per unit of produce are lower in that case.[7] Indeed in the districts where such fields prevail and the percentage share of flax in the total cropland is at its highest (6% in Krasnoye Selo and 7.5% in Nerekhta against 2.2% on average in the entire province) the production costs per unit of produce are well below average. In these districts mechanized operations are more frequently used and traditional skills have been maintained. However, it is exactly here, where flax used to be profitable that the degree of confusion generated by price dynamics and the sales crisis is higher than elsewhere. Because the area under flax has been reduced and because very low paid retirees do much of the harvesting (labor comprises about half of the cost of flax production) flax ought to have remained profitable. But everything rests on prices and sales. For example, in order to harvest all the flax in 1993 it was necessary for farms to hire urbanites. But it was unclear how much of what they produced primary processors would buy, the cultivation costs were likely to exceed sales prices, and subsidies promised did not arrive in time. In such a muddle not to harvest flax seemed better than harvesting it, so half of it remained in the fields. In 1994 the subsidies were promised, to the tune of 400,000 roubles per one ton of fibre. Such subsidies would ensure that the ratio of revenues-minus-costs to costs (the Russian *rentabelnost*) would be 30-60%. On mass produce such as grain this kind of ratio would be just fine, but on something like flax, that requires more efforts than its share of cropland would suggest, it is not very reassuring. So farms that used to earn the bulk of their profits from flax cut back on it even more.

The province's primary flax processing sites can produce 15,000 tons of fibre annually (the largest of them, *Sholokhovsky* has one-third of the total province's capacity), but the flax-sown areas that remain in the province can furnish only about one-tenth of flax straw needed for this figure. It is clear that three processing sites, *Sholokhovsky, Krasnoselsky,* and *Nerekhtsky,* are sufficient. But what to do with the remaining thirteen? Over half of them do not function any longer. Since flax straw is like fluff, hauling it is costly. In 1990, because special subsidies for transporting flax straw were provided, unreasonably long hauls were made. But then it dawned on somebody in Moscow that trips over 60 km are prohibitively expensive. This may be true but who will handle the produce in outlying districts where straw processors have come to a halt? Kostroma is a vast province where distances of hundreds of km are perceived as normal. Also the primary processors themselves cannot survive without subsidies. They complain that flax-mills, the next link in the production chain (whereas primary processors, or so-called *lnozavody*, produce fibre from flax straw, flax-mills turn fibre into fabric), set prices too low. As a result, the flax-mills' share of locally procured fibre has been reduced to as low as 10%. In these ways the whole chain turns out to be truncated: the share of local produce in what local flax-mills consume is too low and the problems of flax-growing are too formidable.

And the flax-mills have grievances as well. There are three powerful mills in the city of Kostroma, one of which, *Bolshaya Lnianaya Manufactura* (Great Flax-Textile Mill), has been in operation since 1866 and used to thrive on local supply. Up until 1992, 30 percent of the fibre that it turned into fabric was from the province; in 1992-93, the share of locally procured fibre dropped to 18%. Though its quality rank does not exceed 10, it is still better than what is transported from the provinces of Smolensk and Tver, where 40% of Russia's fibre is now produced. The flax-mill's share of Byelorussian and Ukranian fibre, which is of higher quality and cheaper than Russian fibre, had grown to about 50% in 1993. But there are problems here as well: the flax-mill does not have enough cash; with local processors of straw it can operate on barter, which is more difficult with Byelorussian and Ukranian sources.

However, the most acute problems that flax-mills face are the same the prior elements of the production chain face: prices and sales. At the flax-mill in Nerekhta the weaving section is idle: sales are in free fall; produce in the form of packing materials and sacks valued at half a billion roubles has accumulated in the warehouse. *Bolshaya Lnianaya Manufactura* has halved its output, which is now mostly composed of fabrics for linen (20% of which is for foreign export). As for prices, the flax-mills' reaction to the grievance of the preceding elements of the production chain are simple: if prices were set to the satisfaction of all these elements, the price of one square meter of fabric would increase by 20,000-30,000 roubles. No one would buy such a fabric, so the flax-mills would go out of business. As of now, however, locally produced flax blankets are being sold in the city of Kostroma for 1800 roubles per square meter. Earlier, mills produced only the same 35 different types of fabrics; now retailers urge them to vary their types every now and then and they bargain for prices. Before, different flax-mills were in close touch with each other, learned from each other, and agreed on the "division of labor." Now each secretly determines its range of goods and is equally secretive about its technological pursuits. All of this is so hard to get used to, mill managers complain.

What Comes Next?

Dismal jokes circulated at the Great Flax-Textile Mill in the city of Kostroma in 1993: "Compared with others we are tough. Those at the heavy cloth mill will go down the drain by 1994-95, while we will still languish for a couple of years." Neither at the flax-mill itself nor at primary processing plants are any long-term supply and sales contracts available, so prospects for the future are totally uncertain.

Actually the mill equipment would allow mills to switch to cotton. But the principal supplier of cotton yarn, the Kostroma Cotton-Spinning Mill (built in 1846, it had processed flax up until this century and then switched to cotton) is also on the verge of halting operations. While severed ties with cotton-growing republics could be restored, the factory lacks the financial means to buy cotton even at 30% lower than the world's prices; moreover it cannot pay for electricity, and its workers' pay checks are long overdue.

Just as in agriculture, the root cause is not the new economic relations as such or severed trading ties but exorbitant production costs that have been put up with for a long time; heavy taxation and inflation take their toll as well. Flax-mills are begging provincial authorities to lower taxes and to subsidize their "social sphere," that is, factory-owned housing stock, day care, recreation sites, etc. Then, they say, they will get themselves out of the mess. Some kind of assistance is indeed highly desirable, especially for the upper element of the production chain, the flax-textile industry, which is a valuable creation of the last 150 years. It is a unique industry with a highly skilled labor force that would be a pity to lose.

And what about the lower elements of the chain? The main factors that slowly but steadily killed the production of flax-fibre were: rural depopulation; the low mechanization level of harvesting operations; and low level of farming technology and management. As in other areas of agriculture, the remedy is not simply increased financial backing; it is abundantly clear that one must first identify the fittest enterprises and channel help to them. Such a strategy has long been against the grain of the tenacious Russian obsession with equity. For the fittest to survive, this obsession has to die first. Perhaps now this death will occur spontaneously, during the course of developments like those in the flax-growing branch of agriculture, and Table 15.1 seems to confirm this.

On the 28-29th of May 1996 a conference was held in Moscow with the title "Stabilization of the Textile Industry In the Commonwealth of Independent States. The Example of Russia's Flax-Producing Composite." Prospective domestic and foreign investors, scientists, and professionals of textile manufacturing took part in the proceedings. At the conference the key elements of the Federal Investment Program, "Development of the Flax Composite of Russia in 1996-2000," were made public. The cost of the program is 8,143,000,000 roubles of which manufacturers and farmers are to receive one half each under the assumption that investment is to be repaid (by competitive produce) in two years. Deep structural changes are proposed, mostly based on a study of the French management experience. Regional holding companies are to be set up in which the interests of flax-mills, primary producers, and flax-growing farmers are to be interlocked and managed harmoniously. The state subsidies will be directed only to these companies. Some foreign investors and Moscow-based banks have offered to participate in this federally overseen program. We will soon see whether it will prove to be effective.

Notes

1. *Finansovye Izvestiya* May 30, 1996, p. 4.
2. Nefedova, T. "Chto Proiskhodit s Kostromskim Lnom." *Vash Vybor* 1994, No. 2, pp. 15-16.
3. Rybnikov, A.A. *Ocherki Razvitiya Krestyanskogo Promyshlennogo Lnovodstva Niechernozemnoi Polosy Rossii.* Kostroma 1912.
4. *Selskoye Khoziaistvo Rossii v 20 viekie.* Moscow: Novaya Derevnya 1923, p. 175.
5. Ibid., pp. 175, 190.
6. *Chislennost Naseleniya Rossiiskoi Federatsii.* Moscow: Goskomstat 1994, p. 4.
7. Markov, N.P. *Organizatsiya Proizvodstva Lna-Dolguntsa v Novykh Usloviyakh Khoziaistvovaniya. Avtoreferat Kandidatskoi Dissertatsii.* Moscow: VASKHNIL 1990.

16

Large Mechanized Farms

American Drumsticks and Russian Hens

A Spring 1996 "chicken war" between Russia and America illuminates one of the most acute problems facing Russian agriculture, the survival of large mechanized farms.

We have already touched upon the issue of food imports (see Chapters 3 and 6). The gist of the international conflict around what Russians nicknamed "Bush's legs" (because it was under George Bush that drumsticks from America first made it to Russian food stores) is as follows. Since 1989 chicken legs from the US have been one of the cheapest and most favorite meat products of different social strata in the largest cities of Russia. They are substantially cheaper than red meats: up till the very end of 1995 their retail price in Moscow and Saint Petersburg did not exceed 10,000 roubles (roughly $2) per kilogram, whereas imported and domestic red meats ranged from 13,000-25,000 roubles.

Custom duties on imported food were first introduced in 1994; for American drumsticks, however, they did not entail price hikes or cutbacks in supply because the contract price went down at the same time. Tariffs raised in 1995 did not impact retail prices or the supply of chicken legs either, because losses were made up for by the introduction of a currency exchange corridor very beneficial to importers. The Spring 1996 scandal was prodded by a Russian sanitary control agency alleging that American drumsticks might be tainted by salmonella. Somebody was evidently interested in an interruption, however temporary, of US chicken imports, which would cause price hikes and thus invigorate domestic production reportedly ill-affected by US competition. To this end some American media articles featuring cases of chicken-infected salmonellosis in America were used as a pretext to halt the importation of US chickens.

Let us take a look at what occurred to Russian domestic producers of poultry that led their lobbyists to resort to this move, a truly desperate one, even in the eyes of the uninitiated, for the simple reason that the Russian domestic sanitary food control agencies had never before suffered from excessive vigilance.

Poultry farms in today's Russia are typical exurban enterprises. They are especially numerous and technologically advanced in the vicinities of Moscow and Saint Petersburg, where 13% of all of Russia's farm birds were held as well as 14% of all the chicken eggs produced in the late 1980s.[1] Only the province of Sverdlovsk (now Ekaterinburg) and piedmont provinces of the Northern Caucasus (Stavropol, Krasnodar, and Rostov) held a commensurable number of chicken. Together these six provinces accounted for one-third of all the nation's farm birds.

There are 29 poultry farms in the province of Moscow and 16 in the province of Leningrad. (Recall that the province of Leningrad [*Leningradskaya oblast*] has retained its name even though the city of Leningrad switched back to Saint Petersburg in 1991). The following discussion is based on material collected in the latter province in the Winter of 1996 under the auspices of *Vash Vybor* magazine.[2]

Russian Holland

Agriculture in the province of Leningrad has long been under the formative influence of two factors. First, the second-largest city in Russia with a population of about 5,000,000 generates huge food demands so that all suitable parcels of nearby land have been converted to food production. Secondly, agriculture in the province had to adjust to low-quality soil and an abundance of forests and marshes: overall, arable land accounts only for 4% of the province's land area, exceeding 10% only in two districts. Both factors facilitated specialization in livestock and a huge concentration of agricultural investment on limited chunks of land predominantly close to the city. For a long time Leningrad agriculture worked well enough to meet the city's demand and it enjoyed the city's financial and logistical support.

The situation changed in the 1990s. On the one hand, the largest cities of Russia fell in love with imported foods. In the case of Saint Petersburg one need not spend very much on transporting food if it is procured in nearby Finland or shipped over the Baltic Sea. On the other hand, a general crisis in domestic agriculture began taking its toll.

In the 1970s and '80s the province was noted for its robust agriculture compared to other areas of European Russia. In terms of output per unit of land it exceeded even the province of Moscow and Kuban' (*Krasnodarsky kray*). It outproduced every area but Stavropol and Belgorod in meat output per one resident of the province (without the city of Leningrad) and was among the top ten producers of milk in Russia. On the verge of the 1990s one agricultural worker in the province fed 1.5-2 times more people than a

TABLE 16.1 Agriculture in the Province of Leningrad Against the Backdrop of Adjoining Provinces and Russia as a Whole.

	Leningrad	Novgorod	Pskov	Moscow	Russia
1994 Meat Output per One Farm Worker in tons					
	2.0	1.4	1.4	1.5	1.3
1994 Vegetable Output per One Farm Worker in tons					
	2.8	1.6	1.4	2.9	1.2
1991-94 Grain Yield in c/ha					
	15.3	7.9	9.0	20.9	15.6
1991-94 Potato Yield in c/ha					
	122	103	97	94	107
1990 Milk Yield per cow in tons					
	4.1	2.4	2.3	3.9	2.7
1994 Milk Yield per cow in tons					
	2.6	1.7	1.6	2.5	2.0
1994 Percentage Share of Cattle in Socialized Farms					
	89	76	71	91	71
Gross Output of Socialized Farms in 1994 as a Percentage of 1990					
	52	51	50	65	59

Source: Siatistichesky Bulletyn 1 (APK). Osnovnye Pokazateli Funktsionirovaniya APK Rossiiskoi Federatsii in 1994. Moscow, Goskomstat 1995.

worker in neighboring Russian provinces even though in, say, Pskov, natural conditions are more favorable. Leningrad's success was attributable to huge farming investments (in terms of the value of fixed assets per unit of agricultural land the province led all but the province of Moscow) and to large-scale production units of the *sovkhoz* type with an average 510 employees in each. Needless to say, both capital provinces, Moscow and Leningrad, had not lost nearly as many rural residents due to depopulation as most of non-Chernozem Russia had. In terms of farm labor productivity in the European USSR, the province of Leningrad was second only to Estonia.

The province was definitely Russia's agricultural success story: even under marginal natural conditions it was able to meet the demands of its entire population, including the city of Leningrad, for milk, eggs, and vegetables, and it met half the area's need for potatoes and meat. Even in 1994 the province retained its high farm labor productivity (Table 16.1), and, up till 1995, it held the top position in the Russian Federation in per capita output of eggs and the third position in per capita vegetable produce.

And yet in the '90s the output of the province's socialized farms has been plummeting more rapidly than the overall average in Russia and than the average in the province of Moscow. For example, in the interval between

1991 and 1994 alone, meat output was reduced by one-third and the output of eggs by 15%.

Are Bush's Legs Really at Fault?

At the Russko-Vysotski poultry farm (in the Lomonosov district, 15 km from Saint Petersburg) not a single chicken was left by the end of 1996. In order to get a feel for what this means let us go back two years and take a look at some of the 16 poultry farms in the province in the early-to-medium stage of crisis (Table 16.2). Russko-Vysotsky was the largest broiler factory not only in the province but recently in all of Russia. It was a real giant in the Russian *Agroprom* with a settlement of 9000 residents with multi-storey apartment buildings. Together, it and the Lomonosov poultry farm (until the latter came to a halt in 1994) produced between a quarter and one-third of the province's chicken meat. Only recently 16 trucks delivered 80 tons of chicken meat to Saint Petersburg food stores daily. Now all 100 sections of the chicken factory, 96 by 18 meters each, are empty. The new multi-storey addition to the farm built in 1980 looks especially disheartening.

What killed this giant and all the other broiler factories of the province? Was it Bush's legs, as many think? Hardly so: they only finished them off. Domestic chicken meat prices did not exceed import prices in recent years. But beginning in the Summer of 1995 a shortage of domestic grain and the resultant cutbacks on grain feed began driving prices up. Broiler production costs (75% of which goes just to feed costs) had grown to 12,000 roubles per kilo by the fall of 1995. In the meantime American chicken legs were still being sold for 10,000 roubles per kilo. A broiler farm cannot resolve the problem of feed supply by itself. The farm Russko-Vysotsky has only 900 hectares of land, so it used to produce only flourmeal, a small portion of the 200 tons of feed required daily for 40-gram broilers to gain up to 1800 grams weight in 40 days. Domestic producers of mixed fodder had by the end of 1995 raised their price to $300 per ton. As a result it became even more beneficial to buy mixed fodder from Finland. The price was the same and the quality higher: per one kg of chicken weight gain one needs 2 kg of Finnish mixed fodder versus 3-4 kg of its Russian counterpart. But working capital is limited and high-interest credit is risky. Raw grain from Russia's south was cheaper; but its delivery would double the price and then processing added to its cost appreciably. So transactions that in the 1980s were being regulated *outside* the farm by Party bureaucrats in Moscow and Saint Petersburg, have become difficult to manage from the farm itself, especially given the highly unstable economic landscape.

The situation of egg-producing farms is somewhat better. They require feed of lower quality which is, therefore, cheaper. They are also less vulner-

TABLE 16.2   Poultry Farms in the Province of Leningrad in 1993.

Farms	Chickens, in Thousands	Number of Employees	Meat, in Centners of Live Weight	Eggs, in Millions	Ratio of Revenue-Minus-Cost to Cost (as a percentage)	
					Without Subsidies	With Subsidies
Russko-Vysotski	2378	745	173732	16	28	46
Lomonosov	2370	766	141004	14	6	20
Lagolovo	936	388	-	112	33	47
Krasnye Zori	879	743	-	119	28	38
Lebiazhye	165	392	-	8	69	70
Siniavsky	2629	941	21951	310	50	63
Severnaya	1363	352	No data	6	-42	-38
Zavodskoi	670	244	-	79	34	45
Skvoritsy	1200	480	-	153	· 27	41
Voiskovitsy	654	422	-	20	101	104
Nievski	1153	671	-	129	13	24
Baltiisky	238	129	-	54	38	49
Roskar	1301	608	21	163	72	85
Primorski	705	298	547	103	35	44
Udarnik	603	240	505	92	49	61
Korobitsyno	285	288	825	22	-54	-21

Source: Osnovnye Pokazateli Deyatelnosti Selxozpredpriyatii Leningradskoi Oblasti v 1993 godu. Saint Petersburg, SPb Statistics Committee 1994.

able to imports: Finnish eggs, for example, are not in high demand in Saint Petersburg because they are less fresh. So there is still a glimmer of life in egg farms. However, they face similar problems, including a shortage of working capital. The replacement of laying hens is far behind schedule, and instead of a normative one-year life-span they are being kept for two years and more, so productivity goes down. The amount of back-up feed is slim. As a result the number of hens goes down as does the number of egg-farm personnel. The farms that have more land of their own and have expanded their own feed base are better off. Still, experts believe that given current fodder price trends, egg-producing farms are also doomed.

Many large poultry farms face one more problem. They used to own and maintain housing for their laborers. Compared to that in outlying collective and state farms of various specialization, the housing of exurban poultry farms is of much higher quality and is more expensive to maintain. Local civil administrations, however, are not in a hurry to take over these costs and support, at best, only cost-sharing.

In this way, the long-famous poultry farms of Leningrad found themselves between a rock and a hard place. They are squeezed out by price hikes on fodder, electricity, and maintenance of the social sphere on one side, and by imports on the other. And it looks like the former factors are decisive. But not only for poultry farms -- these are just the most glaring examples of production units falling into dire straits due to the removal of central planning and subsidies. Pig-breeding farms are also nearing collapse as is animal husbandry as a whole. Compared with the previous year (1994), the output of meat in the province of Leningrad has fallen by 40%; milk by 17%; and eggs by 9%. Multi-purpose farms fare better in the province, while exurban potato and vegetable farms fare best of all.

Are Subsidies a Panacea?

Farm administrators believe they cannot possibly succeed without government aid, especially without at least partial compensation for feed costs. Federal subsidies for animal husbandry in the 1990s paled behind those of the 1980s when the production cost of meat could exceed farm-gate prices by a factor of 2-3. But even in the 1990s subsidies have amounted to 12-25% of production costs. In the province of Leningrad animal husbandry absorbs 70% of all agricultural subsidies. In 1995, however, the situation worsened yet again: only 4-5% of chicken feed costs were compensated from the federal budget, while the provincial budget gave nothing. In this regard each province conducts its own policy. The majority of provinces subsidize poultry farms per one ton of meat with the size of budget aid ranging from 100,000 (Kirov) to 2,000,000 roubles (Voronezh and Penza). Subsidies per 1000 eggs range from 4000 (Kirov) to 40,000 (Kaluga) roubles. In the province of Moscow, the twin brothers of Leningrad poultry farms received, in 1995, 475,000 roubles of provincial budget subsidies per one ton of meat and 14,400 per 1000 eggs; the city of Moscow's share in this aid amounted to 25 billion roubles. Eighteen Russian provinces compensate their broiler farms for up to 50% of their chicken feed costs.

In the region of Saint Petersburg the issue of assigning budget spendings on agriculture is a bone of contention between the city and the province. (Saint Petersburg and the province of Leningrad are separate jurisdictions with separate budgets; an analogous situation exists only in the region of Moscow; all other provincial capitals are part and parcel of their provinces). The Leningrad provincial administration claims it is short of funds. It promises to cancel only 60% of poultry farms' loan debts. However, since the agriculture of the province mainly serves Saint Petersburg, that city must contribute to the partial compensation of feed costs. Saint Petersburg's administration (at least under Mayor Sobchak, who lost the mayoral election in May 1996) believes that it bears no responsibility for other Federation "subjects'" agriculture: in a market economy one buys where it pays to buy; and currently imports are more economical than domestic produce.

According to poultry farm directors, their production units could survive even only a 30% compensation of feed costs plus loan debts cancellation. In this regard they are still counting on a sizeable bailout, even though budgetary aid in Saint Petersburg's exurbia is already 15-20 times larger per one farm employee than in the peripheral districts of the Leningrad province. It looks, therefore, that the problem is too profound to be solved by budgetary aid alone. An effective solution could involve expansion of cropland under fodder, cooperation with other provinces, creation of holding companies controlling the whole feed-meat-processing retail chain, setting up wholesale markets, and looking for capable managers. Some thinking along these lines is taking place but progress is slow.

Yet the economic potential accumulated in exurban agriculture is colossal; farming in the province needs to get back on track and it will. It is hard to imagine that agriculture will fail given the province's existing state-of-the-art and costly farm equipment, lots of skilled people who want to remain where they are, and an enormous market next door. But this market must now be won in competition, not just administratively assigned by fiat as it used to be. The old system has collapsed and the new one is still in the works. Given this, some form of budgetary aid is unavoidable. The Russko-Vysotsky farm mentioned earlier will finally succeed in getting a new low-interest loan, will then buy a consignment of imported eggs, and rise up yet again. This is what happened to the nearby Lomonosov poultry farm which, by the end of 1995, began to roll after several months of idleness and in December delivered 1000 tons of chicken meat to Saint Petersburg.

As for the infamous conflict over Bush's legs, provoking it did not resolve any problem. It is only natural that whenever Russian hens fail for any reason to feed Russia, American ones will fill the void.

Notes

1. *Narkhodnoye Khoziastvo RSFSR v 1987 Godu.* Moscow: Goskomstat 1988, pp. 231- 233.
2. Nefedova, T. "Dotatsiyami Uzhe Nie Oboitis (Problemy Selskogo Khoziaistva Oblasti)." *Vash Vybor* 1996, No 1, pp. 6-9.

Conclusion

When communism fell, it was as if the Roman Catholic Church were hit by absolute proof that Christ was a historical fiction.

--William F. Buckley[1]

A variety of problematic issues facing rural areas in Russia have been discussed in this book. Definitely at a crossroads, today's Russian countryside, and farming, its main activity, lend themselves less easily to clear-cut conclusions than at any time since the 1929-1935 collectivization. Not for the first time the image of a collapsing rural economy looms large, but just as before, the rumors about its premature death may be exaggerated.

At least in part such "rumors" are begotten by analysts' attaching the Western yardstick of economic rationality to things Russian, an operation that does not always make sense even though it provides scores of pundits with an ever-illusive master key that allows them to keep on talking. This is not to say that economic reasoning is entirely alien to the Russian tradition; it is simply that rationality is not a primary characteristic of many Russian practices.

Reluctance to take this into account explains why Russian agriculture suffers when it is subject to an onslaught by a revolution or by reform-minded strategists attempting to change it overnight according to a single-minded master plan. Such radicalism runs aground each time it confronts the powerful bastion of cultural continuity. Yet this failure has happened again and again, confirming that lessons of history are difficult to learn.

That this book has been written by geographers casts a certain perspective on its subject. It allows us to illuminate aspects of Russian village life whose full impact is hardly discernable otherwise. Rarified social space, an east-west land-use intensity gradient, and the growing gap between exurbia and the rural periphery, these indeed formative features of the Russian countryside have become leitmotivs of the book. An over-abundance of space has proved to be a lingering source of Russia's economic backwardness by encouraging what one not very friendly analyst pointedly called "Russia's simultaneous thrust in all directions."[2] This effect was magnified by its crucial conjunction with Russia's vestigial cult of equity deeply ingrained in the village tradition. The desire to suppress ever-recurring departures from equity, the desire to move ahead in close order rather than through pioneering attempts by a selected few, have exerted a drag on Russia's advancement. This is nowhere so vivid as in the non-Chernozem countryside. The tenacious habit to put more land under cultivation than any reasonable economic thinking could condone, has imparted the vivid spatial dimension to the above-mentioned drag.

However, this incessant struggle with space may in fact be nearing its end. Not because the cultural tradition has changed, but because the current demographic situation attaches more rigorous and implacable constraints than economic rationality alone would ever attach to spatial development. As a consequence we expect that Russian agriculture will benefit from the demographically-driven imperative that it shrink its cultivable land expanse, especially in the areas where soil is of marginal quality, and that it employ a smaller but qualitatively improved labor force.

This silver lining in the cloud of Russia's socio-demographic crisis, however, may be insufficient for a shining future -- the issue of land ownership still needs resolving. This would allow not only more flexible land transfers, but would also help bring farming management into harmony with local conditions and ensure that agriculture is a secure sphere for private investment.

Formally speaking, Russia has ceased to be a peasant country. This happened relatively late, in fact, only in the second half of this century; yet this belatedness is not unique. What may indeed make some claim to uniqueness is that though now a country without a large peasant class, Russia nevertheless has not espoused either family farming or agribusiness, or, finally, a cooperative ideal, as the mainstay of its agrarian order.

On the one hand, there are about 20,000 "socialist latifundia" (collective and state farms, including those that have changed their formal status). Many of them are being pilfered by their members and managers alike whose initiative is also strangled by provincial administrations. These "latifundia" have produced a rural proletariat, those 10 million still-remaining farm-laborers, time-workers with individual parcels of land, but they have not produced *proprietors* -- either large or small. Interestingly enough, the average size of one member's land share in socialized farms, between five and ten hectares, is the same as the average peasant landholding over the whole of Russian history since the 16th century. Along with scores of small-town residents, these time-workers-turned-shareholders work as many as 16 million tiny subsistence and semi-subsistence parcels. They mostly use manual labor and produce 30-40% of the nation's agricultural output from about 3-4% of its cropland. Thus the peasant character of Russian agriculture is being de facto reproduced though without a numerous peasant class. As for the third type of farming estates, individual private farms, they are so feeble that only some Western commentators continue to take them seriously. Besides, many private farms are no more than subsidiary plots.

No sense has as yet developed that the bulk of Russia's vast agricultural land actually belongs to somebody. It is only natural that all the attempts to stimulate the accumulation of farming capital on this no-man's land have been a failure. The fixed assets of Soviet-style agriculture are now in a state of rapid decay. Large farms do not have enough means to renew their fleet of machines or to buy fertilizers and fuel, and many of farm buildings and pieces of equipment are becoming deserted landmarks of the Soviet Agro-Industrial Composite, the so-called *APK*. All farm producers complain in chorus about price disparity: what they produce, they say, is priced too low for them to recover their production costs. However, in most cases the lack of profits rests not so much on the price of a unit of produce but on the disastrously low productivity of a unit of land. The circle is thus being closed: it is becoming entirely disadvantageous to work land. In many places the authorities are trying to master the situation by protecting their local markets, subsidizing, and forcibly redistributing working capital in favor of unprofitable production units. As a result, the agrarian economy is decaying even more rapidly. This situation may change when new Russian entrepre-neurial capital begins to look at agriculture as a secure field of investment. This could happen under two inter-related conditions: guaranteed private ownership of land and more consistent restructuring of large farms now stalled somewhere in the triangle of semi-peasant commune (or a com-munity whose aim is collective survival), production cooperative, and private business.

However, Russia is too vast for one to believe that this restructuring can take one and same form everywhere. There is a substantial spatial variance in land, labor, and capital productivity. Partly as a result, patterns of rural life vary considerably along different geographical lines: north-south, east-west, and center-periphery. Not only do yields and profits vary, not only does gravitation to specific corners of the above triangle occur differently in different areas, but also the types of employees and employers are dis-similar in the vicinity of Moscow, in forested Zavolzhye, and in the steppes of Stavropol. These dissimilarities have been largely ignored not only during collectivization and later, but by market reformers as well, which has ex-acerbated rejection of the reforms by the populace.

During the June-July 1996 Presidential elections a solid majority of rural voters rallied behind the communist contender and the party he represents.[3] Even if he does not carry the national vote, of which country folk account only for about one-quarter, the underpinnings and the repercussions of their attitude cannot be overemphasized.

Communism "would not be so dangerous if it were only evil incarnate," wrote Alexander Vasinsky in early June 1996. "Dispossessed from the sur-face of our life, communism is inside and around us.... It would be too easy if it were linked only to GULAGs, a Pol Pot-like cannibalism, lingering poverty, and a mass lobotomy extracting free-thinking abilities from entire peoples... Communism derived not only from utopians but also from ...other sources. Among other things it has always been a reaction to the genuine unfairness of life, the last refuge of hope for millions who are humiliated and rightless."[4]

In a famous 1943 play by Yevgeniy Shwartz, a dragon maintains its grip on a city. Every now and then it selects a beautiful girl to become its wife and each time she dies of disgust in the dragon's chambers but nobody, even the victims and their parents, find the situation outrageous. Moreover, the dragon collects a monthly tribute from the city in the form of one thousand cows, two thousand sheep, five thousand chickens and 36 kg of salt; in summer and fall the city is supposed to add ten beds of lettuce, asparagus, and cauliflower. "He eats you out of house and home," says Lancelot, a stranger who decides to challenge the dragon, to the despair of many loyal citizens. "Oh, no, you don't mean that," replies Charlemagne, one such loyalist. "We don't complain. And how could it be otherwise? After all, while he is here, no other dragon would dare touch us." "All the others have long been slain, haven't they?" exclaims Lancelot. "But what if they have not? Let me assure you," replies Charlemagne, "that the only way to rid yourself of dragons is to have one of your own." In fact Shwartz's dragon embodies a system of governance that, on the one hand, totally subjugates people and, on the other, shields them from uncertainty and relieves them of the necessity of making decisions. A person gets accustomed to his dragon, it suits him. When Lancelot miraculously kills the dragon, nobody in the city is relieved; and its power is quickly captured by the dragon's yesmen, human beings. They claim they slaughtered the tyrant themselves *but introduce a regime even more corrupt, cynical, and cruel than the dragon's.* When Lancelot comes back to the city he is stunned, but he now realizes that it was not enough to kill the dragon. "What I am to do now is going to be tedious, worse than needle-work," says Lancelot. "Now it's on to killing the dragon within each citizen."[5] While the dragon of communism in Russia may have already been defeated, the second part of the task is far from completion. Especially in the countryside.

There is no way to understand Russia without delving into its history and without exploring its countryside. Today when only a minority of Russians are non-urbanites, this may seem less evident than before. But this would be another great illusion for foreign students of Russia. While the countryside may no longer be as fertile an incubator of new national myths as it once was, it is still one of the most popular subjects of these new myths, and the old myths are not about to vanish either. One has to be fully blinded by overt politicking and totally untouched by life experience in Russia to stick to the belief that Russian communism derived from some trend of political and economic thought subsequently adjusted to Russian soil. Today this mainstay of cold-war sovietology seems terribly off the mark.

Notes

1. *We* (a newspaper), 1-14 November 1993:5.
2. Kissinger, Henry A. "Beware: A Threat Abroad." *Newsweek* June 17, 1996:43.
3. In the first round of elections (June 16) Ziuganov carried 43 out of 89 subdivisions of the Russian Federation; however, he garnered a much higher proportion of a rural vote.
4. Vassinsky, Alexander. "Golos iz Bolota." *Izvestia* 4 June 1996.
5. Shwartz, Yevgeniy. *Piyesy.* Moscow, Sovietsky Pisatel 1972:277-351.

About the Book and Authors

Contrary to the viewpoint of many Western scholars, the authors of this penetrating analysis argue that private farming is not a viable option in Russia's future. Instead, a convergence of Soviet-style subsidiary farming with traditional and reorganized collective farms is the most plausible path of evolution in most rural areas.

Grigory Ioffe and Tatyana Nefedova arrive at this conclusion by a careful examination of ongoing reform efforts in Russian agriculture against the backdrop of European and Russian agrarian history and rural spatial development since the late 19th century. The comparisons at the national level are then filled in with consideration of a number of Russian provinces (*oblasti*) and regions (*raiony*). Their research reveals the substantial impact of rural depopulation on the Russian agrarian economy. Seventy original maps richly complement and support the narrative.

Grigory Ioffe is associate professor of geography at Radford University. **Tatyana Nefedova** is senior researcher at the Institute of Geography at the Russian Academy of Sciences.

Index